AF262594

HOW TO READ
THE WILDERNESS

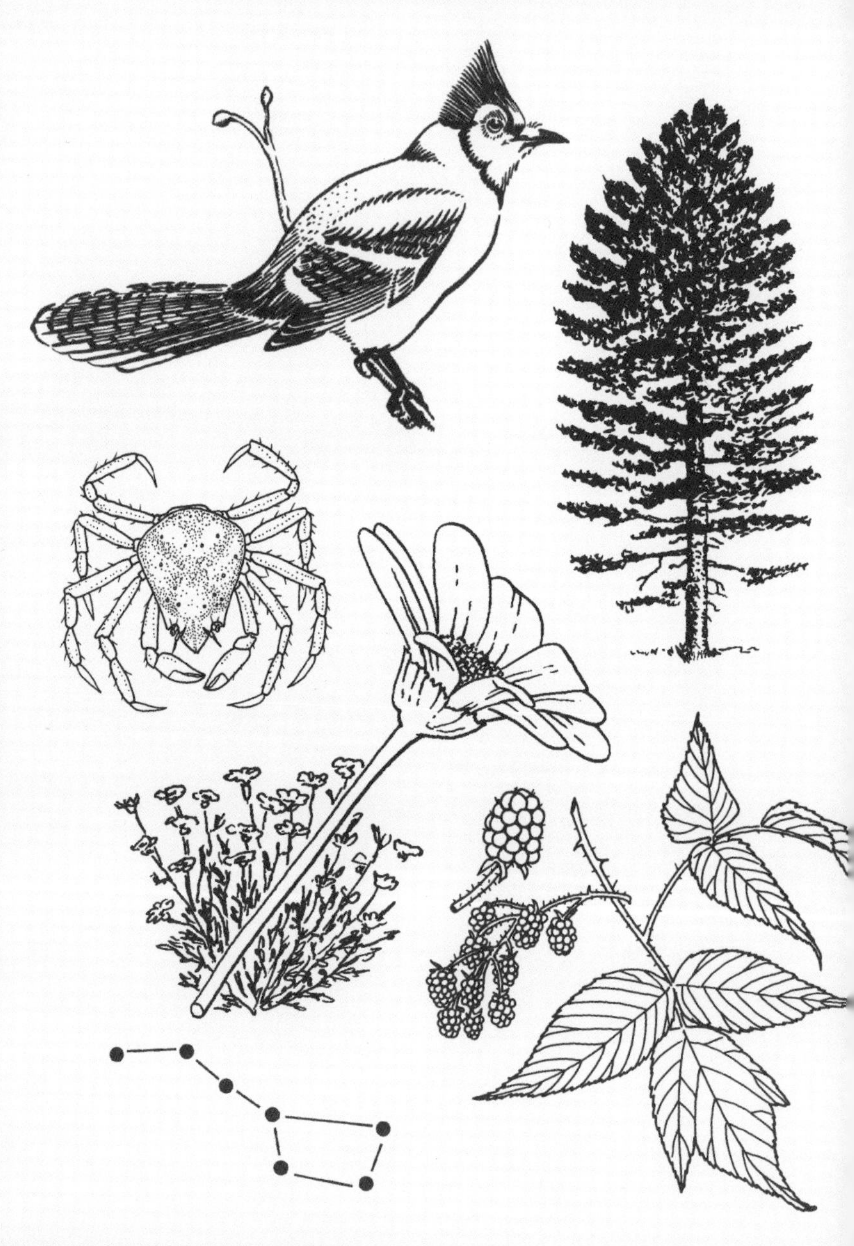

BY THE NATURE STUDY GUILD

HOW TO READ THE WILDERNESS

AN ILLUSTRATED GUIDE TO THE NATURAL WONDERS OF NORTH AMERICA

CHRONICLE BOOKS
SAN FRANCISCO

Library of Congress Cataloging-in-Publication Data

Names: Nature Study Guild (Berkeley, Calif.), author.
Title: How to read the wilderness : an illustrated guide to the
Natural Wonders of North America / by The Nature Study Guild.
Description: San Francisco : Chronicle Books, [2022] |
    Includes bibliographical references.
Identifiers: LCCN 2022012376 (print) | LCCN 2022012377
    (ebook) | ISBN 9781797206868 (hardcover) |
    ISBN 9781797212371 (ebook)
Subjects: LCSH: Natural history--North America--Juvenile
    literature. | Plants--Identification--North America--Juvenile
    literature. | Animals--Identification--North America--
    Juvenile literature. | Stars--Names--Juvenile literature.
Classification: LCC QH102 .N38 2022 (print) | LCC QH102
    (ebook) | DDC 508.7--dc23/eng/20220523
LC record available at https://lccn.loc.gov/2022012376
LC ebook record available at https://lccn.loc.gov/2022012377

Manufactured in China.

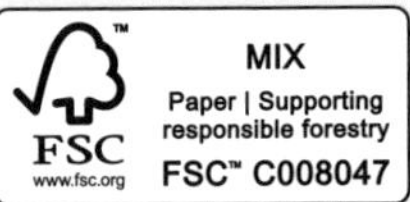

Curation by John Lasack.
Design by Kelley Galbreath.

10 9 8 7 6 5

Chronicle books and gifts are available at special quantity
discounts to corporations, professional associations, literacy
programs, and other organizations. For details and discount
information, please contact our premiums department at
corporatesales@chroniclebooks.com or at 1-800-759-0190.

Chronicle Books LLC
680 Second Street
San Francisco, California 94107
www.chroniclebooks.com

CONTENTS

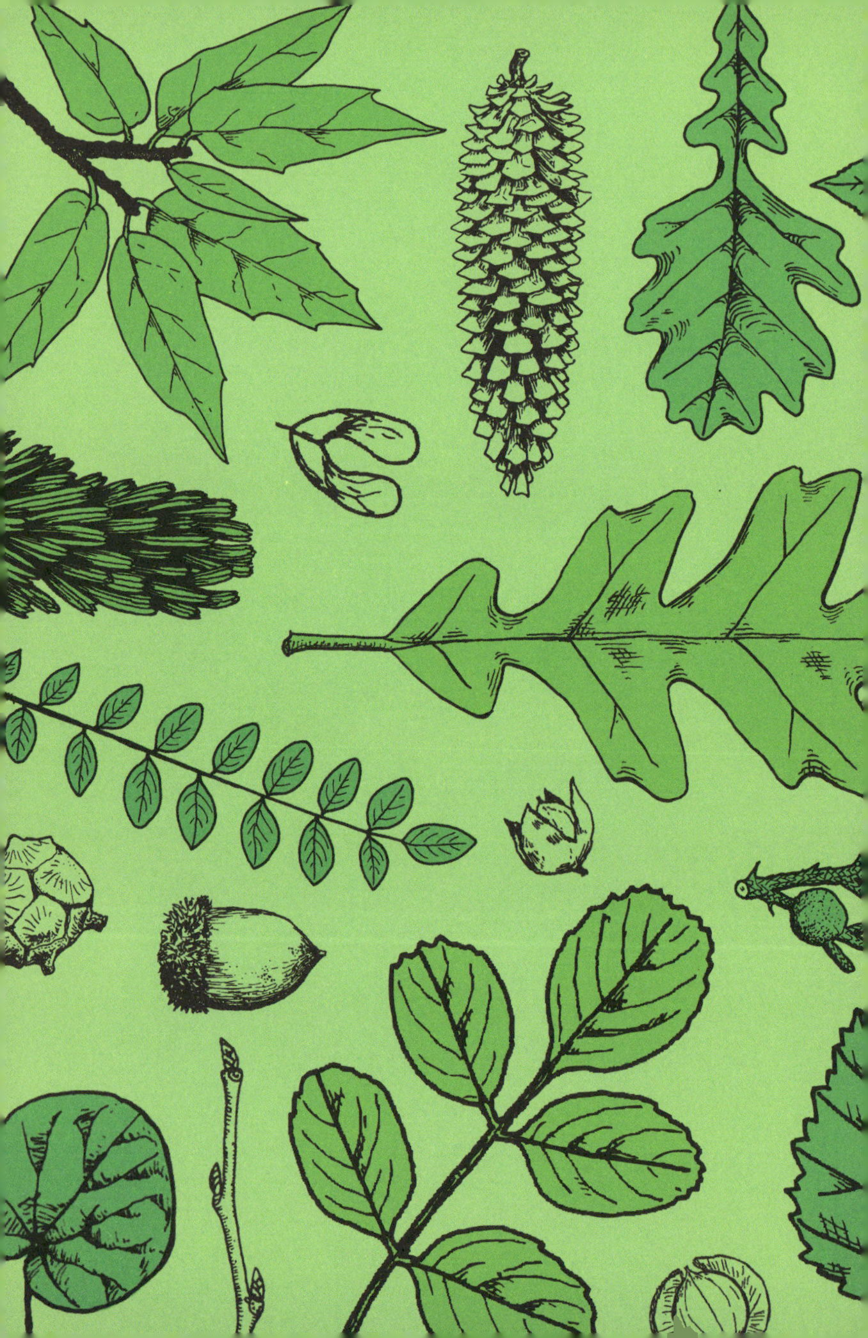

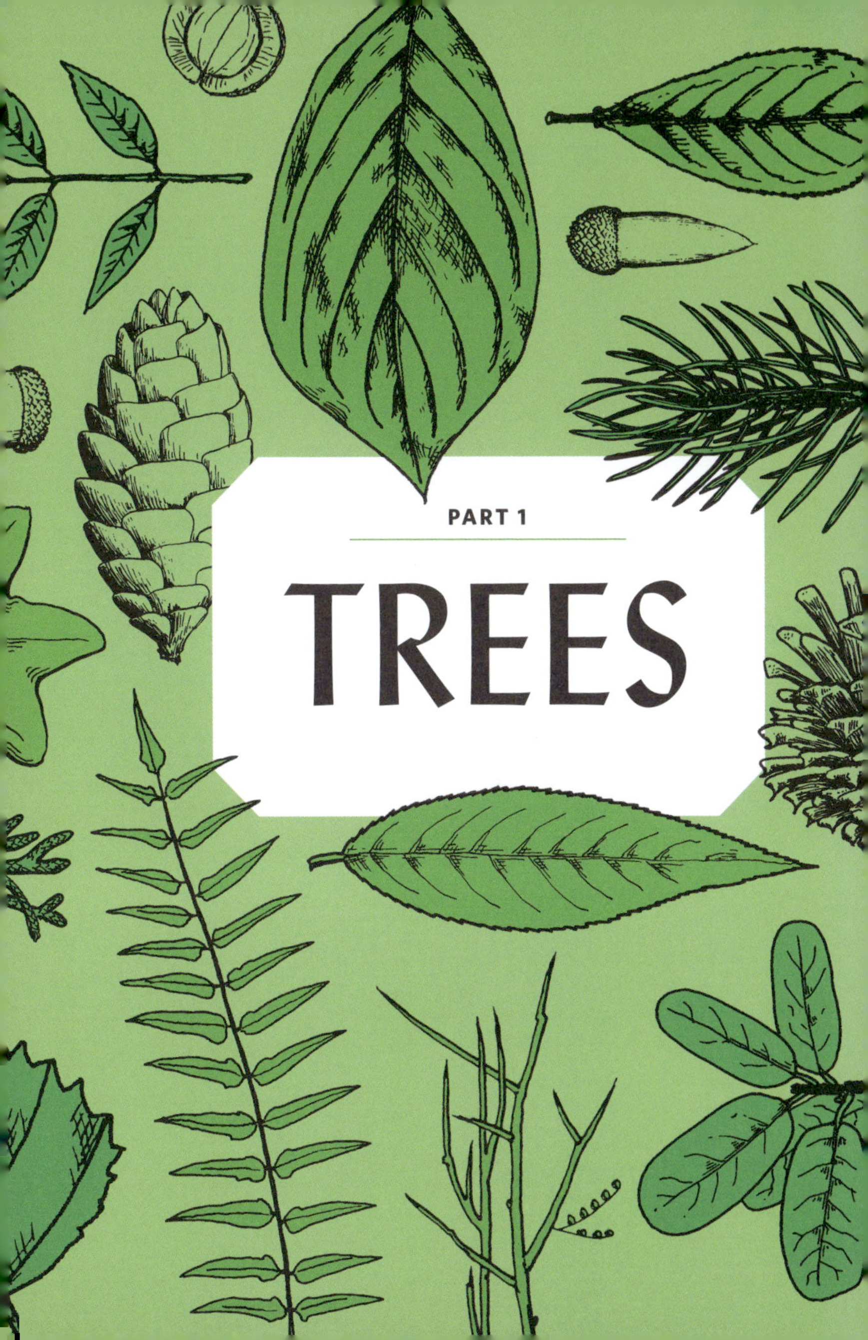
PART 1
TREES

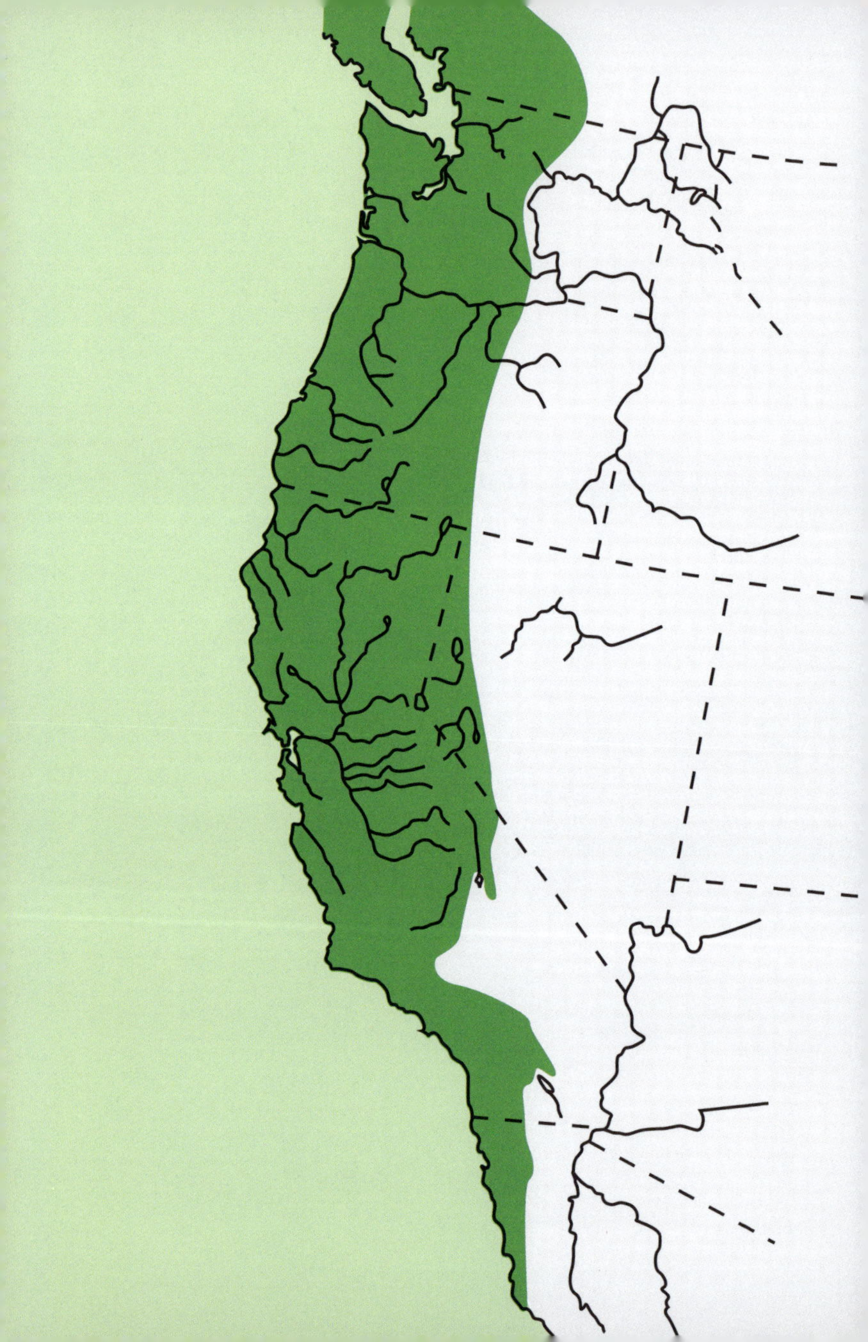

Trees of the Pacific Coast

 where you're likely to find certain types of trees growing:

ON WARM, SUNNY RIDGES

These trees grow slowly with roots in dry, rocky soil, where wind blows away the snow and shapes their resinous foliage. Among them you'll find lairs and lizards.

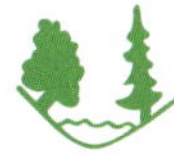

ON STREAM BANKS OR IN SOGGY SOIL

These trees grow fast, have shiny, pliable foliage, easily broken twigs, soft wood, and birds.

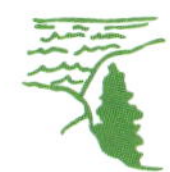

ON THE SEACOAST

These are often trees that grow only on windy slopes facing the sea, usually in sandy soil.

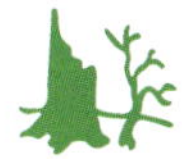

IN BURNED AREAS

Unless fire returns, these trees will eventually be replaced by more shade-tolerant species.

IN ABANDONED FARMYARDS, OLD SETTLEMENTS

These trees grow so well on the Pacific coast they may seem to be native species, but they or their ancestors were planted here.

Other ways of categorizing trees:

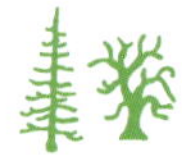

DECIDUOUS TREES
These are leafless in winter
or the dry season.

SHRUBBY TREES
These are trees that often
grow as shrubs and become
trees only on sheltered slopes
and canyons.

The climate of this area has damp, foggy
sea air, a long growing season, and abundant rain. It supports a deep, dark, ferny
forest where trees must compete
for sunlight.

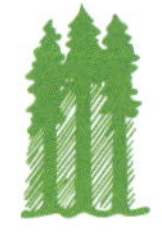

The **DOMINANT TREES** of this
forest grow rapidly, straight
upward toward the light.
Before clear-cut logging,
the shade kills back lower
trunk branches and produces
fine-grained, knot-free lumber.

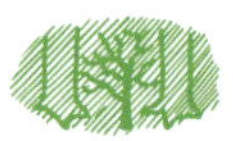

The smaller **UNDERSTORY
TREES** are shade-tolerant
throughout life and often
capture weak light with
thin, broad, horizontally
held leaves. They may grow
sideways toward a patch of
sunlight.

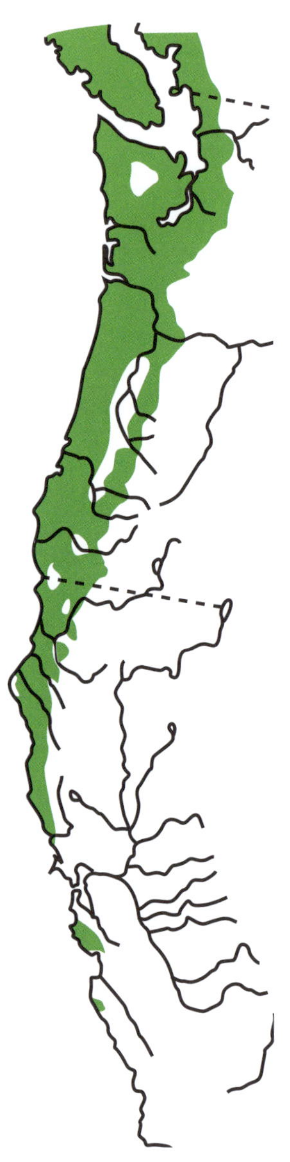

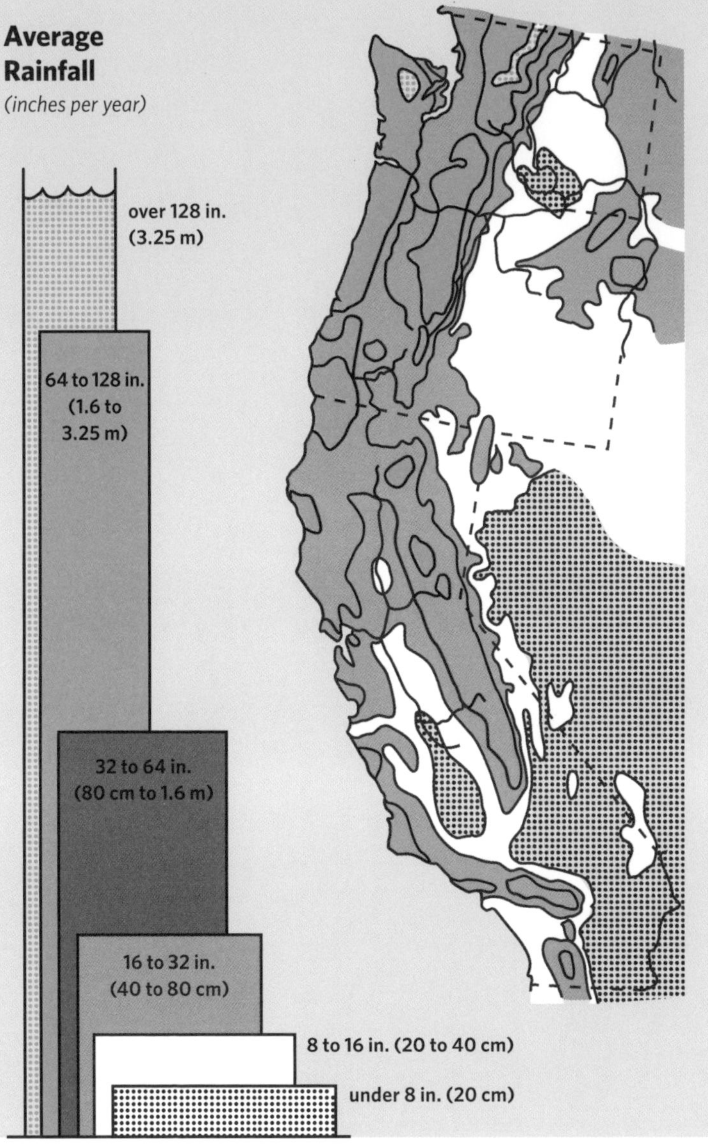

Average Rainfall
(inches per year)
over 128 in. (3.25 m)
64 to 128 in. (1.6 to 3.25 m)
32 to 64 in. (80 cm to 1.6 m)
16 to 32 in. (40 to 80 cm)
8 to 16 in. (20 to 40 cm)
under 8 in. (20 cm)

The Pacific coast climate is drier toward the south and in the low-altitude inland valleys:

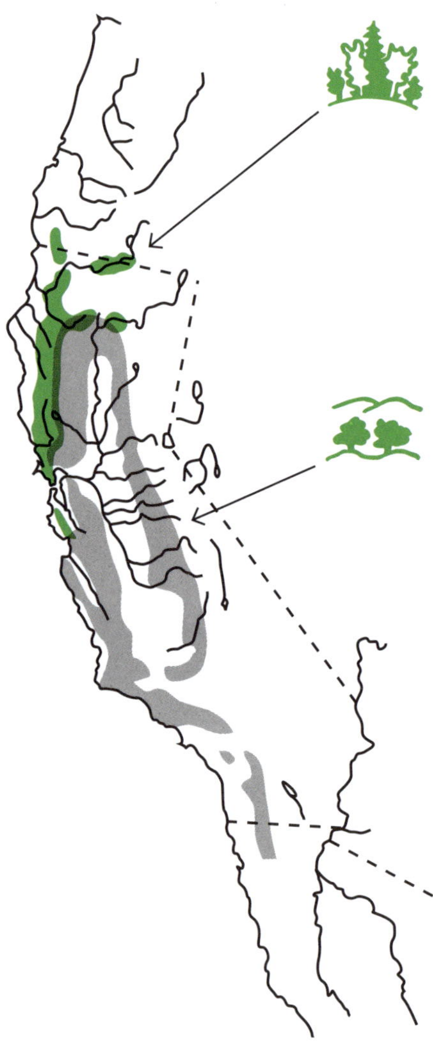

In the **MIXED-EVERGREEN FOREST** of the coast ranges, there are no big redwoods or hemlocks. Small Douglas firs dominate. Leaves here are often thick, leathery, and evergreen. This slows evaporation of moisture during dry summers and takes advantage of mild, wet winters.

In the **CALIFORNIA OAK-WOODS**, short, muscular-looking, deep-rooted trees with small, bristly leaves stand far apart, or else cluster along canyons and north-facing slopes. Even the pines here look stunted.

Much of this area is a mosaic of woods, grassland, and chaparral (stiff, dry, evergreen shrubbery) on hot slopes.

Average Growing Season

(frost-free days)

This map shows why coast-range forests differ from higher-altitude forests of the Sierra Nevada and Cascades. *(see next page)*

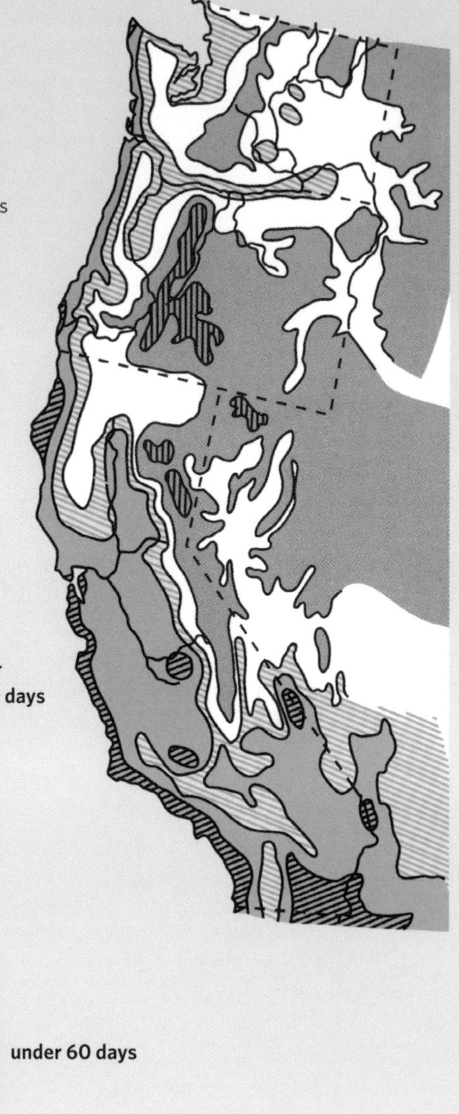

Climbing into high mountains is like going north (and going north is like climbing). The climate turns colder and usually wetter, and vegetation resembles the fir forest of Canada. Further up you see small spire-shaped trees as in interior Alaska. The alpine zone is treeless like the Arctic barrens.

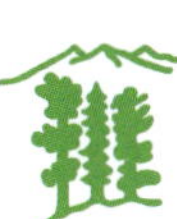

The dry lower elevations of the **MIXED CONIFER FOREST** are often open pine woods (with chaparral). In higher elevations, more rainfall supports more kinds of trees, larger and closer together.

Most adult trees here have thick, corky trunk bark and can survive small ground fires. Accumulations of unburned undergrowth and fallen wood may support larger, killing fires. Hillside trees may be fire-hollowed on the uphill side where logs roll against trunks and later burn.

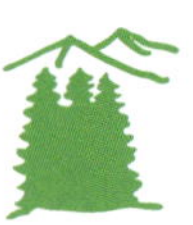

In the **MOUNTAIN FIR FOREST**, branches are shaped to shed heavy snowfall. You'll find curved lower trunks on trees that grew from saplings bent by a thick, sagging snowpack.

Fires are rare here but wreak disaster when they come to the "tinderbox" of twiggy deadwood and lower branches.

The dense fir forest is on the better soils. Pines grow where it's rockier, and in burned areas. Lodgepole pine grows in mountain basins or "flats" where cold night air drainage collects.

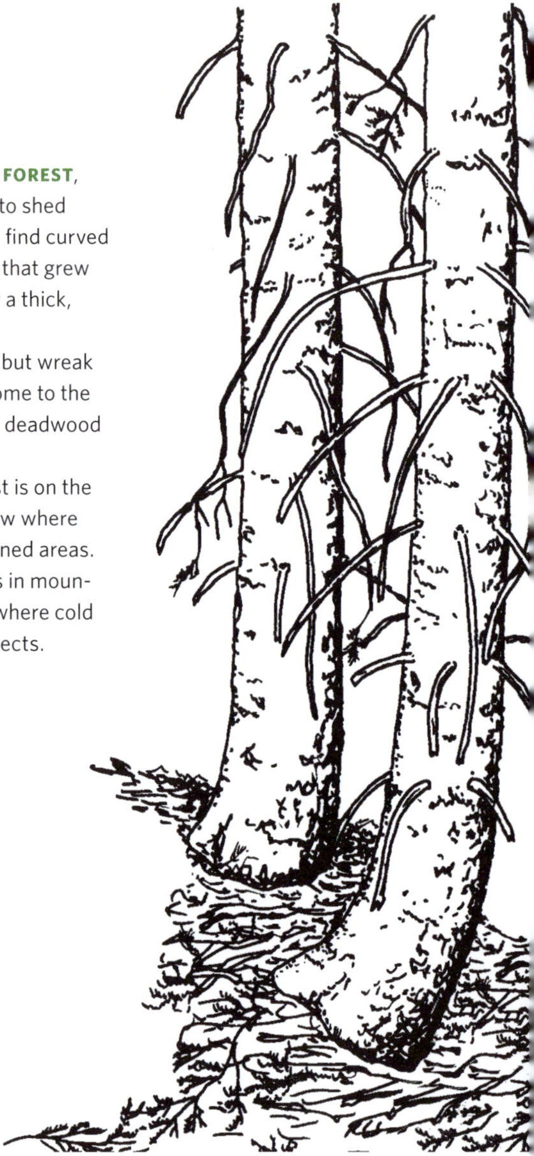

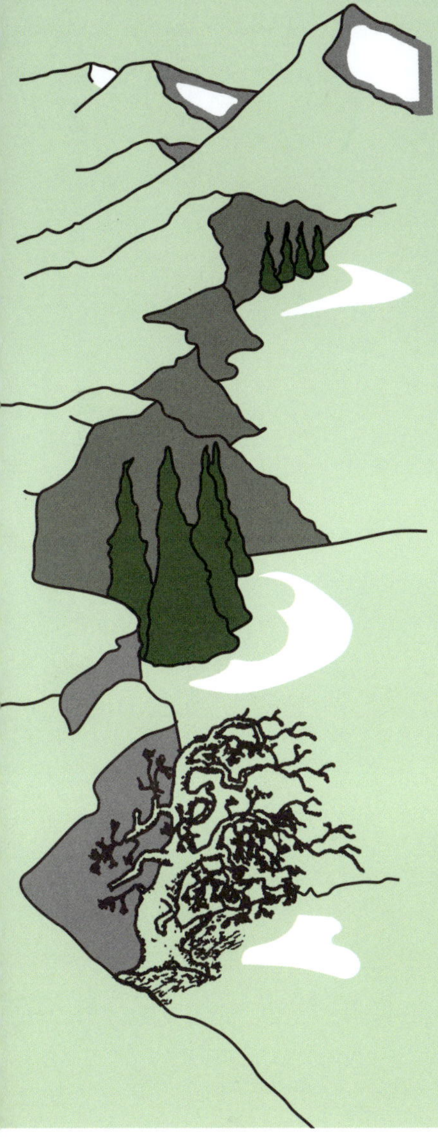

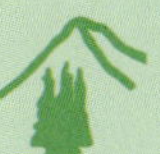

To survive in the **SUBALPINE ZONE**, trees must mature their new growth in a short, unreliable summer where you can make snowballs in July and expect frost in late August.

The higher you go, the smaller the trees. Near the upper limit of this zone, trees can survive only where snowdrift seals out the abrasive, drying wind.

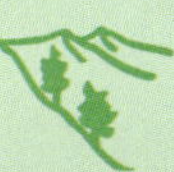

On the dry **EASTERN SLOPES** of high mountains, the trees most likely to grow are Rocky Mountain trees, adapted to a severe cold-dry climate.

Identifying Oak Trees

If the leaf is simple (not made up of leaflets), the main veins branch off from a single large central vein, and the leaf is lobed, it's an OAK. **Go to** ⟶

If the first two are true but the leaf is not lobed, **go to** ⟶

If the lobes are pointed and bristle tipped, **go to** ⟶

If the lobes are blunt and rounded (may have a single point at end of the lobe), **go to** ⟶

If the lobes are shallow, it is **ORACLE OAK**
Quercus ×morehus

If they're deeper, or if the leaves are over 4 inches (10 cm) long, it is **CALIFORNIA BLACK OAK**
Quercus kelloggii

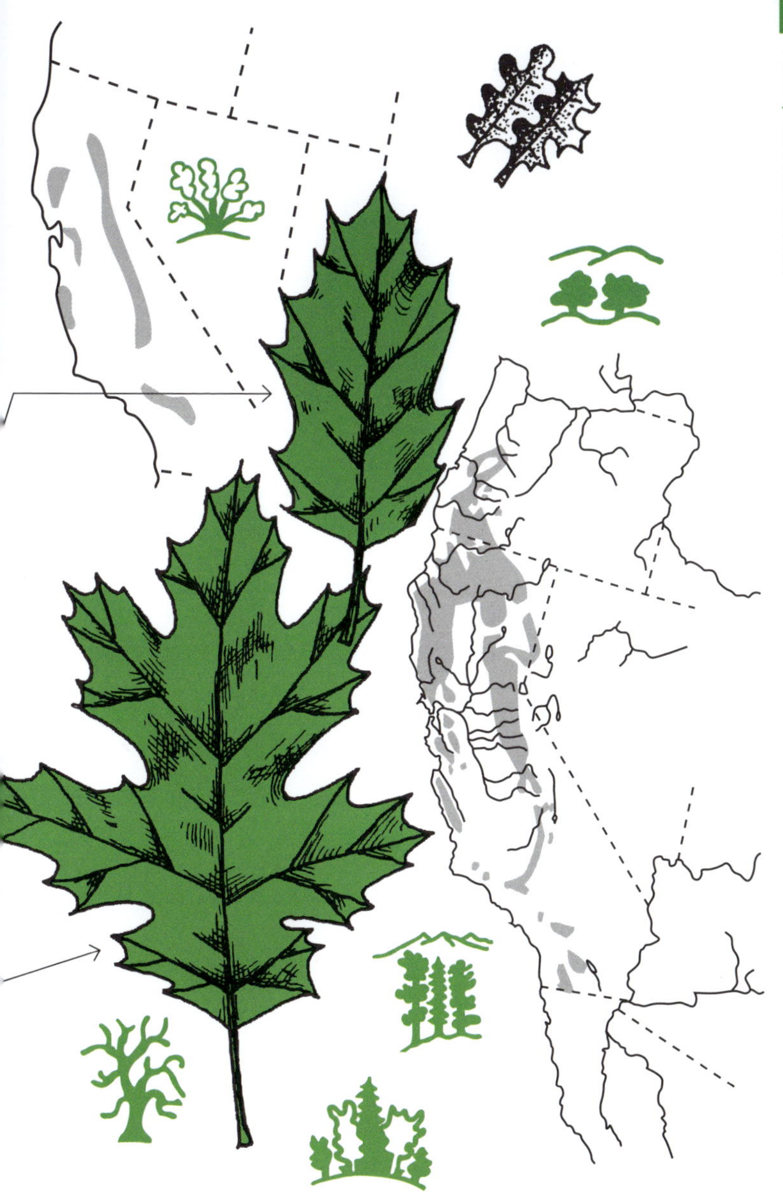

If the leaves have soft hairiness on the topsides and seven to eleven deeply cut lobes; and if the trunk bark is deeply checkered into squarish plates; or if there are long hanging branches or long acorns, it is **VALLEY OAK, ROBLE** *Quercus lobata*

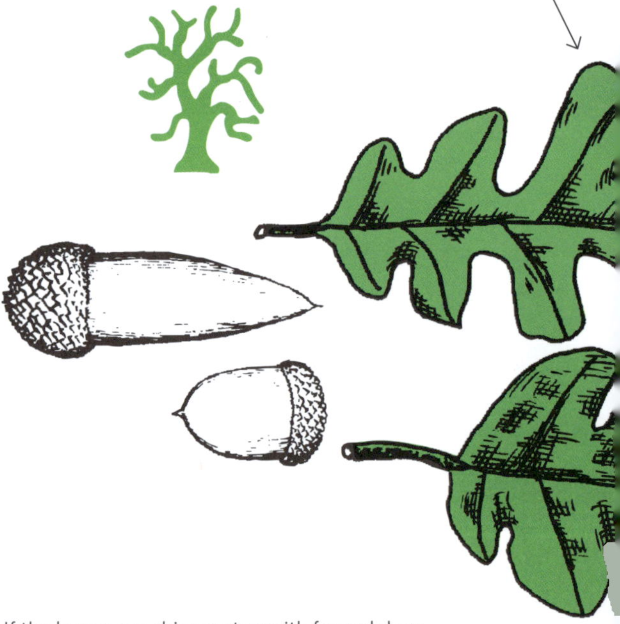

If the leaves are shiny on top with fewer lobes; or if you can rub off whitish-gray bark scales; or if there are stubby acorns, it is **OREGON OAK** *Quercus garryana*

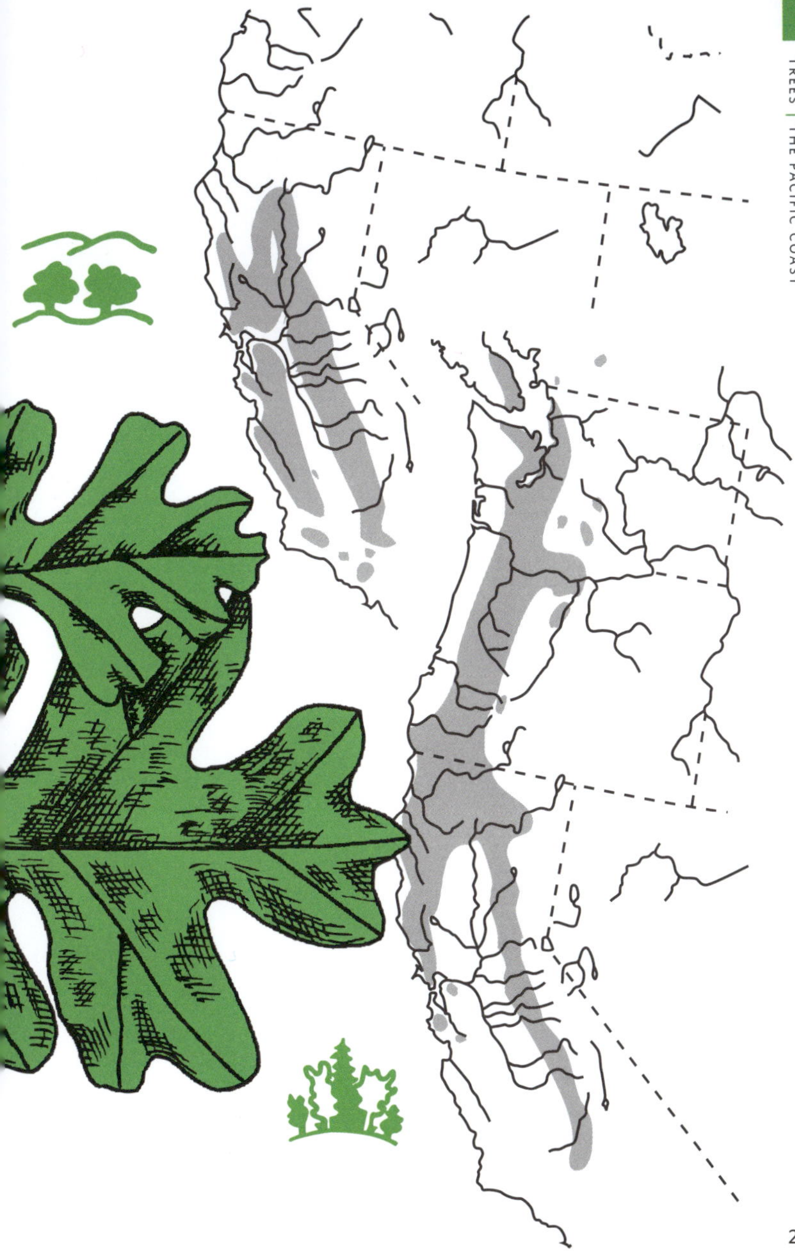

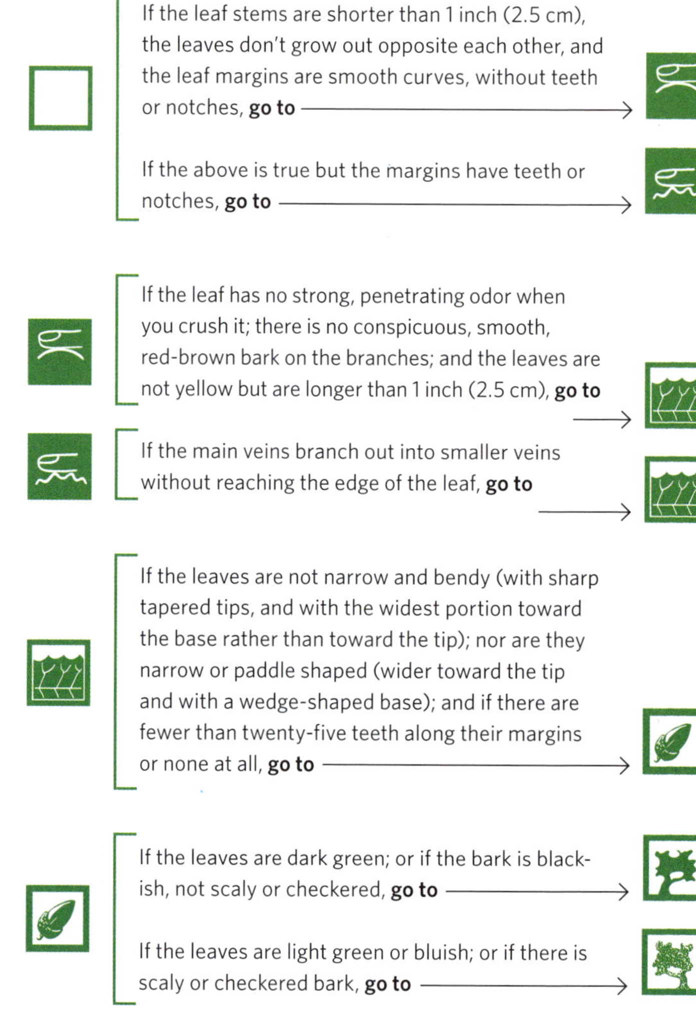

If the leaf stems are shorter than 1 inch (2.5 cm), the leaves don't grow out opposite each other, and the leaf margins are smooth curves, without teeth or notches, **go to** ⟶

If the above is true but the margins have teeth or notches, **go to** ⟶

If the leaf has no strong, penetrating odor when you crush it; there is no conspicuous, smooth, red-brown bark on the branches; and the leaves are not yellow but are longer than 1 inch (2.5 cm), **go to** ⟶

If the main veins branch out into smaller veins without reaching the edge of the leaf, **go to** ⟶

If the leaves are not narrow and bendy (with sharp tapered tips, and with the widest portion toward the base rather than toward the tip); nor are they narrow or paddle shaped (wider toward the tip and with a wedge-shaped base); and if there are fewer than twenty-five teeth along their margins or none at all, **go to** ⟶

If the leaves are dark green; or if the bark is black-ish, not scaly or checkered, **go to** ⟶

If the leaves are light green or bluish; or if there is scaly or checkered bark, **go to** ⟶

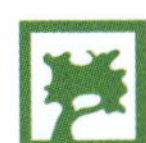

If the leaves are convex on top and have tufts of hair where the veins join on the undersides, it is **COAST LIVE OAK** *Quercus agrifolia*

If the leaves are flatter, hairless, and some are without spines, it is **INTERIOR LIVE OAK** *Quercus wislizeni*

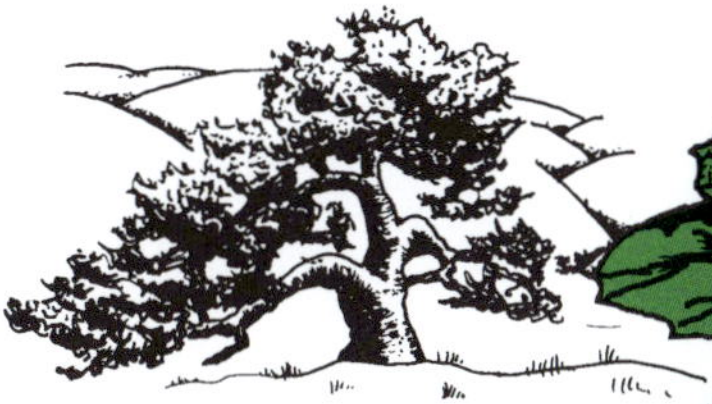

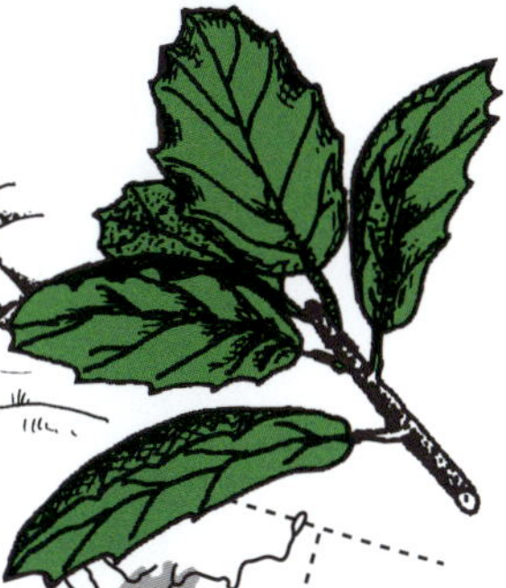

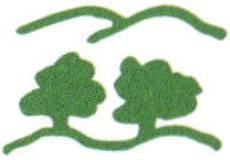

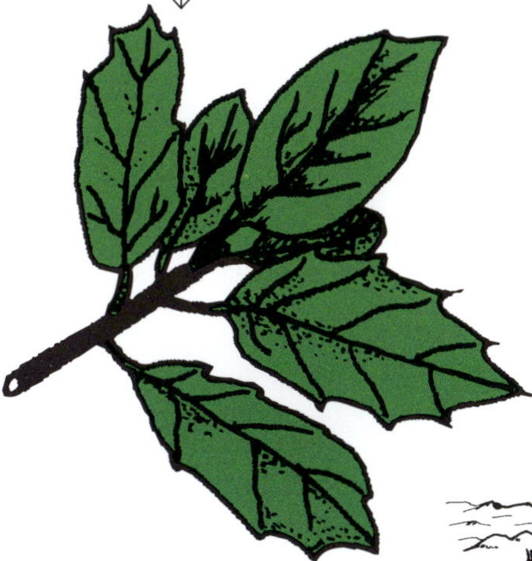

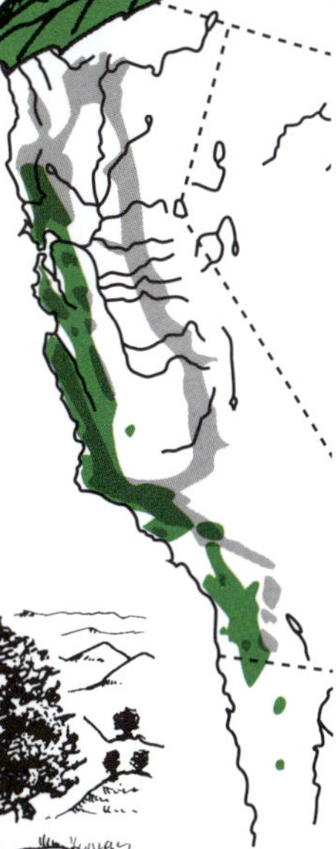

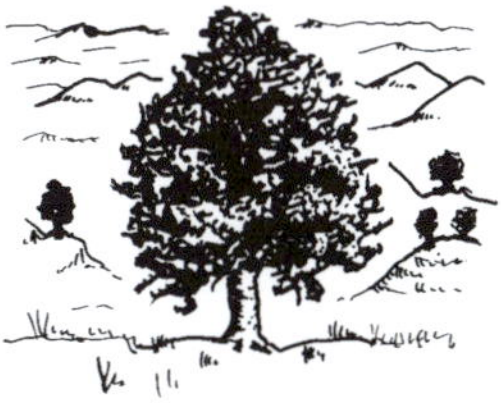

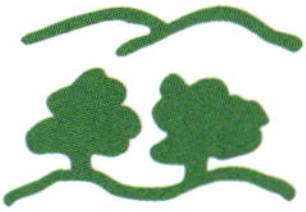

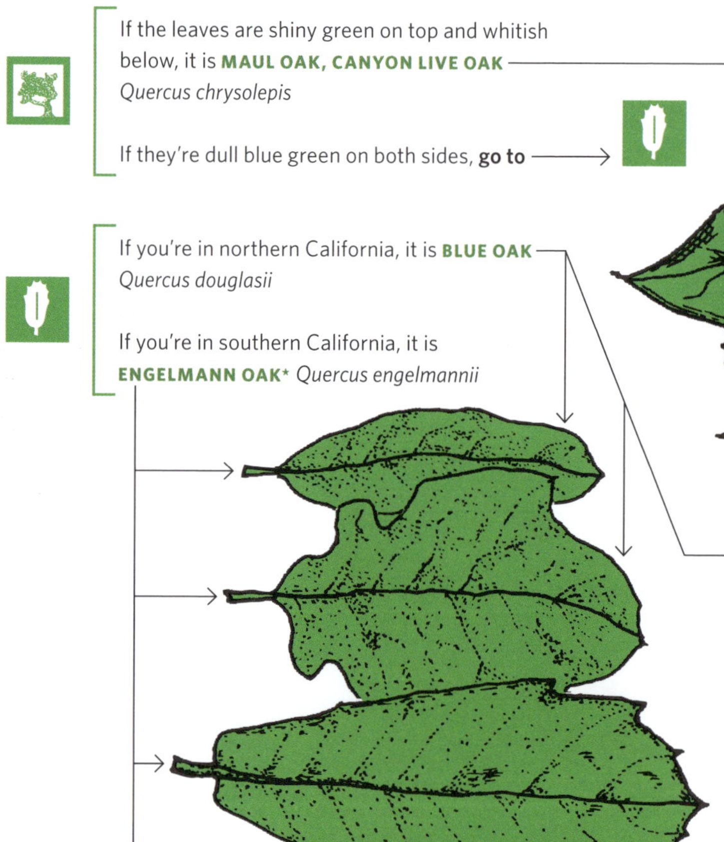

Oak trees and their shrub relatives breed promiscuously. You'll often find hybrid forms with mixed traits of several species. Identifying them is beyond this guide.

*Except at Joshua Tree National Park, where the usually shrubby *Q. turbinella* is treelike

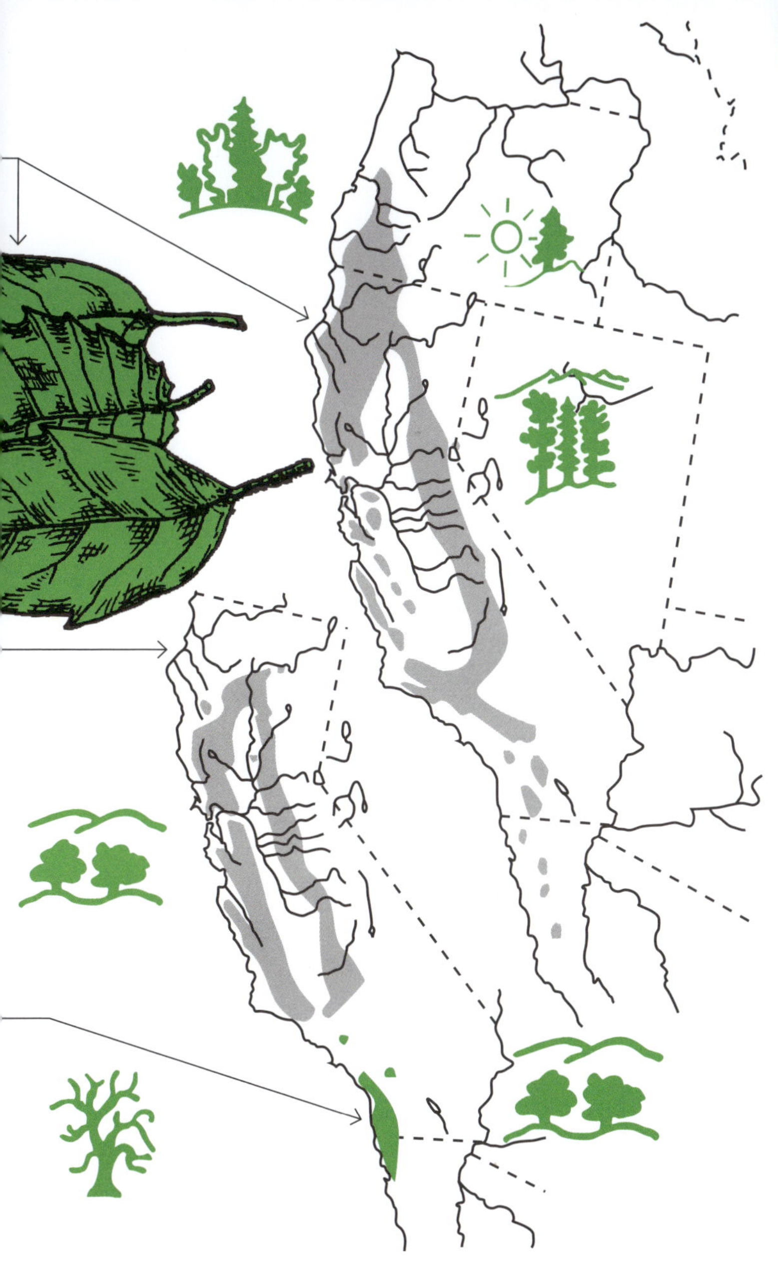

Trees of the Southwestern Desert

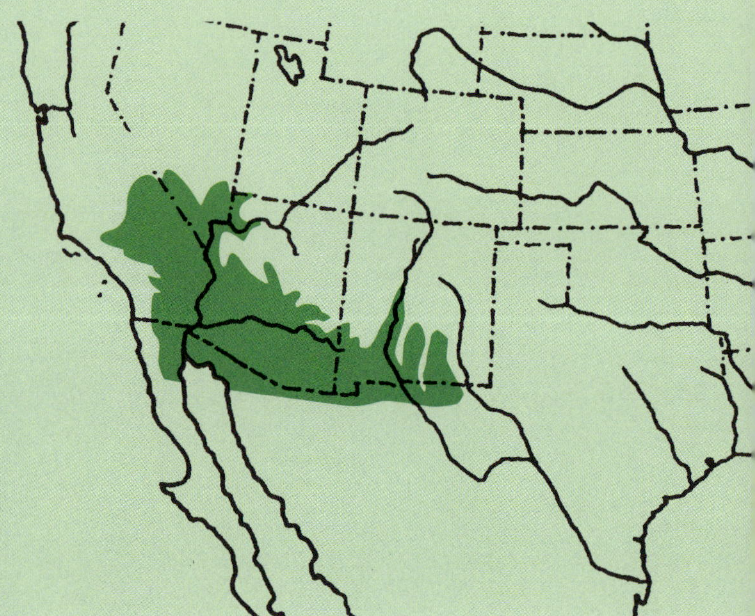

IF THE TREE YOU ARE IDENTIFYING grows among:

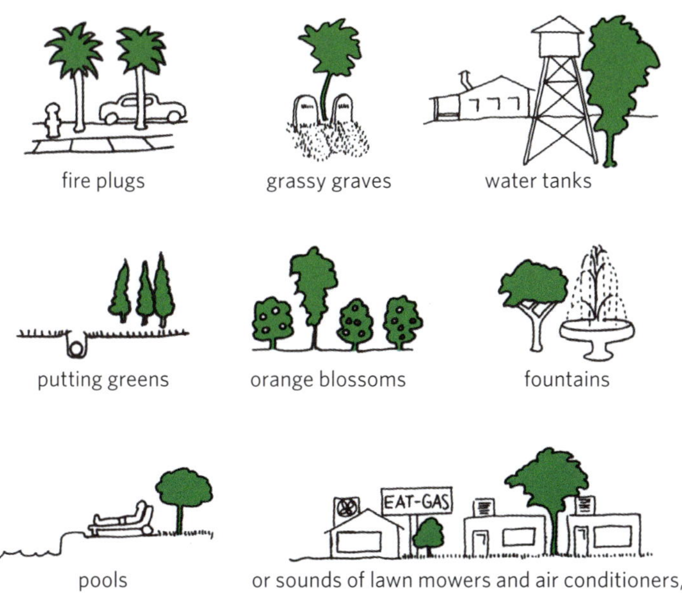

you are in a man-made oasis.

These symbols show how or where desert trees are likely to grow:

 **RIPARIAN TREES:** Their roots are in permanent underground water.

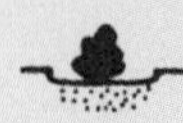 **TREES OF DESERT WASHES:** They grow where flood waters gather after a thunderstorm.

 **EVERGREEN TREES:** In leaf all year long.

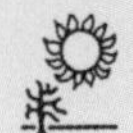 **TREES LEAFLESS IN DRY SEASONS:** They leaf out only after a good rain.

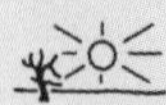 **TREES LEAFLESS** in winter only.

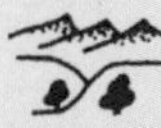 **TREES OF THE OAK-JUNIPER WOODLAND:** These trees of higher altitudes also grow in canyons and cool slopes at the edge of the desert.

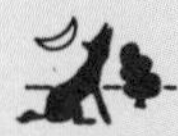 **TREES INTRODUCED BY HUMANS** that have escaped from cultivation and are now growing wild.

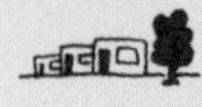 **DOMESTICATED TREES OF THE DESERT,** planted in parks, yards, and cemeteries.

Here are some ways trees survive in the desert.

Trees have long roots

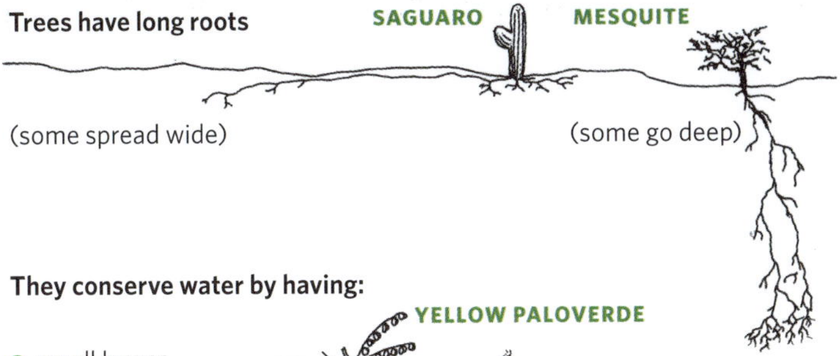

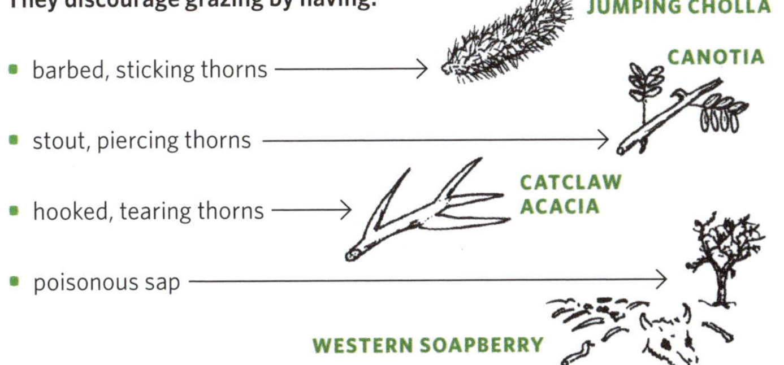

(some spread wide)

(some go deep)

They conserve water by having:

- small leaves →

- thick-skinned leaves →

- leaves that drop in the dry season →

- hairy leaves and twigs →

- thick, fleshy trunks for water storage →

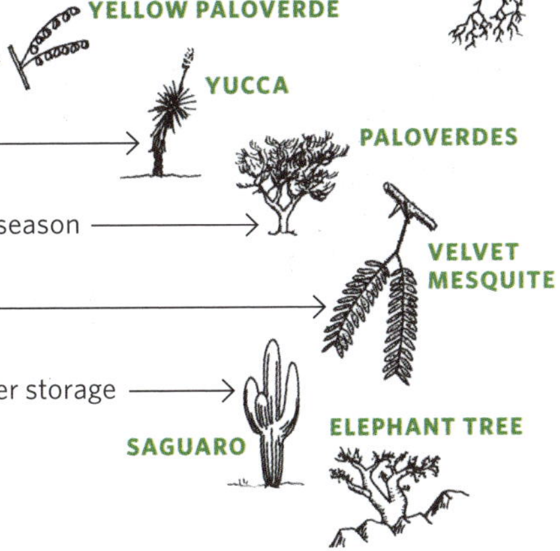

They discourage grazing by having:

- barbed, sticking thorns →

- stout, piercing thorns →

- hooked, tearing thorns →

- poisonous sap →

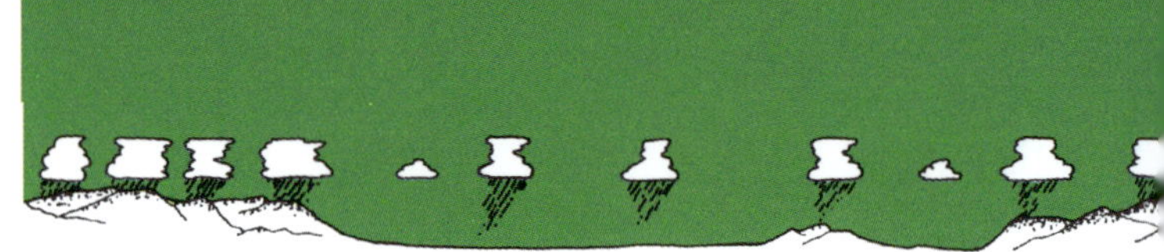

Rainfall

Most of the rain in the desert falls in the cool higher altitudes of the surrounding mountains. Lower altitudes are warmer, but the rainfall there may dry up in midair. The high-rainfall areas on this map also show where the high mountains are:

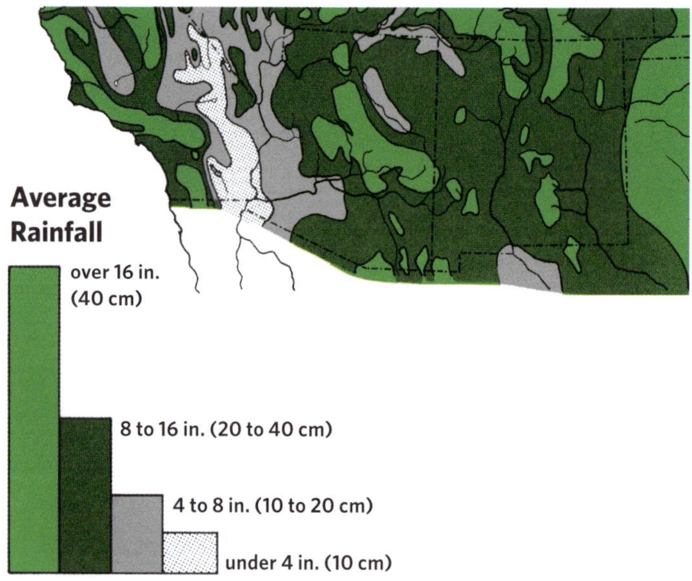

Where the average yearly rainfall is under 16 inches (40 cm), it makes a great difference to trees how much the amount of rain varies from year to year and what season the rain falls—matters which depend on seasonal winds from the Atlantic and Pacific Oceans.

Winter Rain

Weakened remnants of the winter storms that drench the California coast can bring sparse, gentle rain as far east as the Continental Divide. But many desert trees can't benefit from rain in winter, when it's too cool for their roots to grow or seeds to germinate. Winter rains seldom reach the Chihuahuan Desert. It has dry, frosty winters.

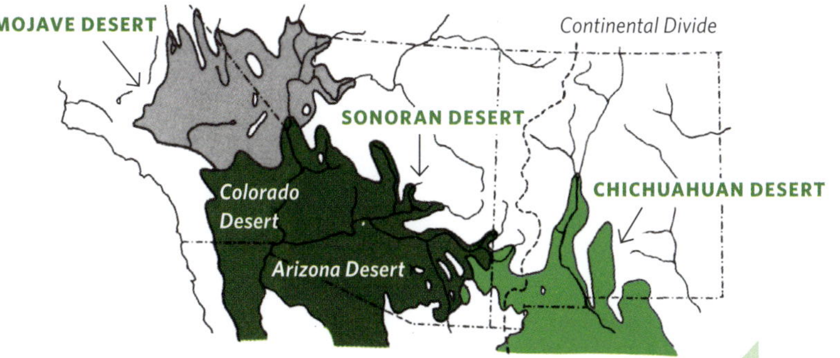

Summer Rain

Monsoon winds can bring unstable, tropical air from the Gulf of Mexico. The resulting thundershowers soak unevenly into the soil, but they come at the best season for trees. Summer rains are most reliable in the eastern end of the Sonoran Desert (the desert with the most abundant trees). To the west, the rains thin out and may skip a year. The trees are smaller and farther apart too.

Desert Soils

Desert mountains stand knee-deep in their own debris—eroded rock, gravel, sand, and clay— which forms long bajadas (slopes) leading down to an intermountain basin or valley. In the desert dryness, soil changes along these slopes can mean life or death for trees. Violent, splashing rain causes sheet erosion, which makes a network of temporary watercourses emptying into desert washes. The erosion carries

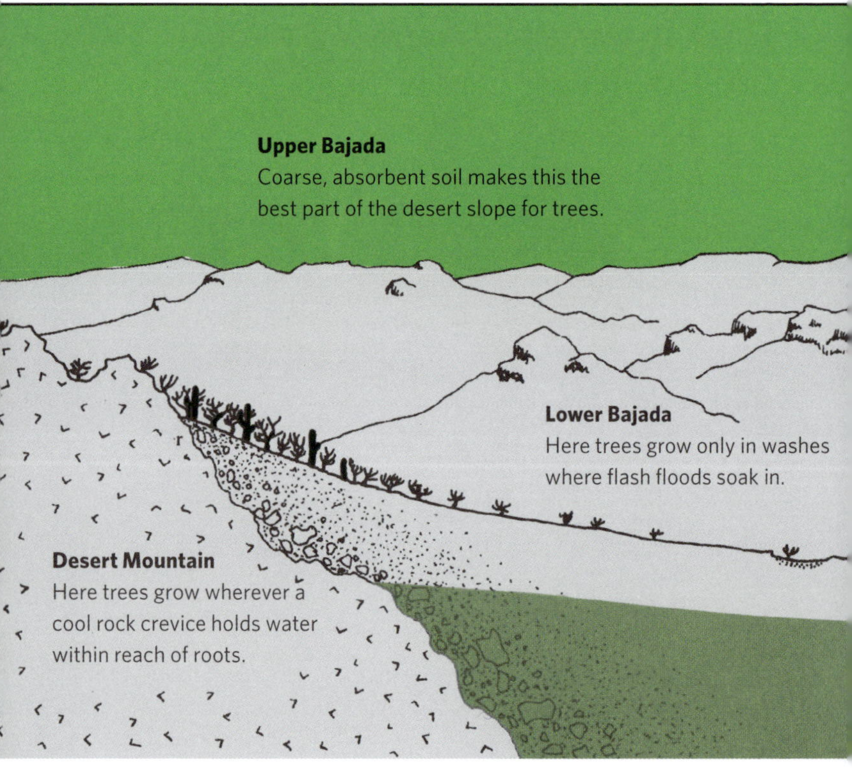

the finer soil particles farther downslope than the coarser sand and gravel. So as you descend a bajada, the trees thin out because the soils become finer, less porous, less able to absorb the quick rains that run off and collect in the washes. Trees grow in the washes, especially where a layer of windblown sand collects in the depressions and holds moisture in the soil.

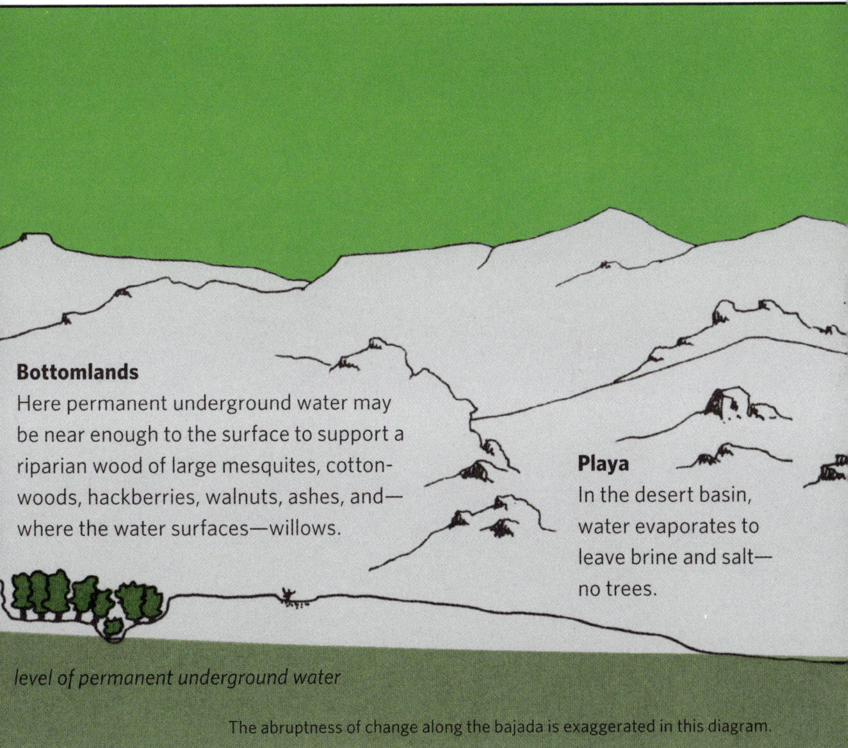

The abruptness of change along the bajada is exaggerated in this diagram.

Identifying Cholla, Saguaro, and Yucca

If the tree is a cactus, fleshy outside, woody within; and bears clustered spines, **go to** 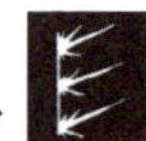

If it's not a cactus, **go to** 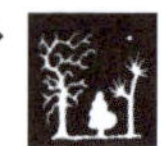

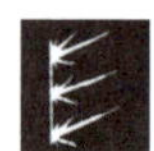

If there are many jointed-looking branches that divide into smaller branches pointing in all directions, it's a CHOLLA. **Go to** 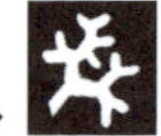

If the branches are few, or if they all come up from the ground, **go to** 

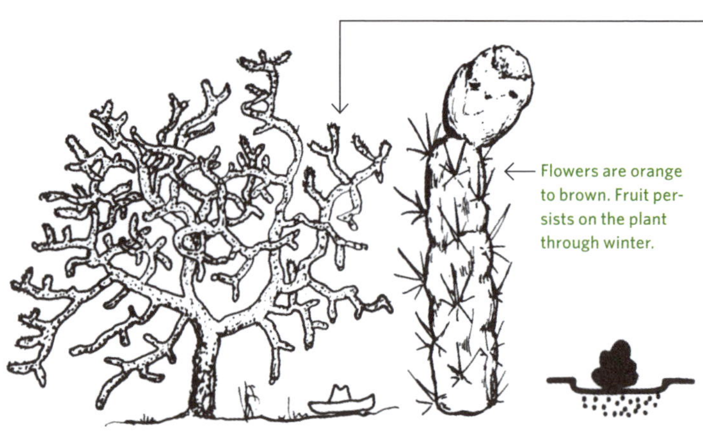

Flowers are orange to brown. Fruit persists on the plant through winter.

If the jointed branches are densely covered with straw colored spines, the trunk bark is black, and the fruits dangle in long clusters, it is **JUMPING CHOLLA** *Opuntia fulgida*

If the spines are dark, of uneven length, and the trunk bark is lighter, it is **STAGHORN CHOLLA** *Opuntia versicolor*

The joints, which can sprout a new plant, detach or "jump" at a touch. Barbed spines discourage a second touch.

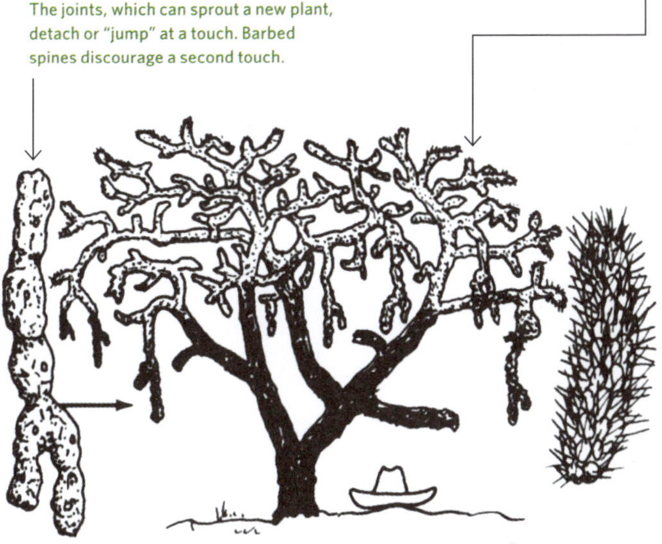

If a single trunk rises from the ground, it is
SAGUARO *Cereus giganteus*

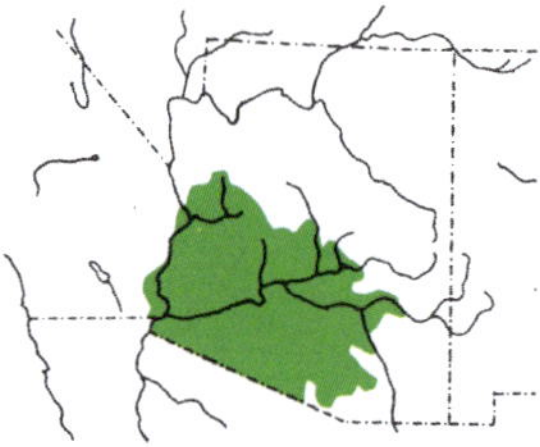

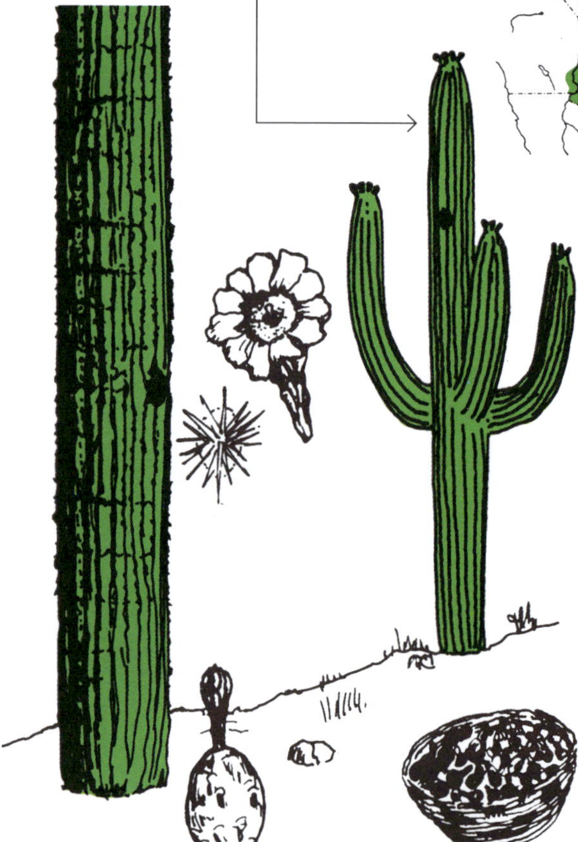

Saguaro branches are crowned with waxen, night-blooming flowers in April, and edible red fruit in June (harvested by the Tohono O'odham). After rain, trunk corrugations flatten out as flesh absorbs water and swells.

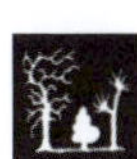
If the leaves are clustered at the tip of a single trunk (or at the tips of branches), and if the leaves are stemless, spear-shaped, it's a YUCCA. **Go to** ⟶ 

If there are only a few branches, or just a main trunk; and if the leaves are edged with peeling fibers, **go to** ⟶ 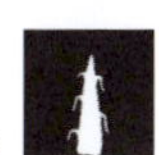

If the tree has many branches, and if there are small, sharp teeth along the leaf edges, it is **JOSHUA TREE** *Yucca brevifolia*

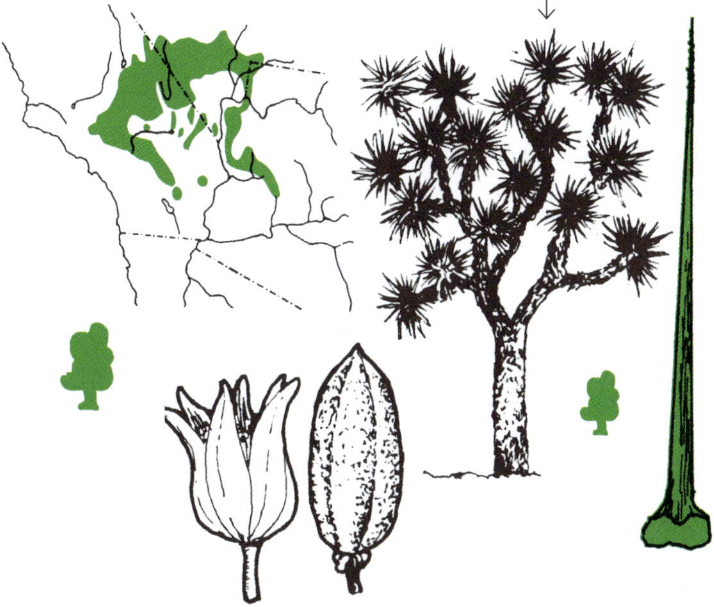

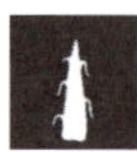

If the leaves are stiff, daggerlike; and dead leaves cover the trunk almost to its base, **go to** ⟶ 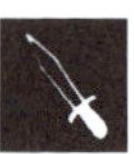

If the leaves are flexible, grasslike, it is **SOAPTREE YUCCA, PALMILLA** *Yucca elata*

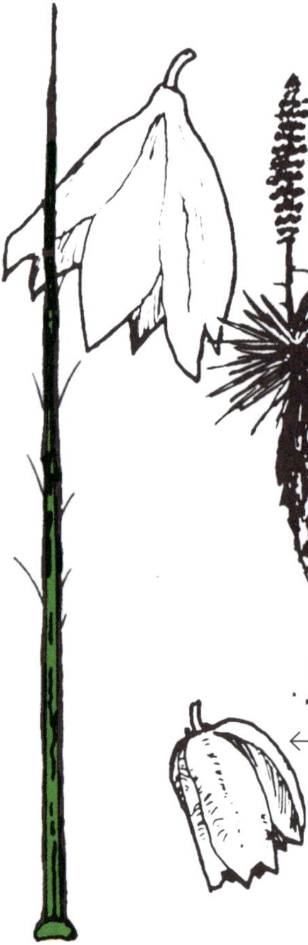

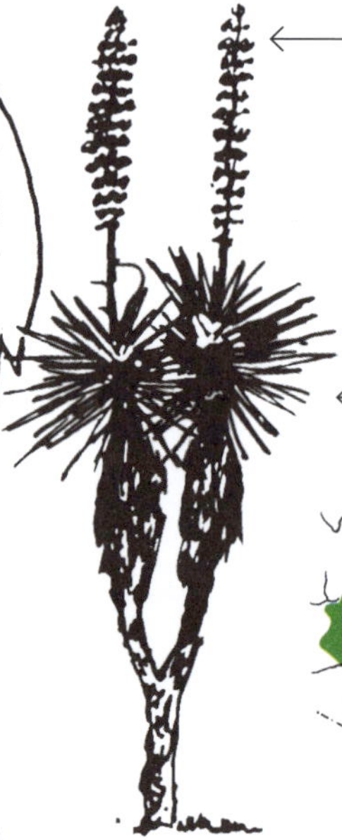

The flower stalk is long, bearing many flowers on its upper part, but naked on its lower part. Seed capsules open in three parts to release seeds and remain on the stalk through the winter.

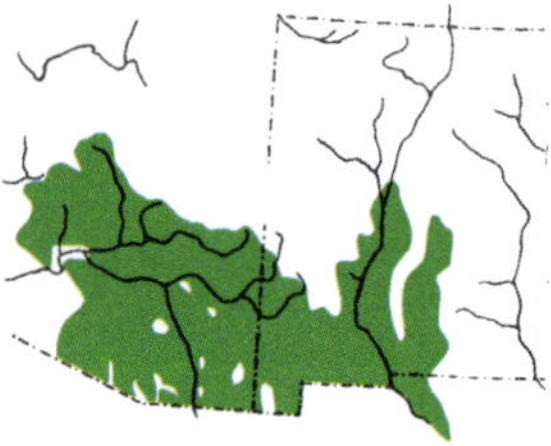

On dry yucca seedpods you'll find a small hole made by the escape of a larva that fed on some of the seeds. It grew from an egg laid by a female yucca moth, whose instinct then made her pollinate the yucca flower, thereby ensuring seed formation. This probably happened at night, when desert creatures do business after hiding from the hot daytime sun.

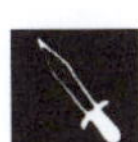

If the leaf has many whitish threads along its edge, and flower clusters are 3 to 4 feet (1 to 1.2 meters) long, it is **TORREY YUCCA** *Yucca torreyi*

If the leaf is edged with coarse, splinter-like fibers, and the flower cluster is shorter, it is **MOJAVE YUCCA** *Yucca schidigera*

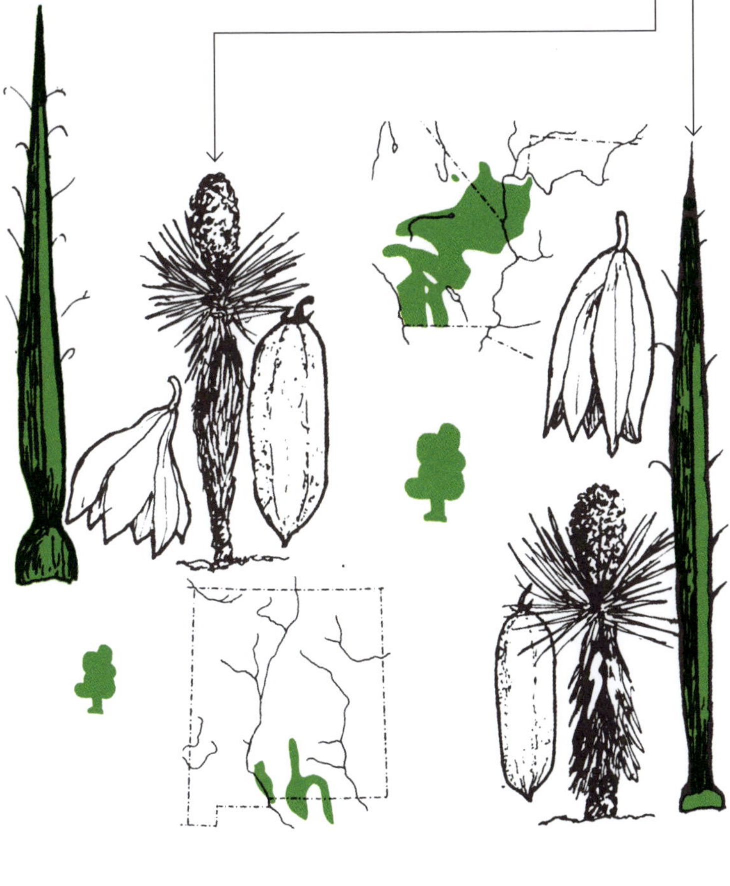

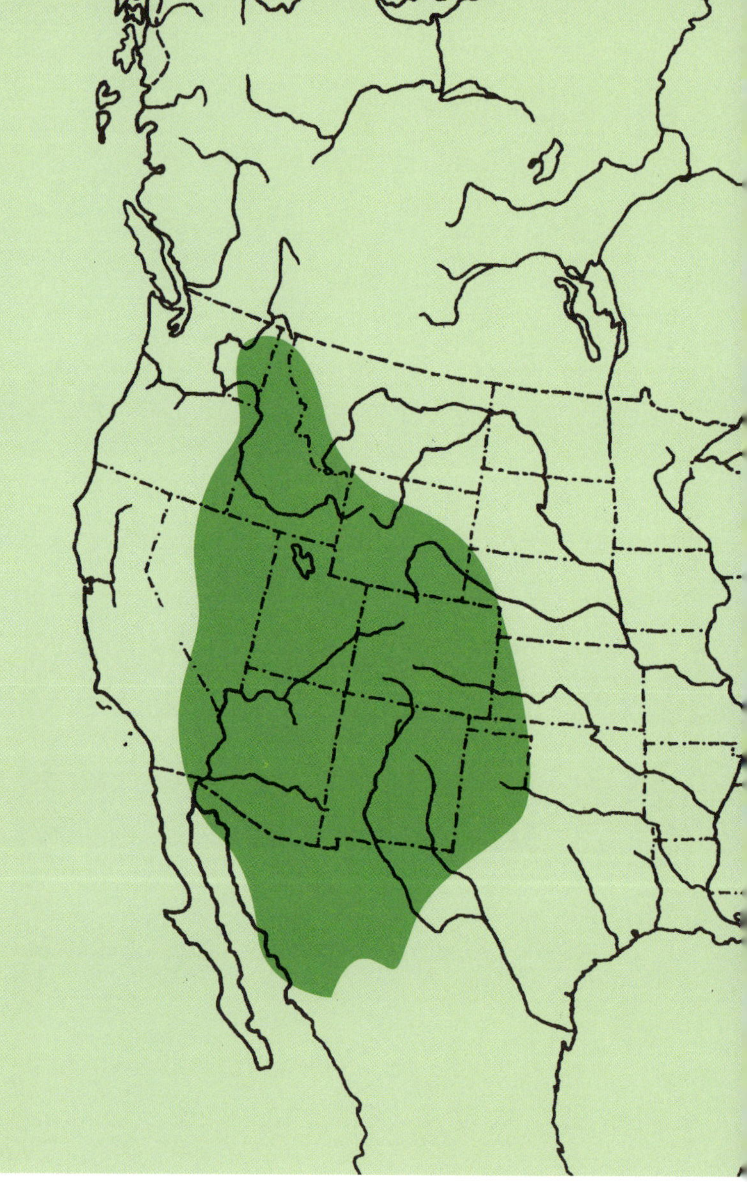

Trees of the Rocky Mountains

Rocky Mountain Tree Distribution

Climate limits where trees can grow. The very things people like about the Rocky Mountain climate—cool summers and deep snow at high altitudes, year-round dry sunshine in the low altitudes—are hard on trees. The Rockies have abundant water and warmth, but not in the same places. Trees here must adapt to extreme temperatures, intense sunlight, and drying winds.

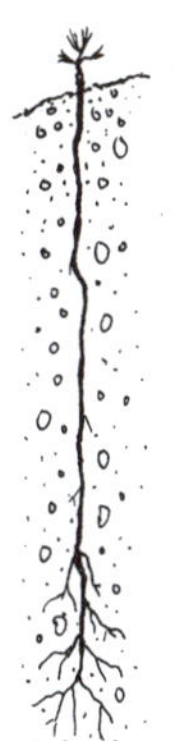

Some trees are adapted to hot, dry foothills: A 3-inch (7.5-cm) ponderosa pine seedling may have roots 2 feet (60 cm) long. It can take soil surface temperatures of 180°F (82°C), but a heavy frost can kill it.

Other trees can survive short summers and bitter winters: The new needles of subalpine fir resist frost, and its branches shed heavy snow. But its roots are shallow and could not survive the dry heat of the foothills.

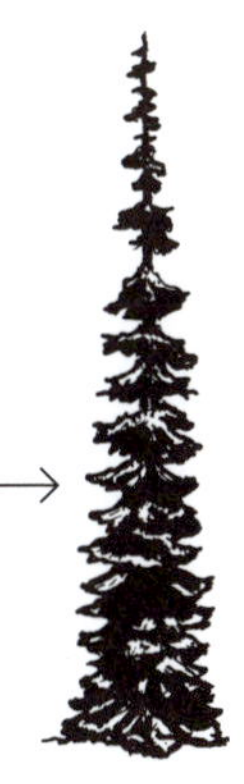

Each tree's range has an upper limit, the altitude above which it's too cold for the tree to grow, and a lower limit, below which it's too dry.

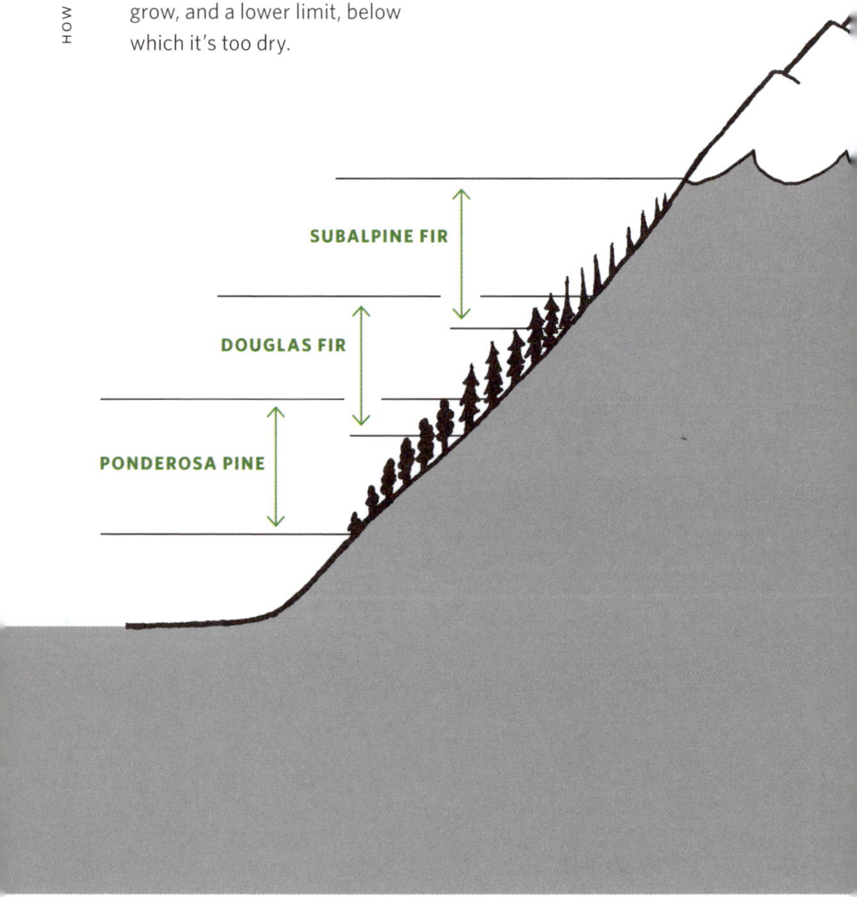

These are the five **LIFE ZONES** found at different altitudes in the Rocky Mountains. Some trees grow in several zones, but if you know what zone you're in you can usually tell what trees you'll find.

Plains Zone

Upper Sonoran Zone—annual rainfall 10 to 20 inches (25 to 50 cm)

Grasses, sagebrush, or dry shrubbery dominate this zone, the driest in the Rocky Mountains. To survive here, trees must have roots that reach moisture.

SMALL TREES grow with wide-spreading roots in loose, porous soil or the cool crevices of a rocky outcrop:

- Pinyons
- Gambel oak
- Junipers
- Mountain mahogany

The only **LARGE TREES** are ones rooted in subsurface water along a stream bed:

- Cottonwoods
- Hackberries
- Boxelder
- Willows

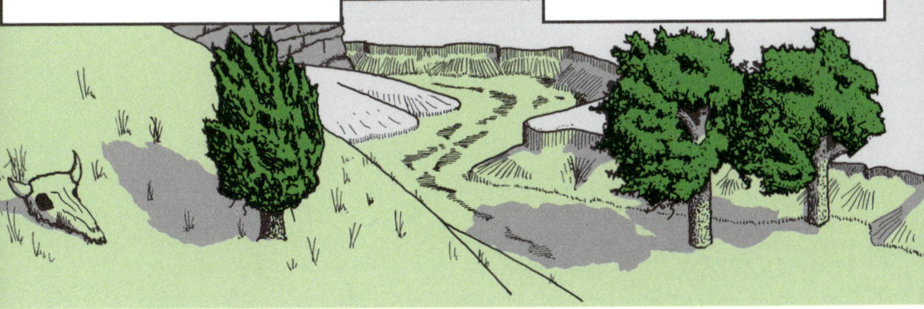

Foothills Zone

Transition Zone—annual rainfall 20 to 25 inches (50 to 65 cm)
Deep-rooted ponderosa pine dominates this zone, except for fine-soiled sagebrush plains in the northern Rockies.

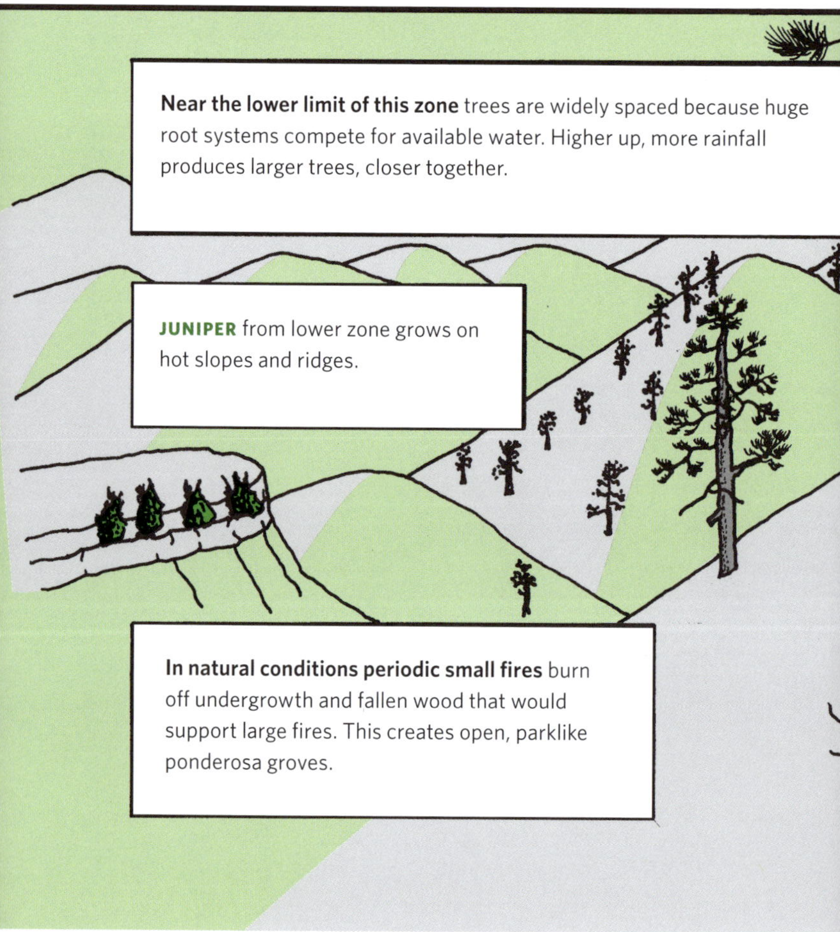

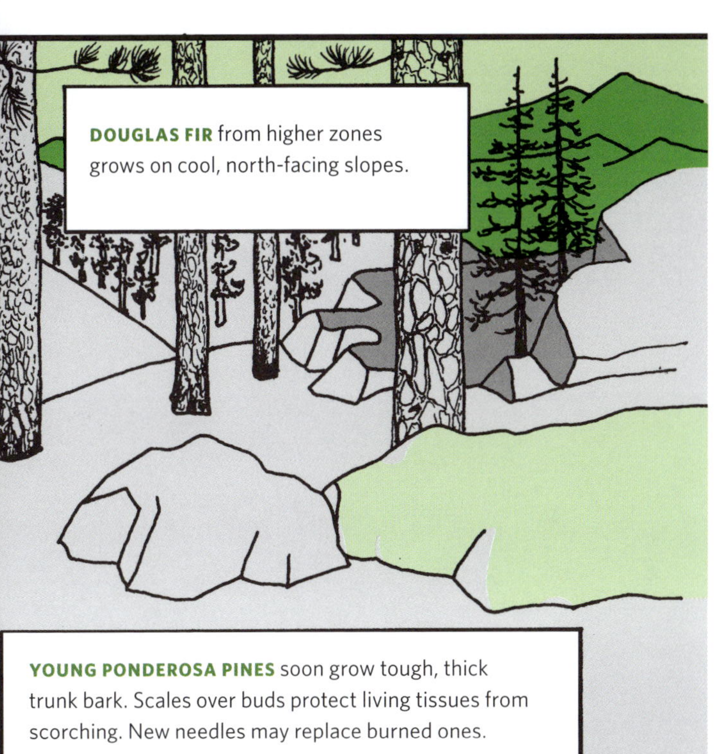

YOUNG PONDEROSA PINES soon grow tough, thick trunk bark. Scales over buds protect living tissues from scorching. New needles may replace burned ones.

Montane Zone

Canadian Zone—annual rainfall 25 to 30 inches (65 to 75 cm)
This zone has enough rainfall to support a dense, shady forest.
Douglas fir dominates except in areas of repeated burning in the
Great Basin and northern Rockies.

Mountain Parks
Flat, treeless "parks" occur in this zone where fine water-deposited soils impede the growth of tree roots. Adjacent slopes with better aerated, porous soils are wooded.

PONDEROSA PINE occupies warm slopes and burned areas in the lower part of this zone.

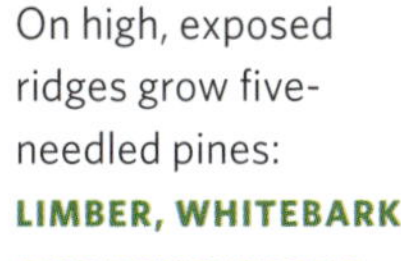

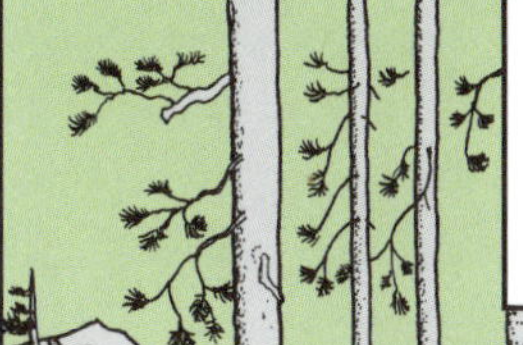

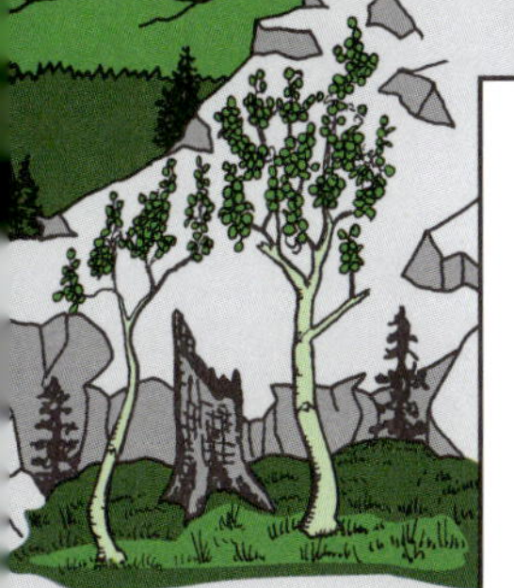

Fire Succession

In this and the next higher zone, **ASPENS AND LODGE-POLE PINES** grow in sunny, burned areas—aspens on the better soils, pines on the poorer. Closely packed, even-aged stands indicate that all trees began growing the same year, after a fire.

Unless fire returns, these trees will eventually improve the soil to the point where seedlings of shade-tolerant trees can grow. Thus pines and aspens are shaded out and replaced by a Douglas fir forest (or spruce-fir forest in the zone above).

Also important in the Montane Zone are white fir (in central and southern Rockies) and white spruce (in the Black Hills of South Dakota, where it replaced Douglas fir).

In western Montana and Idaho, trees of the Pacific coast region—western white pine, western larch, western hemlock, western red cedar, and grand fir—grow in this zone.

Subalpine Zone

Hudsonian Zone—annual precipitation over 30 inches (75 cm)
Engelmann spruce and subalpine fir dominate this zone. To survive
here, trees must resist late and early frosts, and mature new growth
in a short summer.

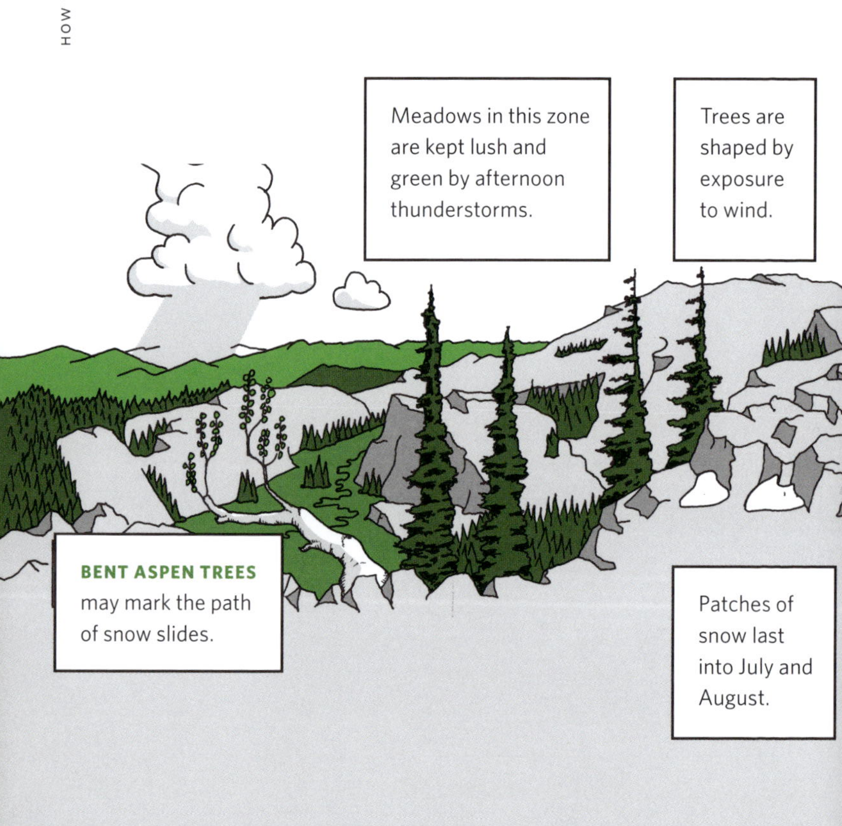

Alpine Zone

No trees here. Only tundra—compact, mounded, or matted vegetation hidden most of the year under snow—can survive.

Fire Damage

The upper Subalpine Zone is too cold for aspens or lodgepole pines. Without them, it takes much longer for a burned spruce-fir forest to replace itself.

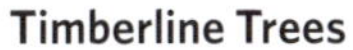

Timberline Trees

Near the upper limit of tree growth, the cold, the drying effects of high-altitude wind and sun, and the abrasion of wind-driven snow kill back all tree growth not protected under snowdrift. An ancient **ENGEL-MANN SPRUCE** may grow into, but never beyond, the drift space in the lee of a rock or ledge.

prevailing wind direction

The place you're mostly likely to find each kind of tree is shown in this chapter by these symbols:

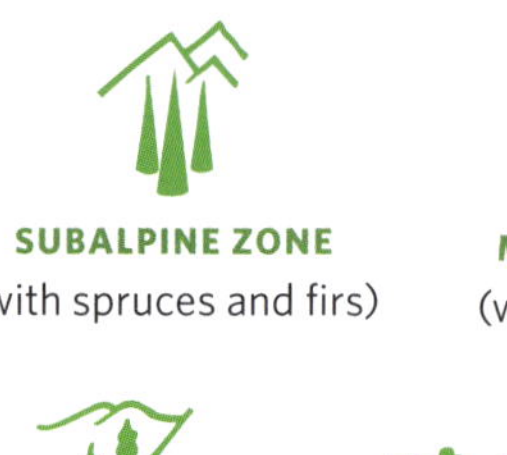

SUBALPINE ZONE
(with spruces and firs)

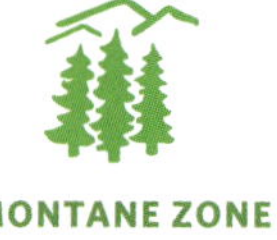

MONTANE ZONE
(with Douglas firs)

FOOTHILLS ZONE
(with ponderosa pines)

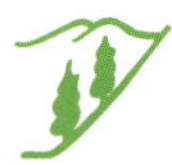

PLAINS ZONE
(among
junipers)

PLAINS ZONE
(along
streambeds)

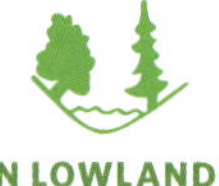

**IN LOWLANDS
OR NEAR WATER**

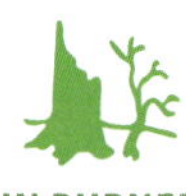

**IN BURNED
AREAS**

Life zones occur at higher altitudes in the southern Rockies than in the north.

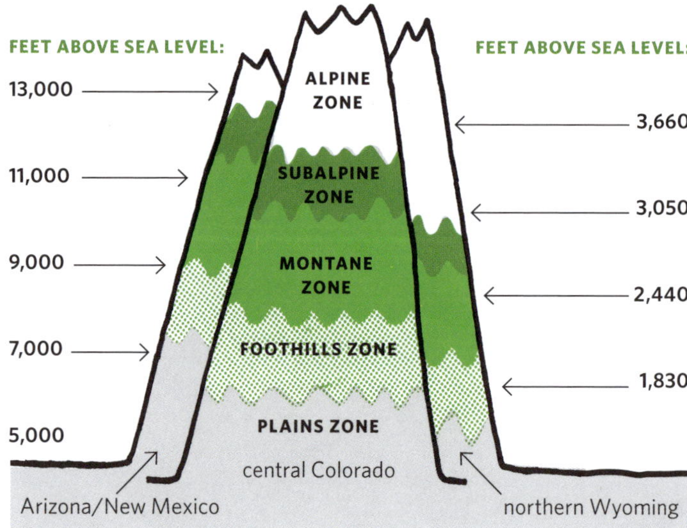

Altitude of the life zones depends also on local climate (microclimate). Southwest-facing canyon slopes and mountainsides are warmed by exposure to full sunlight during the hottest part of the day. Life zones occur at higher elevations there. Northeast-facing slopes are cooler, so life zones extend to lower levels there.

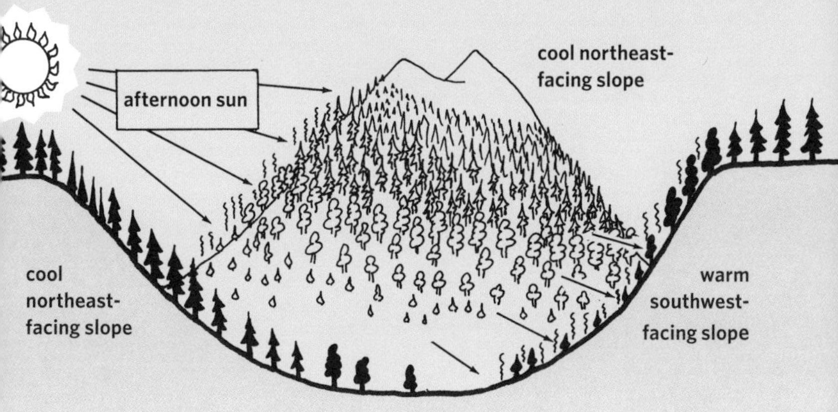

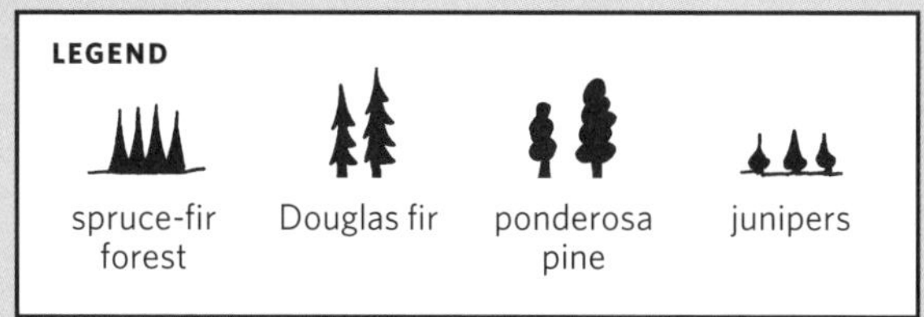

Identifying Fir, Aspen, Cottonwood, and Poplar

If the tree has needles or scalelike leaves, **go to** ⟶ 

If the tree has ordinary leaves, **go to** ⟶ 

 If the tree has needles that are not bundled together, and you can't twirl a single needle because it has a flattened shape, **go to** ⟶ 

 If the needles are thick at the base, it is

ROCKY MOUNTAIN SUBALPINE FIR *Abies bifolia*

Some botanists call this tree *A. lasiocarpa* var. *lasiocarpa*. In the southern Rockies (AZ, NM, southern CO), you may find a closely related tree with thick, corky trunk bark: **CORKBARK FIR** (*Abies arizonica* or *A. lasiocarpa* var. *arizonica*).

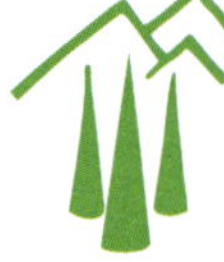

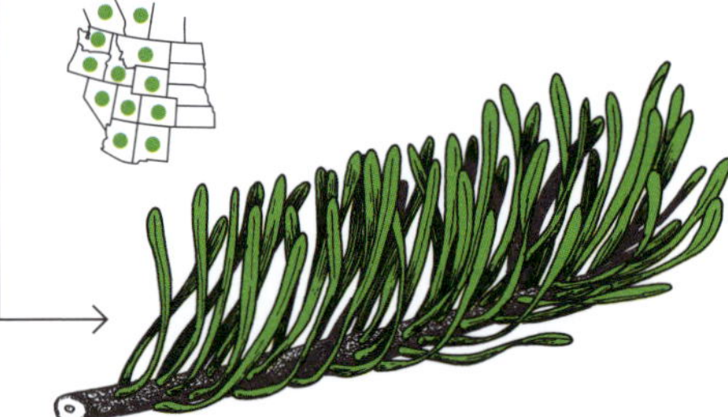

If the needles narrow to a stalk at the base, **go to** ⟶ 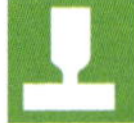

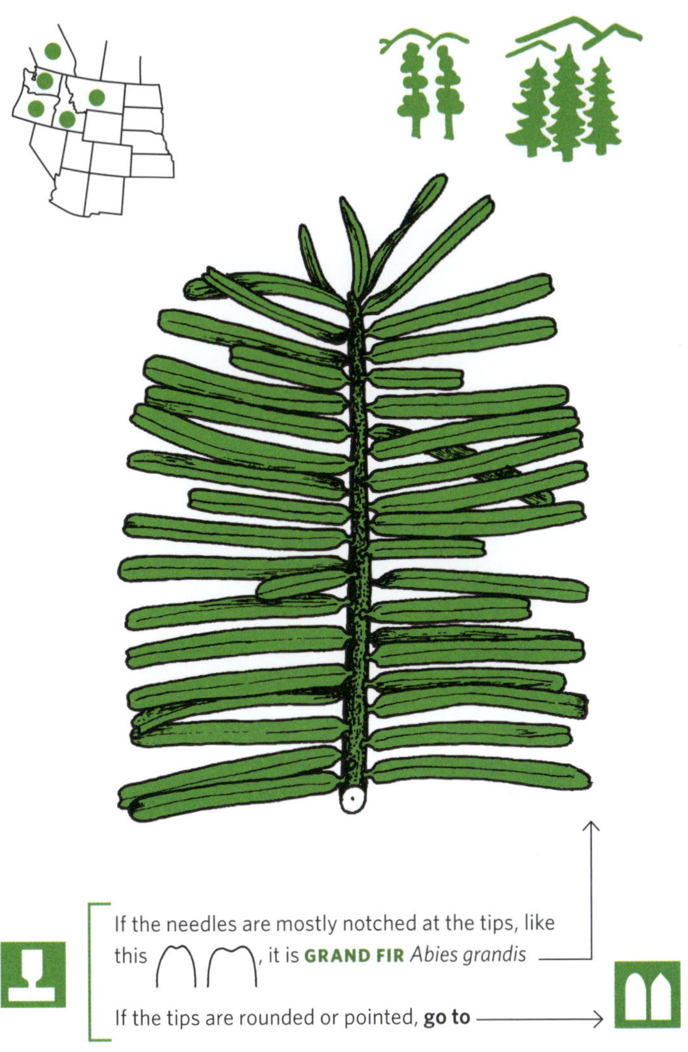

If the needles are mostly notched at the tips, like
this, it is GRAND FIR Abies grandis

If the tips are rounded or pointed, go to

If the needles are flat on top, or grooved, and rounded or blunt at the tip; and if they are longer than ½ inch (12 mm), **go to** ⟶ 

If the topsides of the needles have a single small groove instead of a band, and pulled-off needles have a bit of twig hanging on; or if the tree has dropped cones, it is **ROCKY MOUNTAIN DOUGLAS FIR** *Pseudotsuga menziesii* var. *glauca*

If the topsides of the needles have a delicate, whitish band, and if their bases resemble tiny green suction cups; and the cones do not drop whole, but fall apart on the tree, it is **WHITE FIR** *Abies concolor*

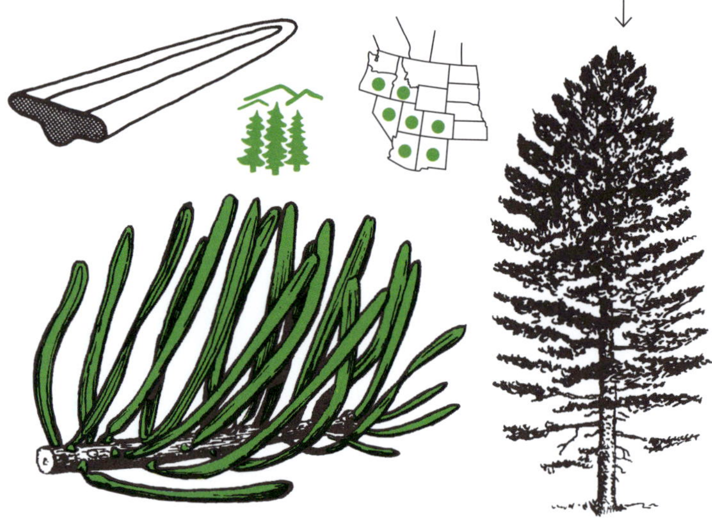

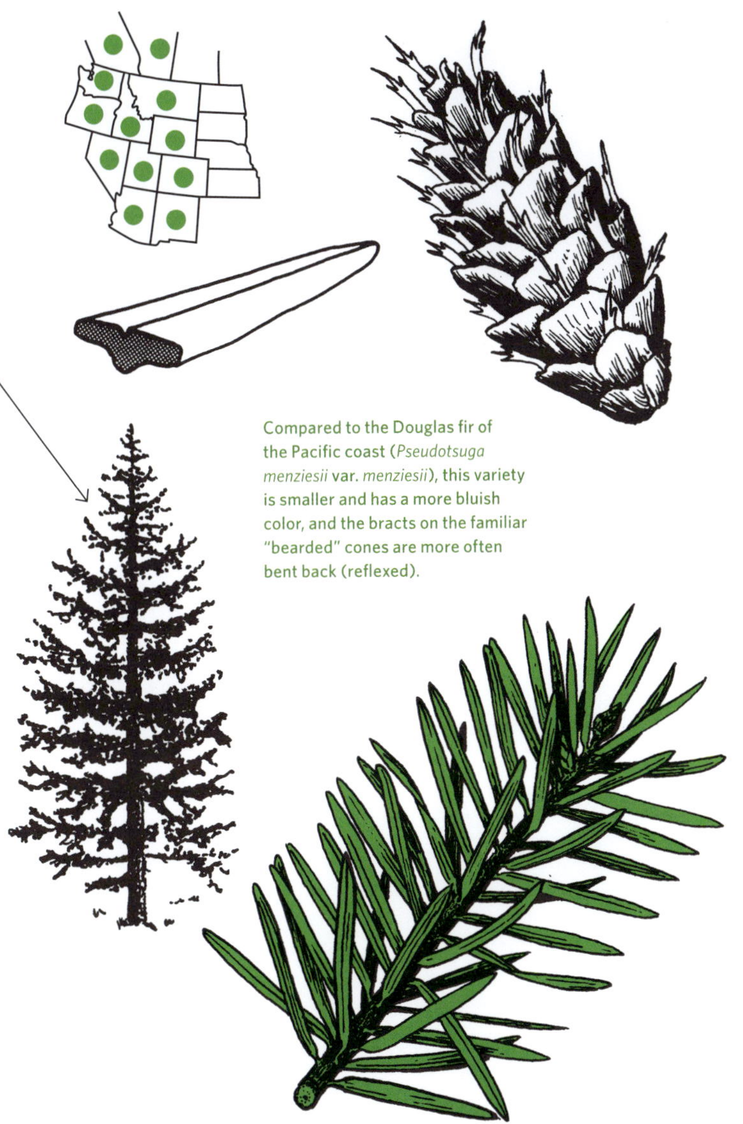

Compared to the Douglas fir of the Pacific coast (*Pseudotsuga menziesii* var. *menziesii*), this variety is smaller and has a more bluish color, and the bracts on the familiar "bearded" cones are more often bent back (reflexed).

If the leaves are simple, not made up of several leaflets, **go to** ⟶

 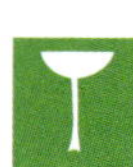

If the leaves do not grow paired on opposite sides of the twigs and aren't lobed; and if the leaf stalks are longer than 1 inch (2.5 cm), **go to** ⟶

If the leaves flutter in a gentle breeze because their stems are flattened where they join the leaves, like this, **go to** ⟶

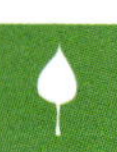

If the stem is not flattened, **go to** ⟶

If the leaf is almost round, it is **QUAKING ASPEN**
Populus tremuloides

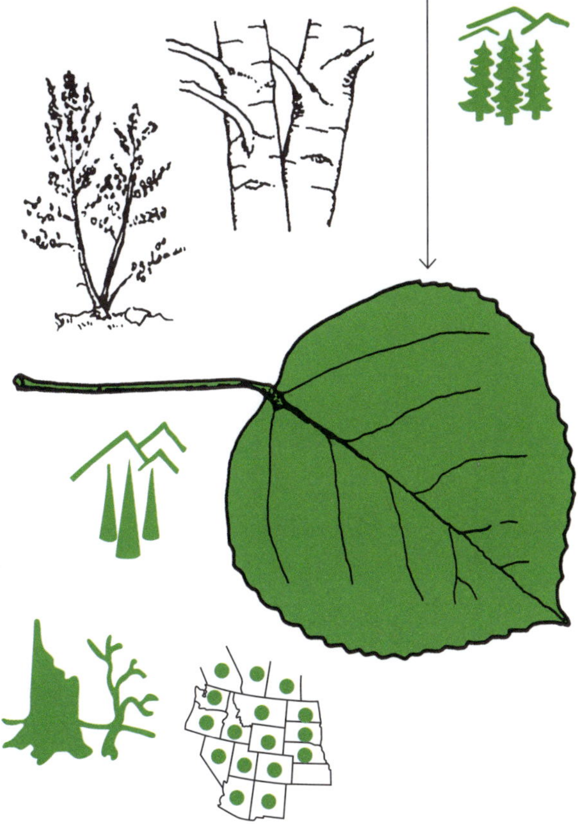

If the leaf is roughly triangular, **go to**

If the leaf blades are over 3 inches (7.5 cm) long, and usually longer than wide, it is **PLAINS COTTONWOOD** *Populus deltoides* ssp. *monilifera*

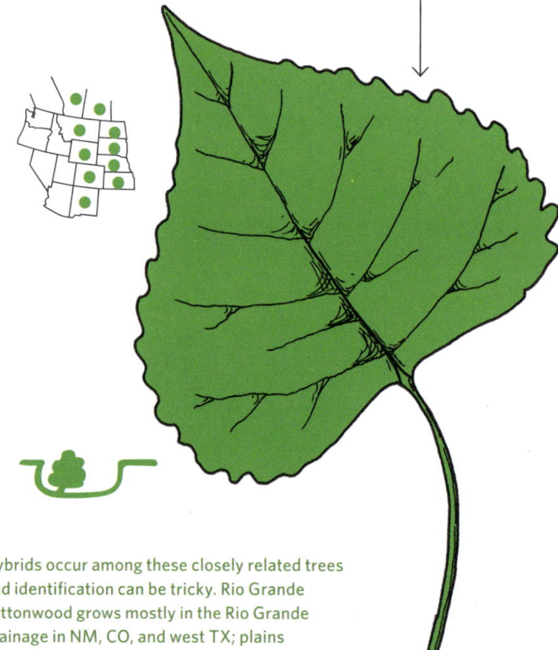

Hybrids occur among these closely related trees and identification can be tricky. Rio Grande cottonwood grows mostly in the Rio Grande drainage in NM, CO, and west TX; plains cottonwood east and north of the Rio Grande; and Fremont to the west, in the Colorado River drainage in CO, UT, and AZ.

If the leaf blades are shorter and mostly wider than they are long, **go to** ⟶ 

If the stalks of seed capsules are shorter, or if the buds are without hair, it is **FREMONT COTTONWOOD** *Populus fremontii* ssp. *fremontii*

If you can find seed capsules that have stalks longer than ½ inch (12 mm), or if the buds on the twig are slightly hairy, it is **RIO GRANDE COTTONWOOD** *Populus deltoides* ssp. *wislizeni*

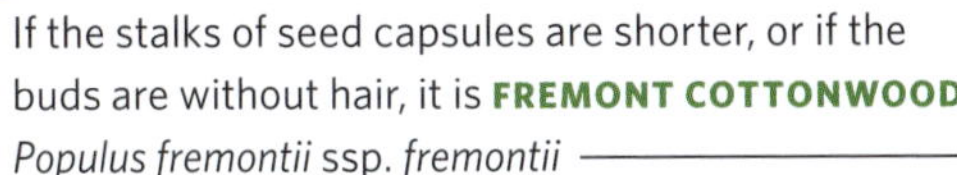

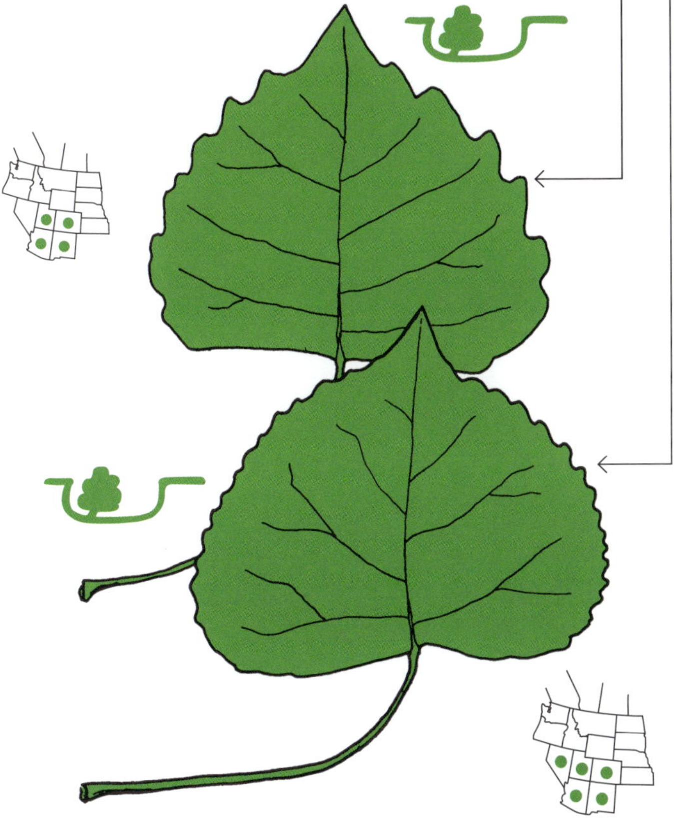

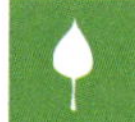

If some of the leaves (stems included) are over 4 inches (10 cm) long, **go to** ⟶ 

If there are no leaves that long, and the bark on young trunks is nearly white, it is **LANCELEAF COTTONWOOD** *Populus × acuminata*

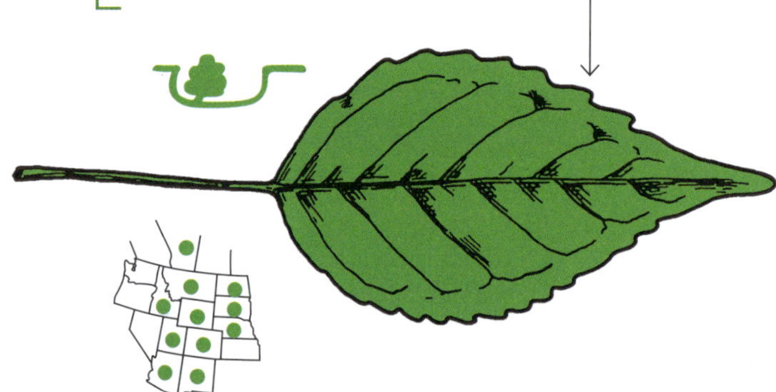

If the leaf stems are over 2 inches (5 cm) long, it is **BALSAM POPLAR** *Populus balsamifera* ssp. *balsamifera*

If they're shorter, it is **WESTERN BALSAM POPLAR** or **BLACK COTTONWOOD** *Populus balsamifera* ssp. *trichocarpa*

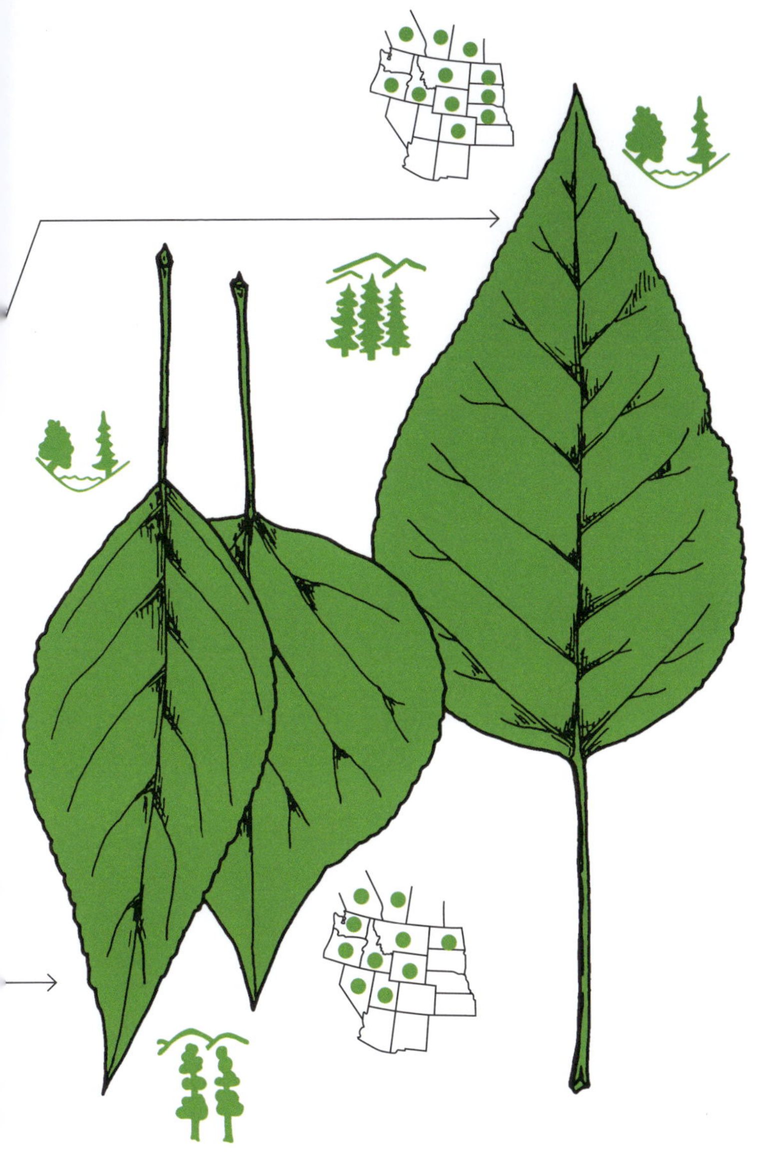

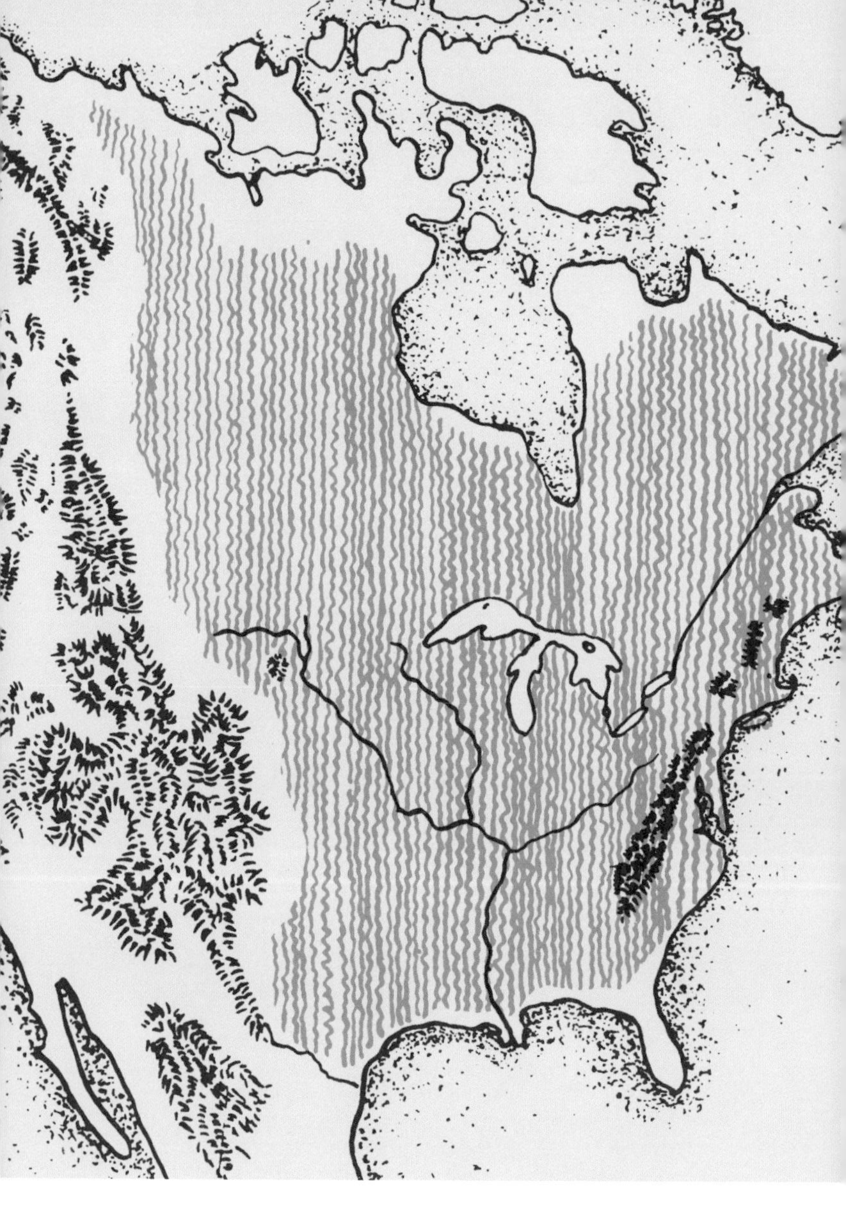

Trees of the Central and Eastern United States and Canada

THE AREAS SHOWN ON THE small green maps beside the trees in this chapter are those in which the tree grows wild. Some of these trees are planted by people over a much wider area.

Some trees have only a narrow distribution.

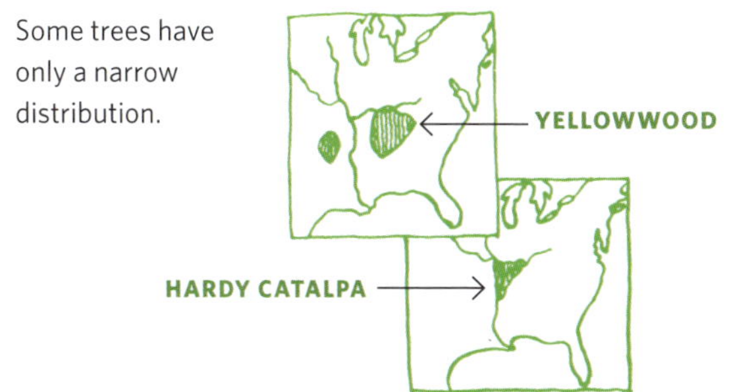

Some trees
have a wide
distribution.

**EASTERN
COTTONWOOD**

Some northern
trees extend
southward along
the mountaintops.

RED SPRUCE

Some southern
trees extend
northward up
the river valleys.

BALD CYPRESS

Some trees
follow the
coast.

LIVE OAK

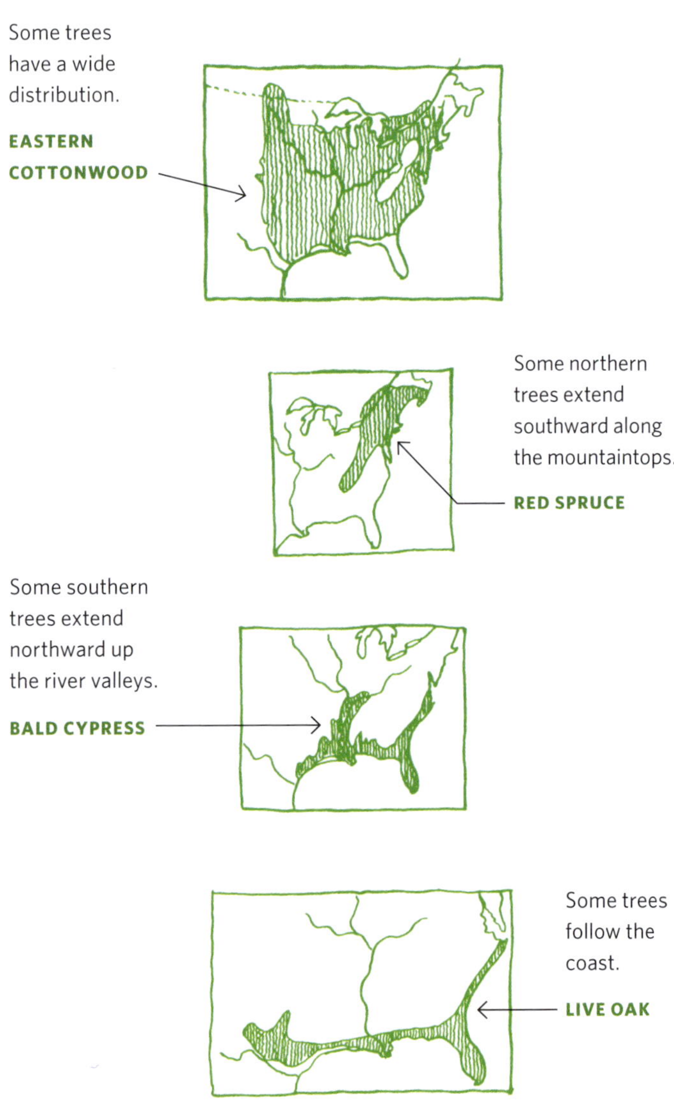

The **HABITAT** of a tree, the place where it is likely to grow naturally, is indicated by the following symbols beside the leaf of each **NATIVE TREE**.

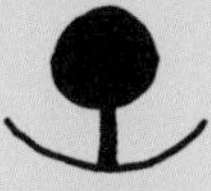

LOWLAND

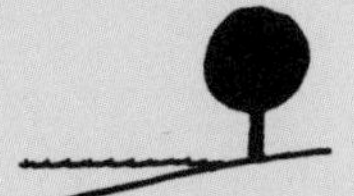

BANKS OF STREAMS AND LAKES

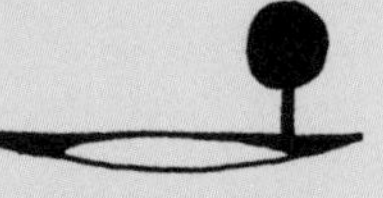

BOGS

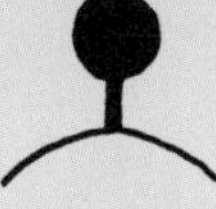

UPLAND

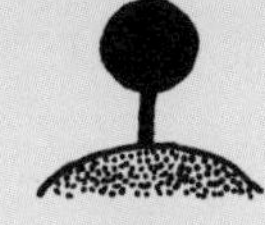

SAND AND GRAVEL

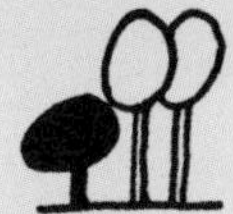

THE EDGE OF THE FOREST

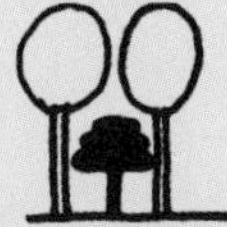

SMALL TREE IN THE SHADE OF TALL TREES

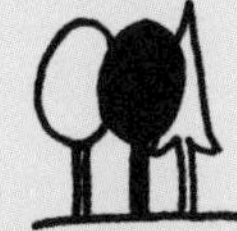

MIXED FOREST OF EVERGREEN AND DECIDUOUS TREES

EVERGREEN FOREST

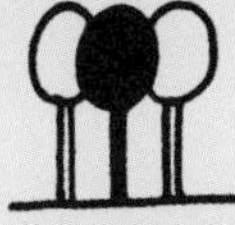

DECIDUOUS FOREST

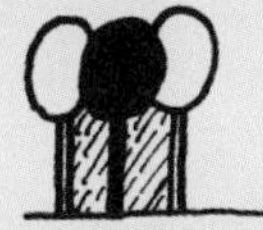

TOLERATING MODERATE SHADE

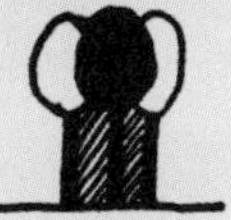

TOLERATING HEAVY SHADE

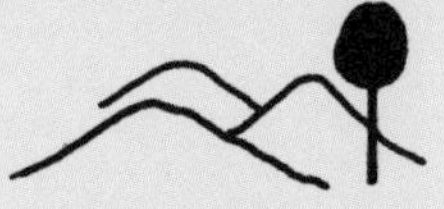

EASTERN MOUNTAINS

65

The PLACE of a tree in association with PEOPLE is shown beside the leaves of introduced trees and some native trees.

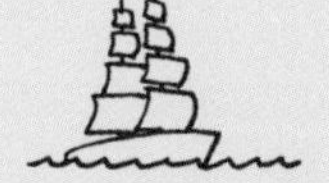

brought from the SEA

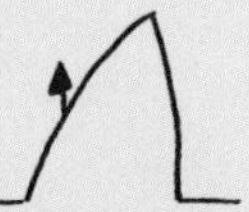

brought from the WESTERN MOUNTAINS

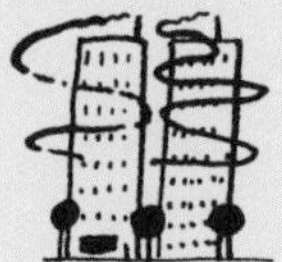

tolerating the conditions of CITIES

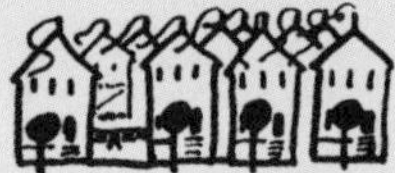

ROADSIDE AND FENCEROWS

(the same kinds of trees are found in fencerows, pastures, and at the edge of the forest)

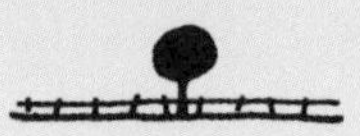

PIONEER IN DIS-TURBED AREAS

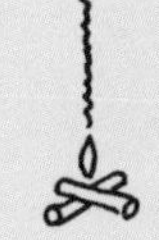

PIONEER AFTER FIRE

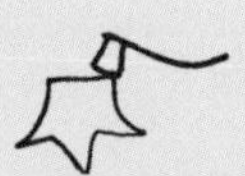

PIONEER AFTER CUTTING OF FOREST

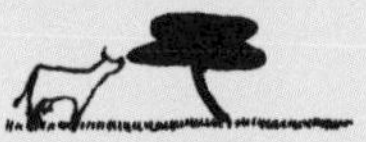

PASTURE PIONEER

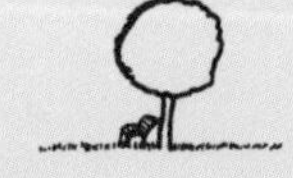

PLANTED IN PARKS, parkways, yards

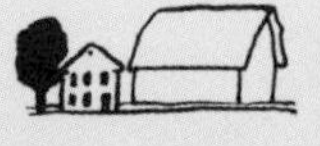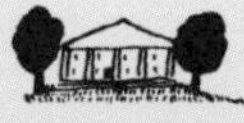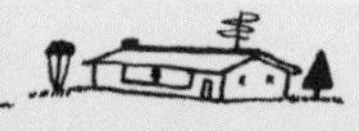

PLANTED ABOUT HOUSES of one sort or another as the times and styles change

Most of the habitats indicated in this chapter are a combination of natural places and places associated with people.

Shapes

The tree shapes shown in this guide are the shapes of mature trees. Tree shape can change with age. Below are shown six stages in the life of an American elm.

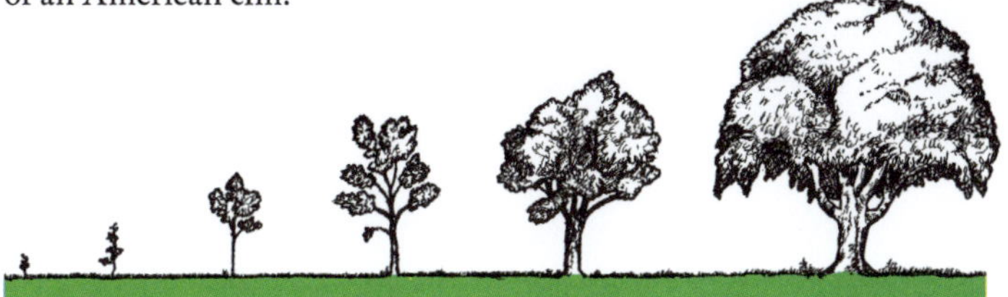

The tree shapes shown below have been modified by:

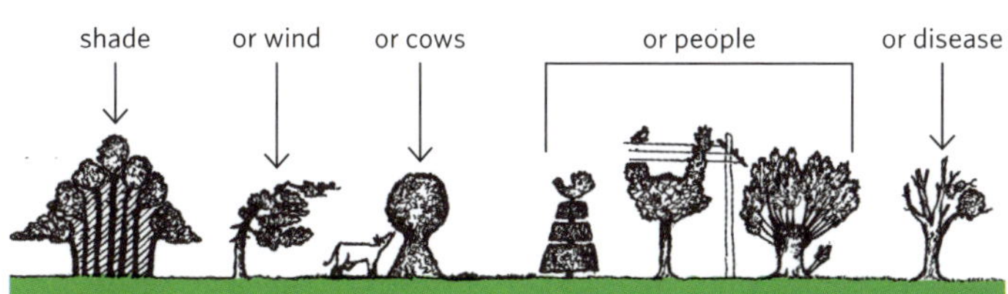

The leaf shape may depend, in part, on its position on the tree:

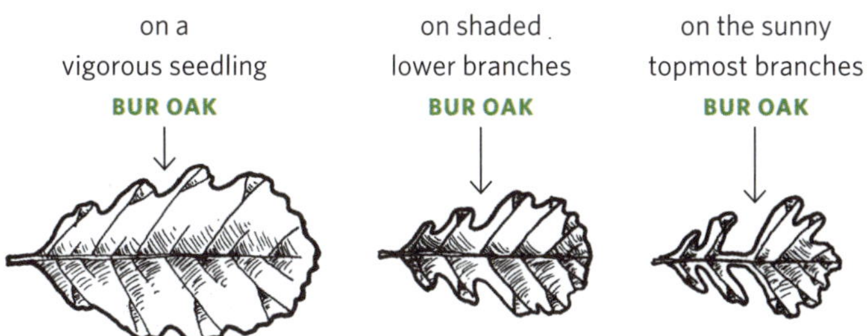

Identifying Spruce

If the tree has long needles, ½ inch (12 mm) to 18 inches (45 cm); and the needles are borne singly, **go to** ⟶ 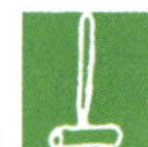

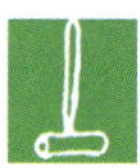 If the needles are stiff, sharp, four-sided, can be twirled between the thumb and finger, and leave the twig rough when they fall off, **go to** ⟶ 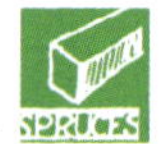

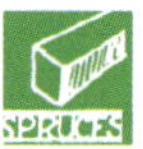 If the needles are not very sharp, nor the branches noticeably horizontal, **go to** ⟶ 

If the needles are *extremely sharp*, and the branches form a flat, horizontal spray, it is **COLORADO SPRUCE** *Picea pungens* ⟶ 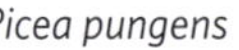

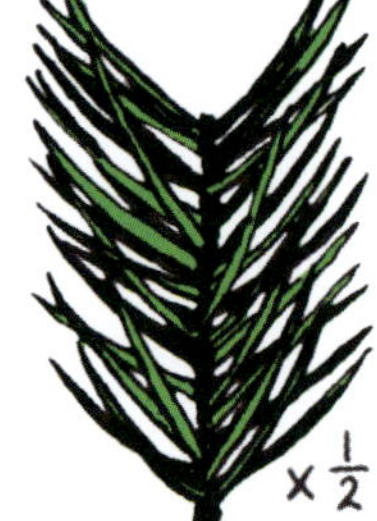

x ½

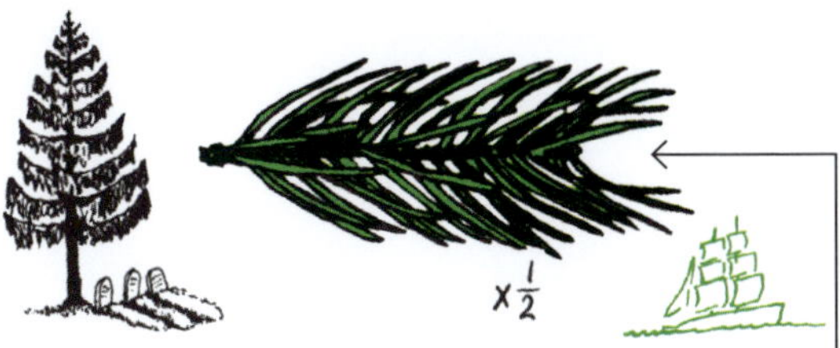

If the branchlets droop and the cones are 4 to 6 inches (10 to 15 cm) long, it is **NORWAY SPRUCE** *Picea abies*

If the branchlets do not droop, and the cones are shorter, **go to** 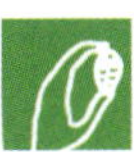

If the needles are short, less than ½ inch (12 mm), and the buds and twigs are hairy, it is **BLACK SPRUCE** *Picea mariana*

If the twigs and buds are hairy but the needles are ½ inch (12 mm) or more, it is **RED SPRUCE** *Picea rubens*

If the needles are blue green, and the twigs are hairless, it is **WHITE SPRUCE** *Picea glauca*

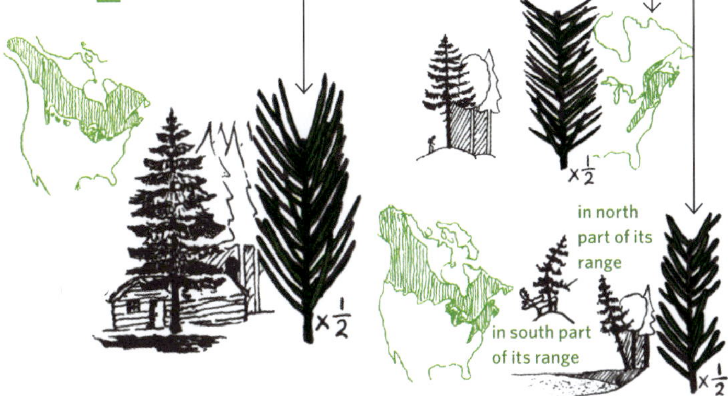

Identifying Willows

If the leaves or buds grow alternately (singly and evenly staggered along the twig), and the leaves are simple, not composed of leaflets, **go to** ⟶ 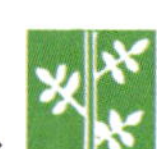

 If the leaf has teeth of any kind (or a wavy margin, or lobes), and the leaf is not evergreen, **go to** ⟶ 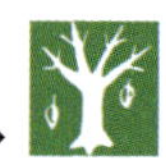

 If the tree has no thorns or thornlike twigs, the leaf margin is toothed (or doubly toothed) continuously (or almost continuously), and if the leaf is not lobed, **go to** ⟶ 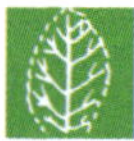

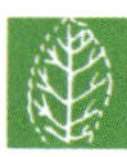 If the teeth are all of about the same size, are more numerous than side veins, and do not terminate the side veins, **go to** ⟶ 

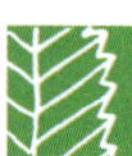 If the leaf stem is not long (shorter than half as long as the blade), the teeth are not somewhat blunted, and the blade is not wide; and the leaf does not have a firm texture and meshed veinlets, **go to** ⟶ 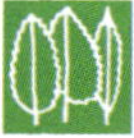

 If the side veins are all of about the same importance (or thickness), **go to** 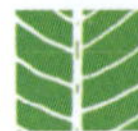

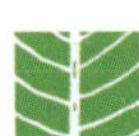 If the leaves are long and narrow; many-veined, tapering gradually and steadily to a long point; and the twigs are slender and limber, with only one scale covering each bud, **go to** 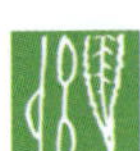

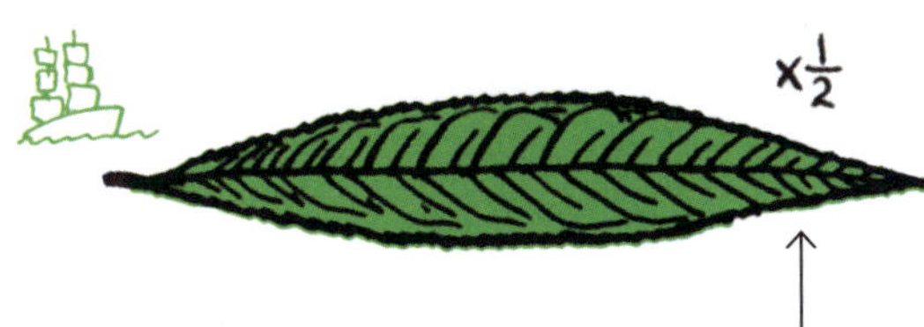

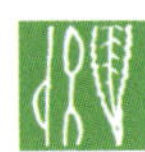 If the leaf has white, silky hairs, and tapers to both ends, it is **WHITE WILLOW** *Salix alba* 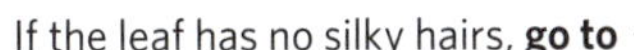

If the leaf has no silky hairs, **go to** 

 If the leaf and twigs are drooping, it is **WEEPING WILLOW** *Salix babylonica*

If the twigs are not drooping, **go to** 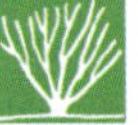

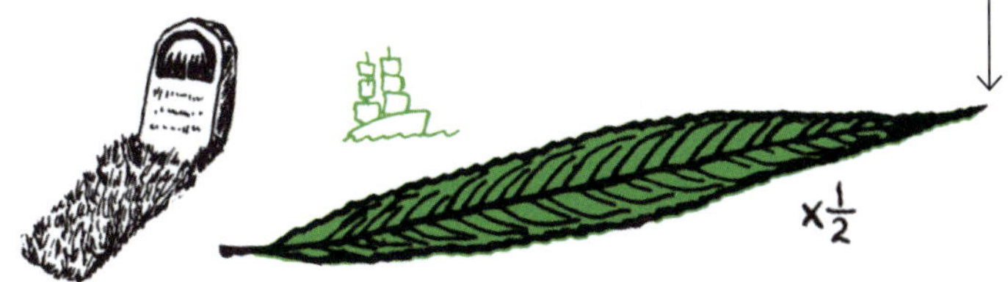

If the leaf is narrow, deep green on both sides, often sickle-shaped, with a downy stem, and a rounded base (vigorous sprouts have leaflike appendages, called stipules, at the base of the leaf stem), it is **BLACK WILLOW** *Salix nigra*

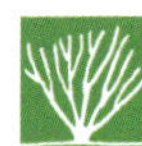

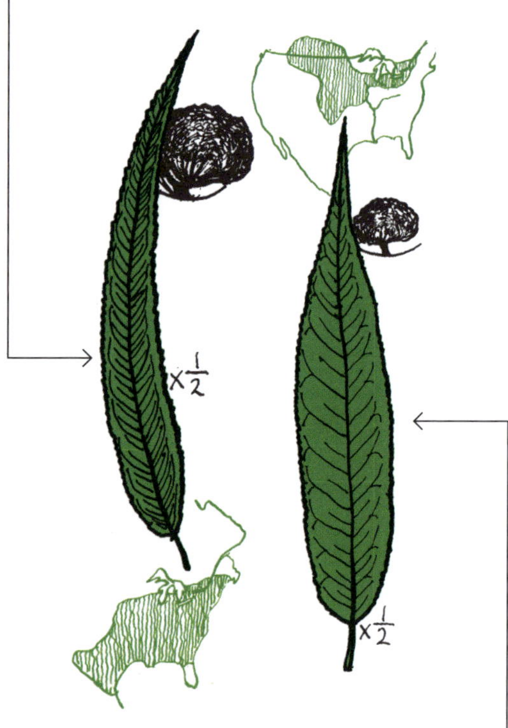

If the leaf is wide at the middle, pale green above, paler beneath, and drooping, it is **PEACH-LEAVED WILLOW** *Salix amygdaloides*

Identifying Deciduous Trees in Winter

The Parts of a Twig

(Twig illustrations in this section are about two-thirds life-size.)

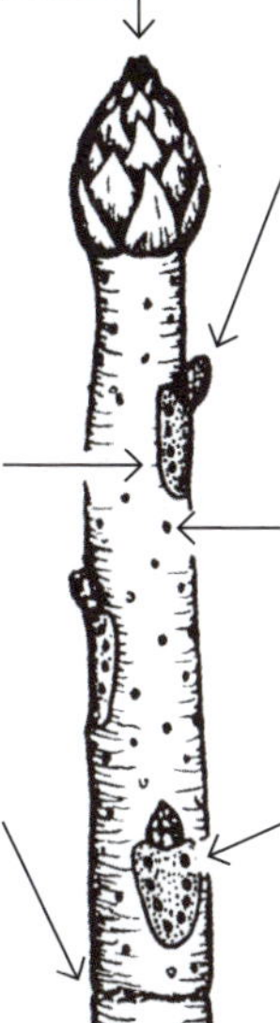

TERMINAL BUD

From this bud the twig will grow longer in spring. It is often much larger than the lateral buds but is absent from the twigs on some trees.

LATERAL BUD

From this bud a side branch will grow—shorter than the growth from the terminal bud.

LEAF SCAR

A leaf was attached here last summer. Leaf scars have many shapes and sizes.

LENTICELS

These cork-filled pores permit the green, living inner bark to breathe.

BUD SCALE SCAR

From this scar the scales of last winter's terminal bud fell last spring. From the tip of the twig to the first bud scale scar is one year's growth. From the tip back to the second scar is two years' growth, and so on. The scale scars encircle the twig.

VEIN SCAR

These dots on the leaf scars are the broken-off, cork-filled ends of the tubes that supplied water to the leaf.

PITH

This is the soft, inner core of the twig.

Here are the kinds of places where certain trees are likely to grow—the habitat:

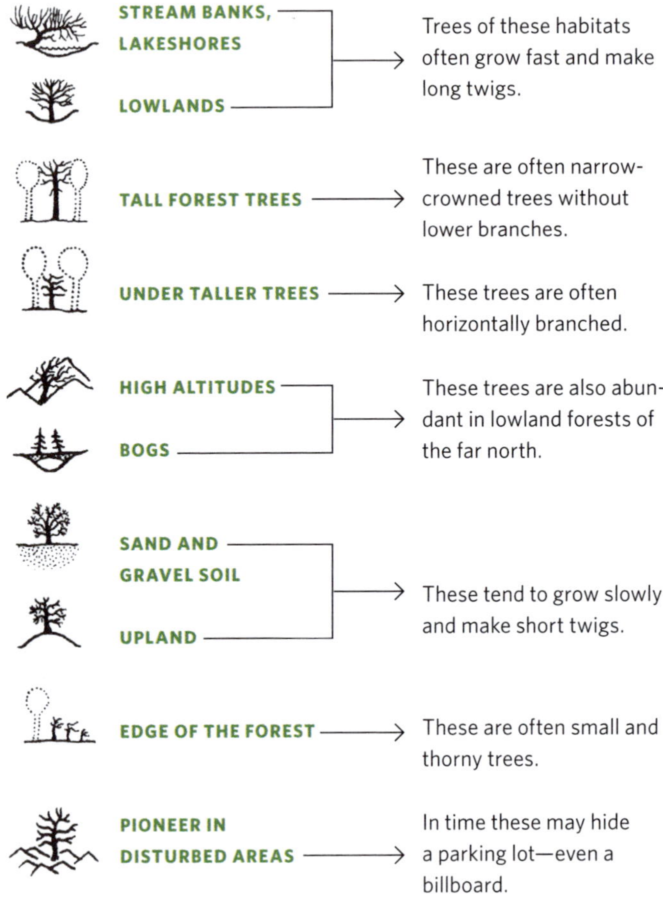

The place where some trees grow depends on what people have done.

Trees are planted in parks, yards, and around houses

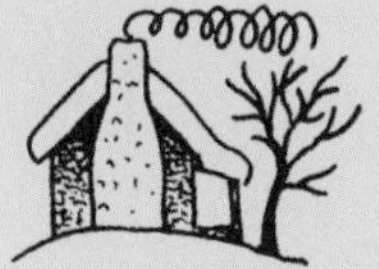

. . . where winters are cold,

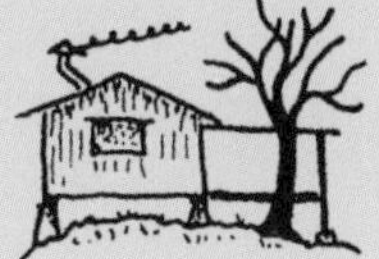

. . . where winters are mild.

Some trees came
from across the sea.

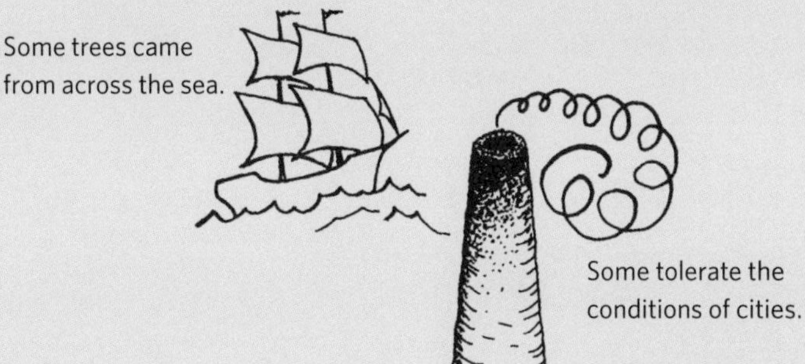

Some tolerate the
conditions of cities.

Pioneer trees begin to grow in a
place after soil has been disturbed,

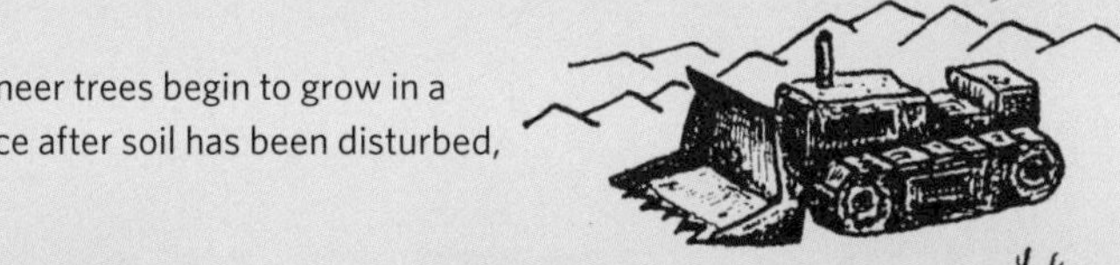

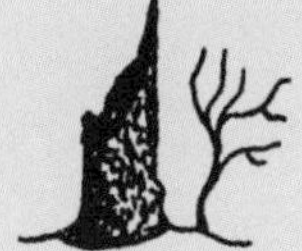

after a fire,

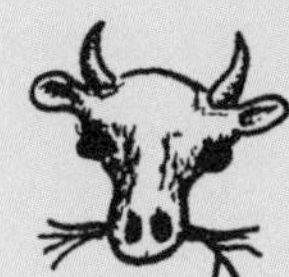

after cattle have grazed,

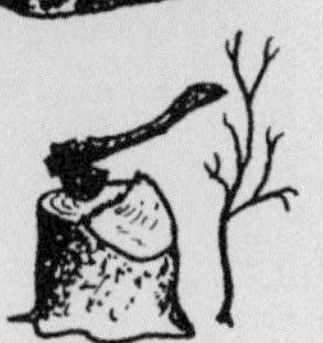

after logging.

Sometimes a conspicuous feature or the location of a tree makes identification easier.

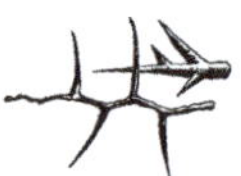

Smooth, conspicuous thorns?

Try: Honey Locust, Hawthorns

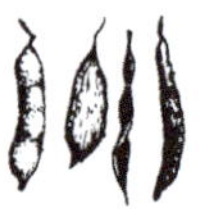

Pods?

Try: Legumes

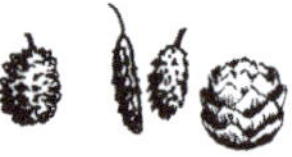

Cones?

Try: Alder, Birch, Larch

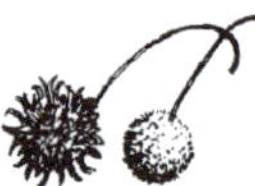

Hanging balls?

Try: Sycamore, Sweet Gum

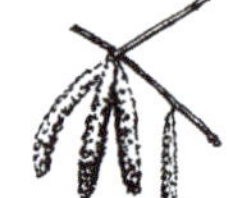

Dangling catkins?

Try: Alder, Birch, Ironwood

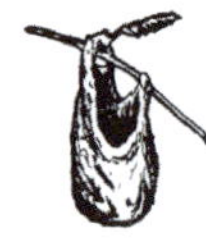

Baltimore oriole's nest?

Try: Elm

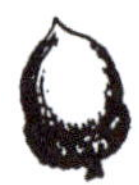

Acorns?

Try: Oaks

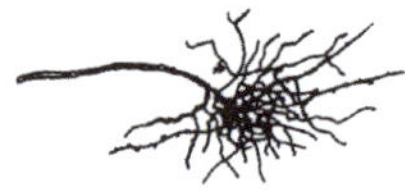

Tangled twigs (witches'-brooms)?

Try: Hackberry

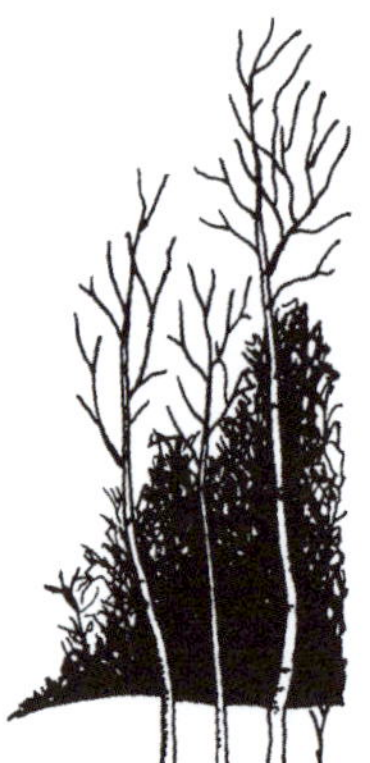

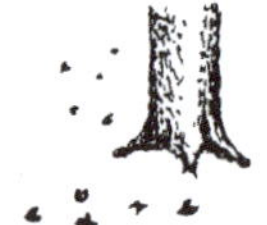

Empty nut husks under the tree?

Try: Hickory, Buckeye, Horse Chestnut

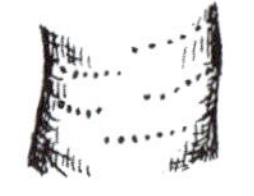

Yellow-bellied sapsucker holes?

Try: Sugar Maple, Mountain Ash, Linden

Slender trees with white or gray bark?

Try: Birch, Aspen

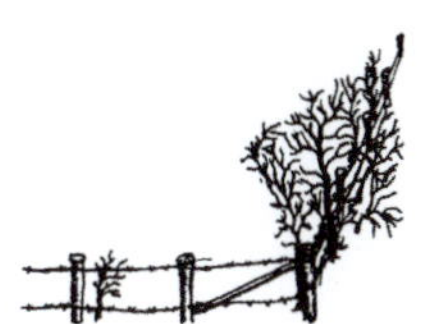

In a fencerow thicket?

Try: Hawthorn, Plum, Sumac, Cherry, Osage Orange

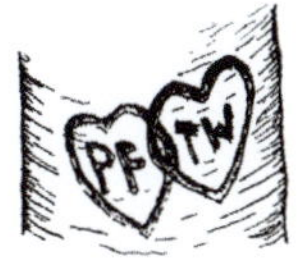

Smooth bark on trunk?

Try: Birch, Beech, Aspen

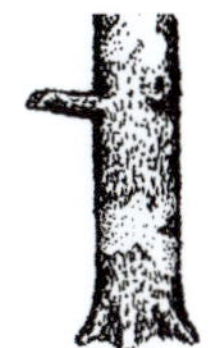

Patches of rubbed-off-looking bark on trunk?

Try: White Oak

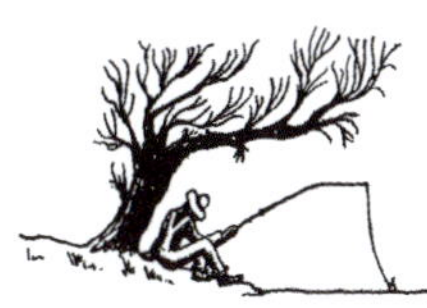

Does the tree reach out over a: fisherman, muskrat, rowboat, canoe?

Try: Willow

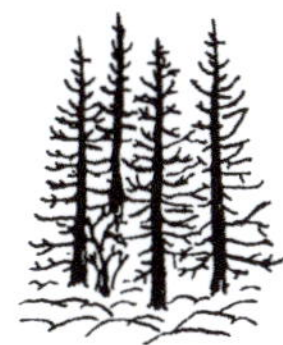

Boggy, unstable, shaky ground?

Try: Tamarack, Poison Sumac

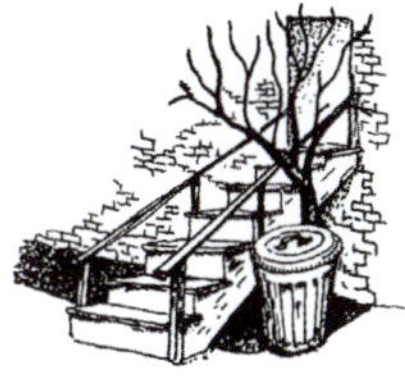

Polluted urban air?

Try: Ailanthus, Mulberry, Sycamore, Willow

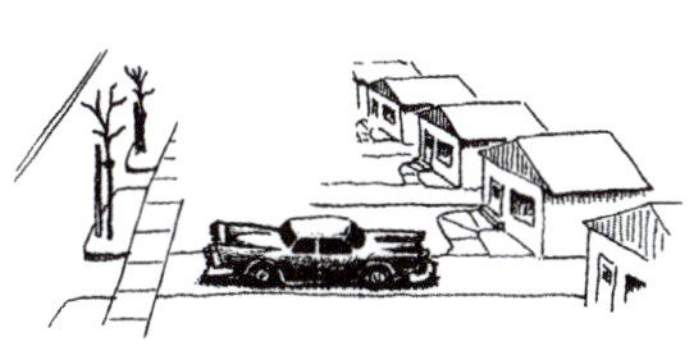

In a ticky-tacky tract?

Try: Thornless Honey Locust, Pin Oak, Magnolia, Birch, Chinese Elm

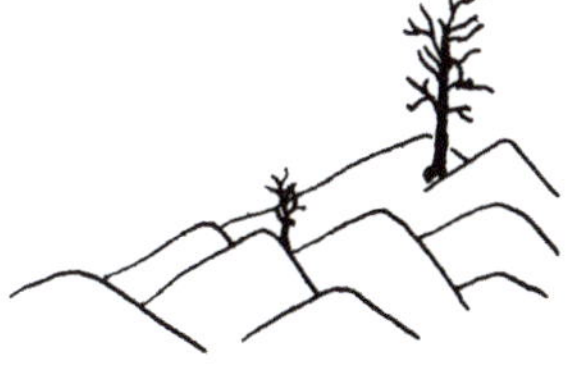

After strip mining, bulldozers, army engineers, urban "renewal"?

Try: Cottonwood, Box Elder

Identifying Maples in Winter

If the tree is not a conifer, has twigs with leaf scars opposite each other,

and if there are no more than two leaf scars around the twig at the same level, **go to** ⟶

If the twig is less than ¼ inch (6 mm) thick; and the terminal bud is not oval and conspicuous, **go to** ⟶

If the terminal buds are not rough and dry; and the leaf scars are narrow, inconspicuous, with three bundle scars, **go to** ⟶
(If you find winged seeds, they will not be symmetrical.)

If the leaf scars are somewhat V-shaped, with three bundle scars; and the end bud is egg shaped or cone shaped, it is a MAPLE. **Go to** ⟶

If the buds are red; and the new growth on the twigs is red or red-brown, **go to** ⟶

If the buds and the new growth are not red, **go to** ⟶

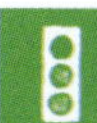

If the tree is not shrub-like, and the flower buds are globular and conspicuous, **go to** ⟶ 

If the tree is shrub-like, and is an understory tree of the forest, with densely hairy twigs (often with fruits that hang on into the winter), it is **MOUNTAIN MAPLE** *Acer spicatum*

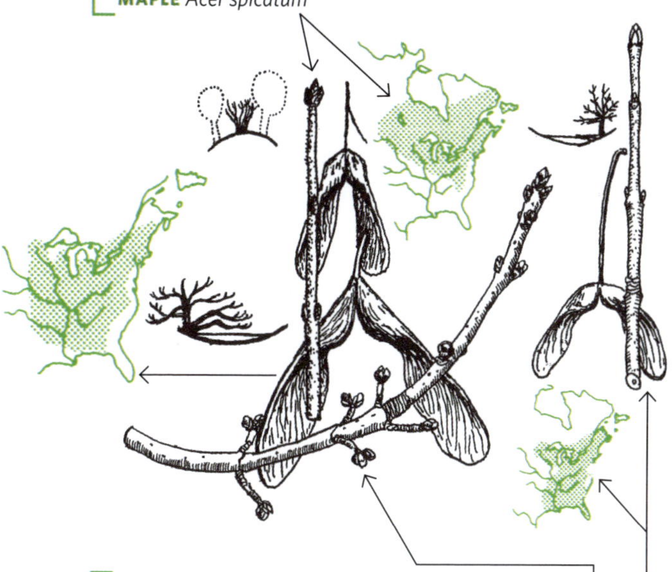

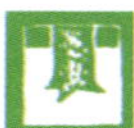

If the twigs give a rank smell when broken; and the bark on old trunks peels in great shaggy flakes; and the bud scales are pointed, it is **SILVER MAPLE** *Acer saccharinum* ⎯⎯⎯⎯

If the younger trunks are smooth, very pale gray, with darker markings; and if the twigs do not have a rank smell when broken; and if the bud scales are rounded, it is **RED MAPLE** *Acer rubrum* ⎯⎯⎯⎯

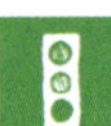

If the buds are whitish and woolly; and the twig area purplish or greenish; and the leaf scars from opposite sides of the twig meet at their tips, forming a toothlike point, it is **BOX ELDER** *Acer negundo* ———

If the buds are not whitish and woolly, **go to** ——————→ 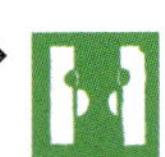

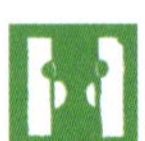

If the twigs are stout, with end buds making a broad, low triangle that is smooth, green, or partly green, **go to** ——————→ 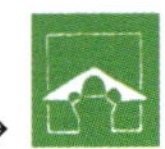

If the buds are longer-pointed, and brown to reddish brown, **go to** ——————→ 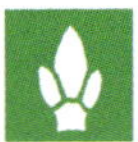

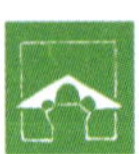

If the buds are marked with green and some red-brown; and the fruits that may still be clinging to the tree are joined in pairs at a wide angle so that they resemble a miniature coat hanger, it is **NORWAY MAPLE** *Acer platanoides* ———
(In early spring the tree is conspicuous with clusters of yellow-green flowers. At that time one can easily identify this tree by its milky juice.)

If the end buds are big and green, it is **SYCAMORE MAPLE** *Acer pseudoplatanus* ——————
(No milky juice.)

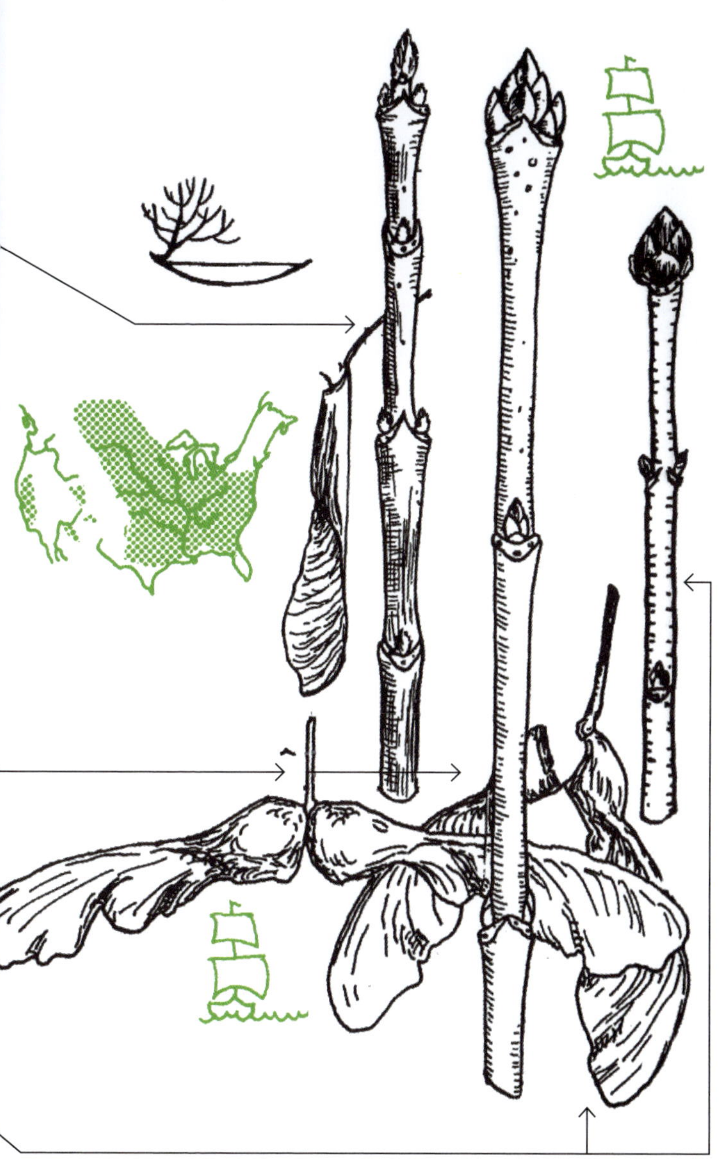

If the tree is small, shrub-like (usually on sandy soil); and the bark is striped with light lines; and the buds have short stalks; and the end bud has two scales that meet at their edges but do not overlap; and the bud scales are keeled, it is **STRIPED MAPLE** *Acer pensylvanicum* ———

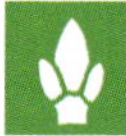

If the tree is not shrub-like; and if the buds are brown to grayish brown; the twigs slender; the bark stone gray with large scales peeling off sideways on the old trunks (usually marked by horizontal rows of yellow-bellied-sapsucker holes), it is **SUGAR MAPLE** *Acer saccharum* ———

If the twigs are rough-hairy; and the buds are dark gray to black; and the bark is very dark, almost black, it may be a tree similar to sugar maple called **BLACK MAPLE** *Acer nigrum* ———

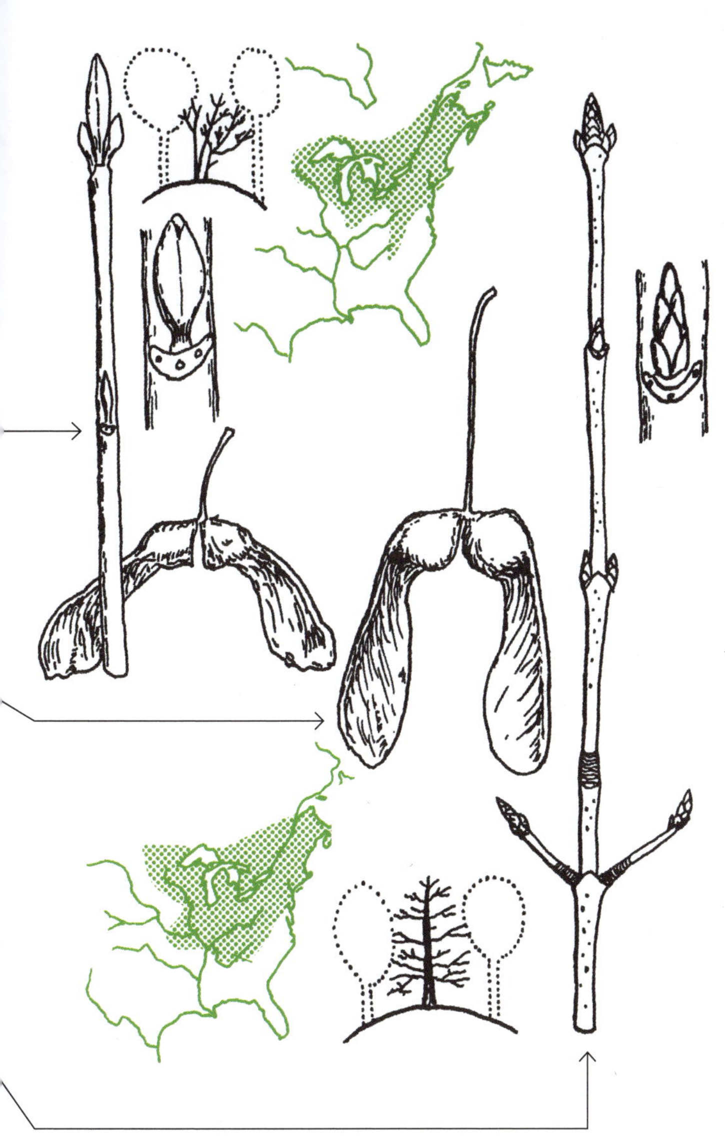

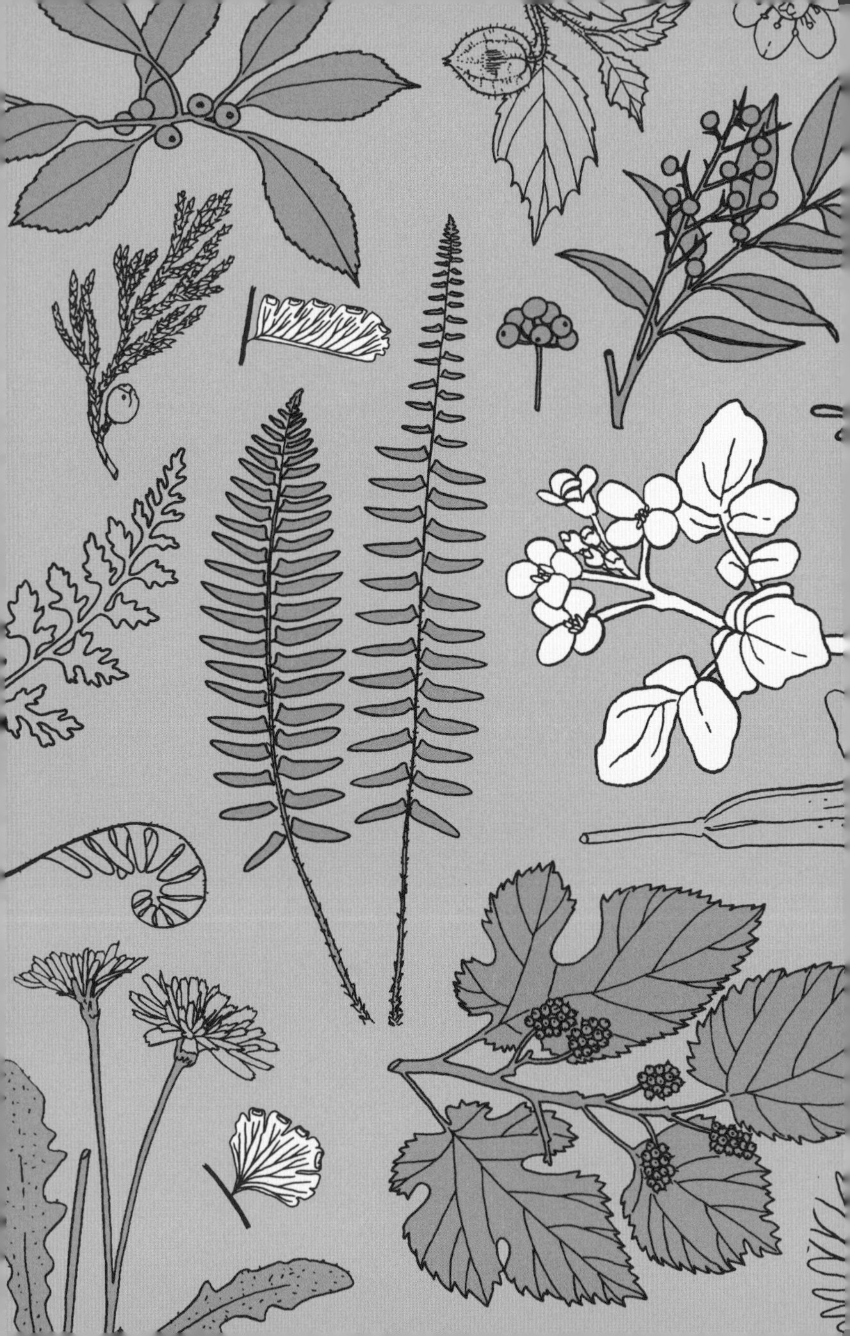

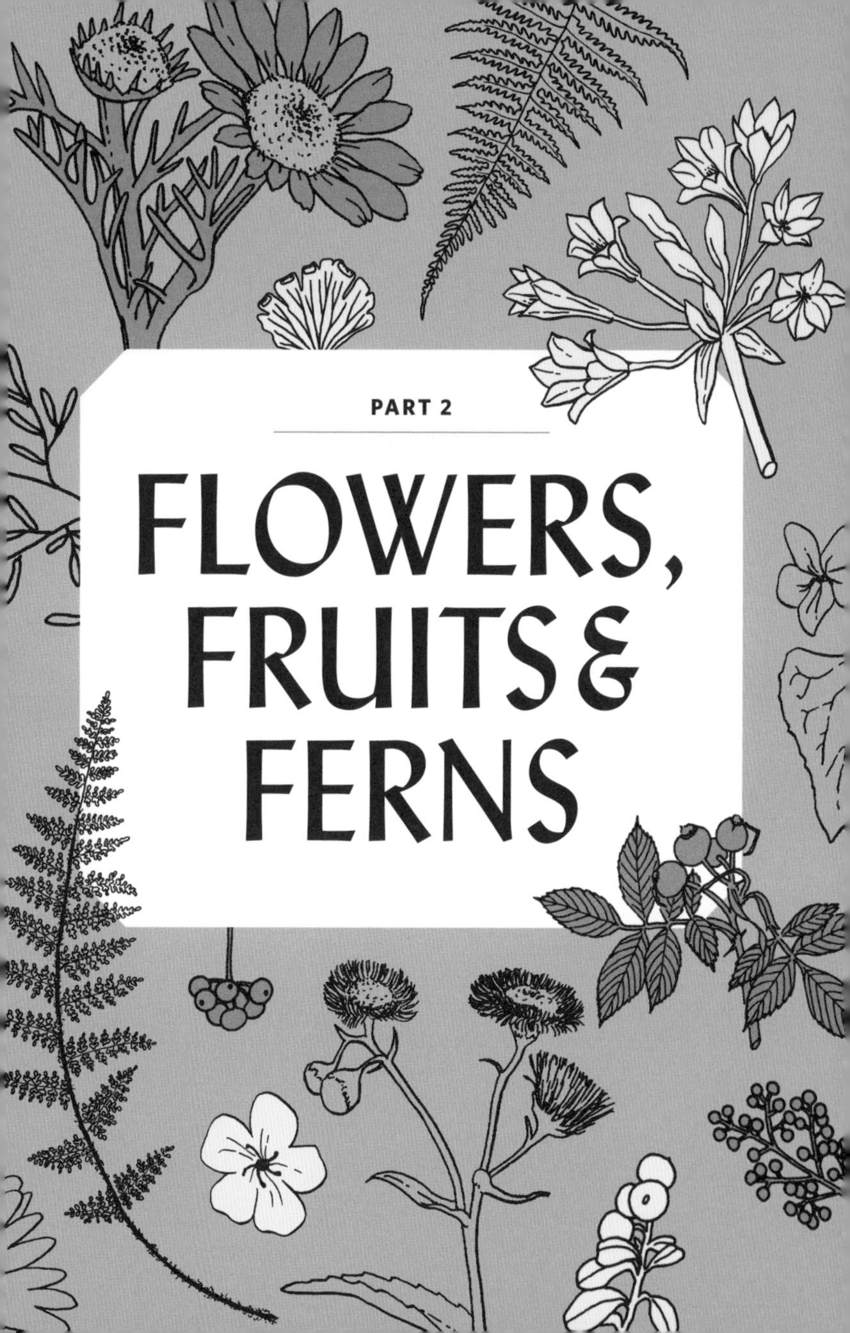

PART 2

FLOWERS, FRUITS & FERNS

Flowers

Terms That Describe Flowers

Flower Parts

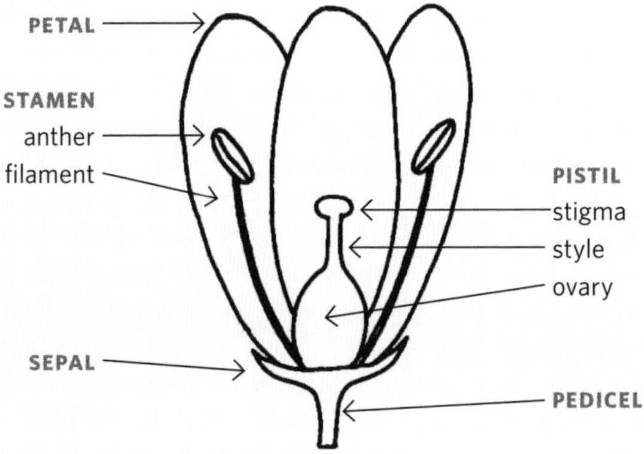

All sepals = CALYX
All petals = COROLLA
Sepals plus petals = PERIANTH

If petals are united to form a tube, corolla can be removed from flower in one piece. If petals are separate, you can remove them one at a time.

Flowers may be alone, solitary, or grouped into one of the following kinds of inflorescence:

- crowded tightly together, unstalked, in a **HEAD** $\longrightarrow$ (usually with bracts under the head)

- along a stem, with stalks, in a **RACEME** $\longrightarrow$

- along a stem, unstalked, in a **SPIKE** $\longrightarrow$

- in a branched arrangement, a **PANICLE**

- clustered with all stalks arising from one point, an **UMBEL** (An umbellet is an umbel within an umbel.)

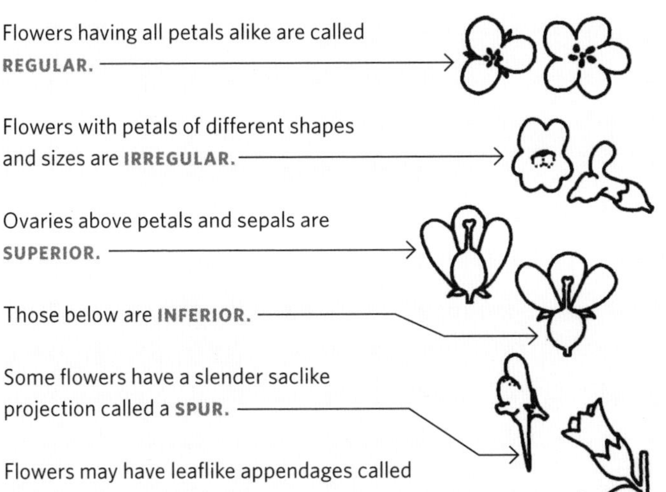

Flowers having all petals alike are called **REGULAR.** $\longrightarrow$

Flowers with petals of different shapes and sizes are **IRREGULAR.** $\longrightarrow$

Ovaries above petals and sepals are **SUPERIOR.** $\longrightarrow$

Those below are **INFERIOR.** $\longrightarrow$

Some flowers have a slender saclike projection called a **SPUR.** $\longrightarrow$

Flowers may have leaflike appendages called **BRACTS.** $\longrightarrow$

Terms that describe leaves

Leaves have

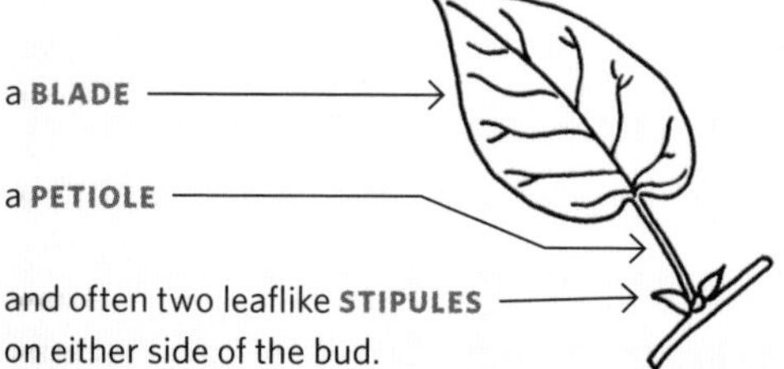

a **BLADE**

a **PETIOLE**

and often two leaflike **STIPULES**
on either side of the bud.

*(Sheathing stipules are united around
the stem.)*

Leaf veins can be

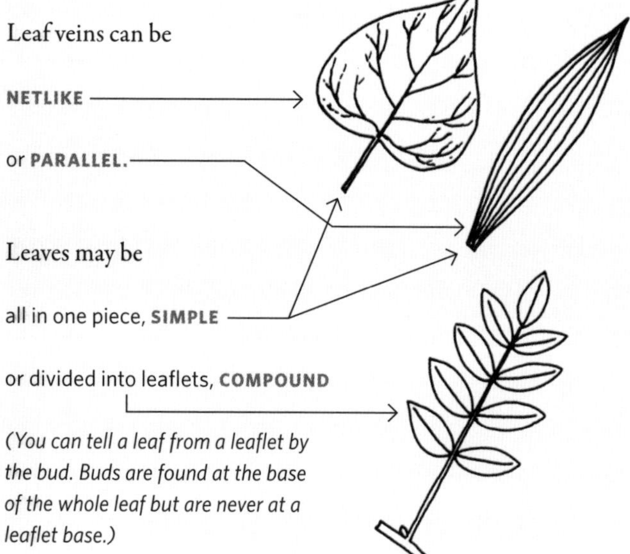

NETLIKE

or **PARALLEL.**

Leaves may be

all in one piece, **SIMPLE**

or divided into leaflets, **COMPOUND**

*(You can tell a leaf from a leaflet by
the bud. Buds are found at the base
of the whole leaf but are never at a
leaflet base.)*

Leaflets may be arranged

PALMATELY

or **PINNATELY.**

(Tendril may replace leaflet.)

Leaves are attached at nodes
and may be

OPPOSITE

ALTERNATE

WHORLED

or **BASAL.**

Leaf edges may be

ENTIRE

TOOTHED

or **LOBED.**

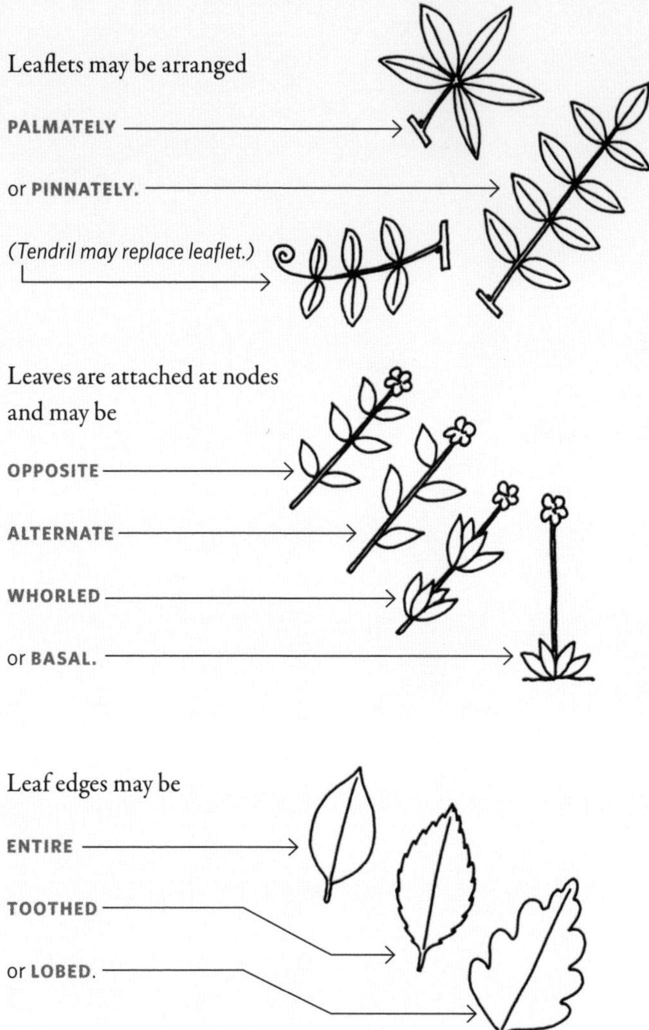

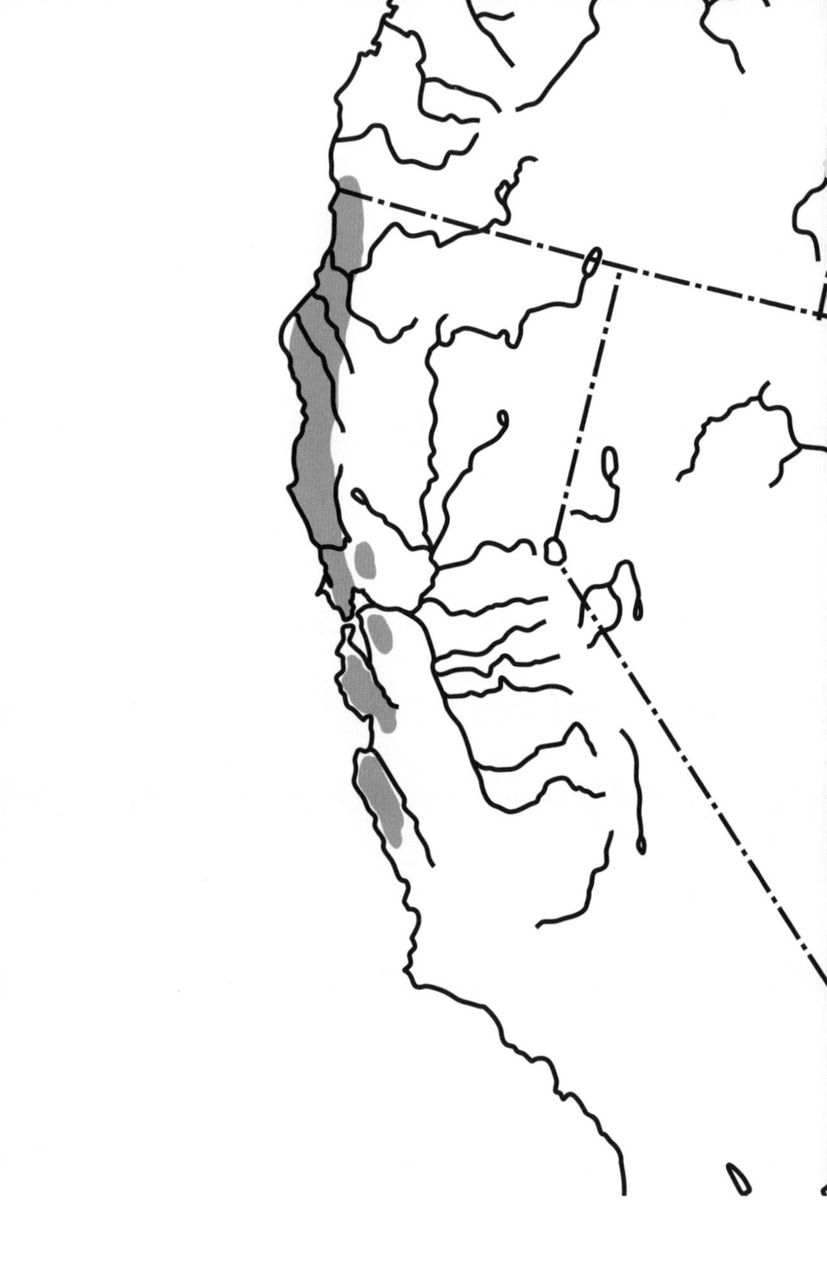

Flowers of California's Coastal Fog Belt

THESE HABITAT SYMBOLS show the kinds of places within the redwood region where you're most likely to find each flower.

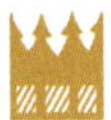

SHADY FOREST

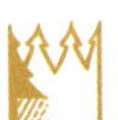

EDGE OF FOREST

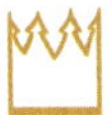

OPEN PLACES

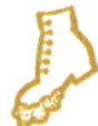

WET PLACES

Identifying the Sunflower Family

If the plant has green parts; the main veins form a netlike pattern; and there is a mass of tightly packed stemless, small flowers, which altogether look like one flower, it is from the **SUNFLOWER FAMILY** *Asteraceae*, a very large family, with flowers borne on heads (see diagram). Pappus takes the place of sepals. Each ovary bears one seed. ⟶ 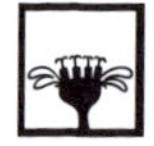

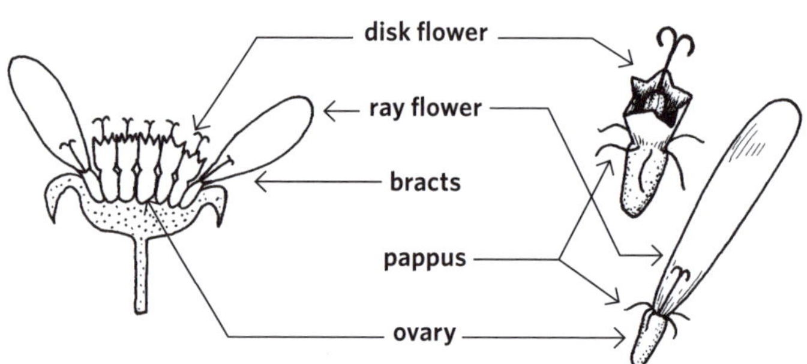

If disk and ray flowers are both present, **go to** $\longrightarrow$

If ray flowers only, **go to** $\longrightarrow$

If disk flowers only, **go to** $\longrightarrow$

If flowers are yellow, and leaves are toothed, it is one of many species of *Agoseris* or *Microseris*. The one pictured here is **WOODLAND DANDELION** *Agoseris apargioides*

If the flowers are white and leaves are entire (not toothed), it is **WHITE HAWKWEED** *Hieracium albiflorum*

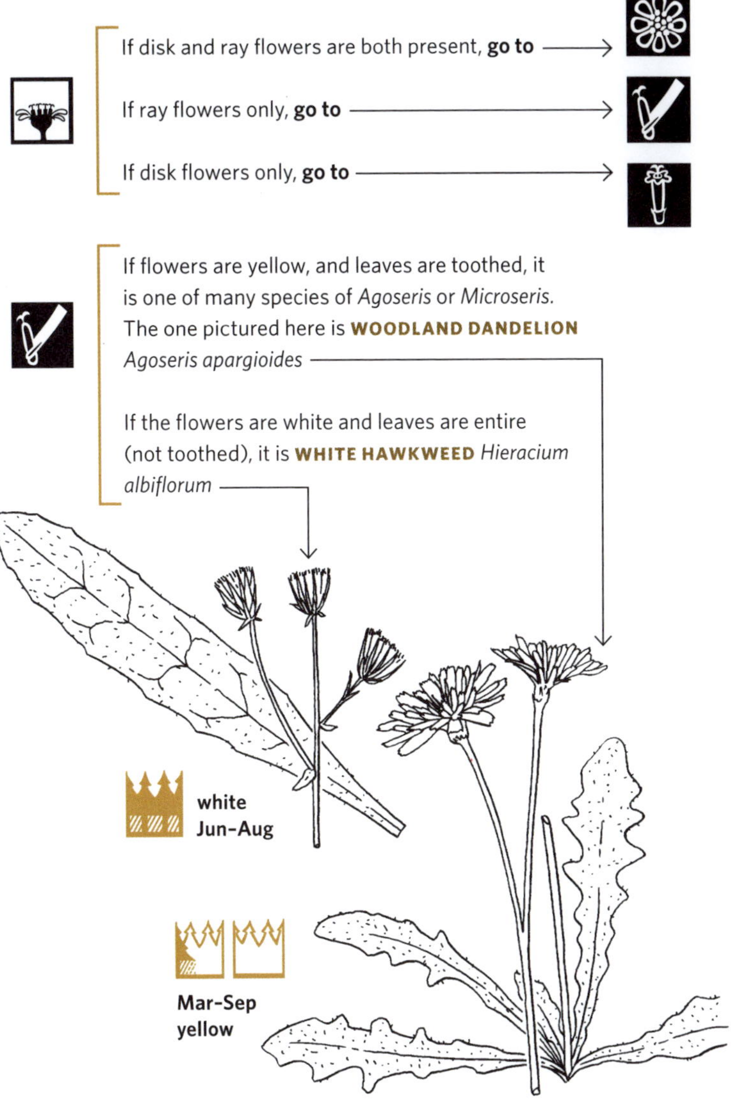

white
Jun–Aug

Mar–Sep
yellow

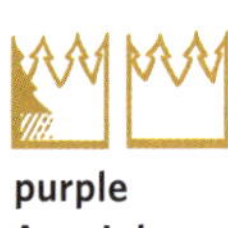

**purple
Apr–Jul**

If leaves and bracts are spiny, it is one of many species of *Cirsium*. The one pictured is **THISTLE** *Cirsium occidentale* and features purple flowers. ⟶

If there are no spines and the leaves are

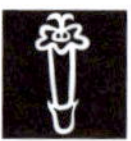

large, palmately lobed and veined, it is **WESTERN COLTSFOOT** *Petasites frigidus* var. *palmatus* (Flowers are white or pink and may precede leaves in early spring.) ⟶

narrow, entire, it is **PEARLY EVERLASTING** *Anaphalis margaritacea* (Yellow flowers are surrounded by papery, white, "everlasting" bracts.) ⟶

triangular, not lobed, it is **TRAIL MARKER PLANT** *Adenocaulon bicolor* (Flowers are tiny, greenish.)

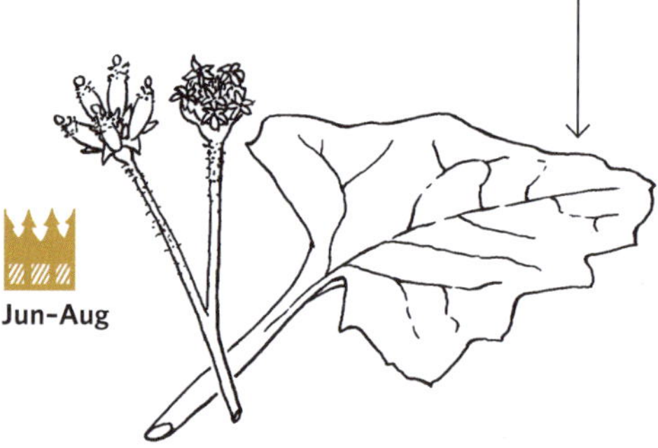

Jun–Aug

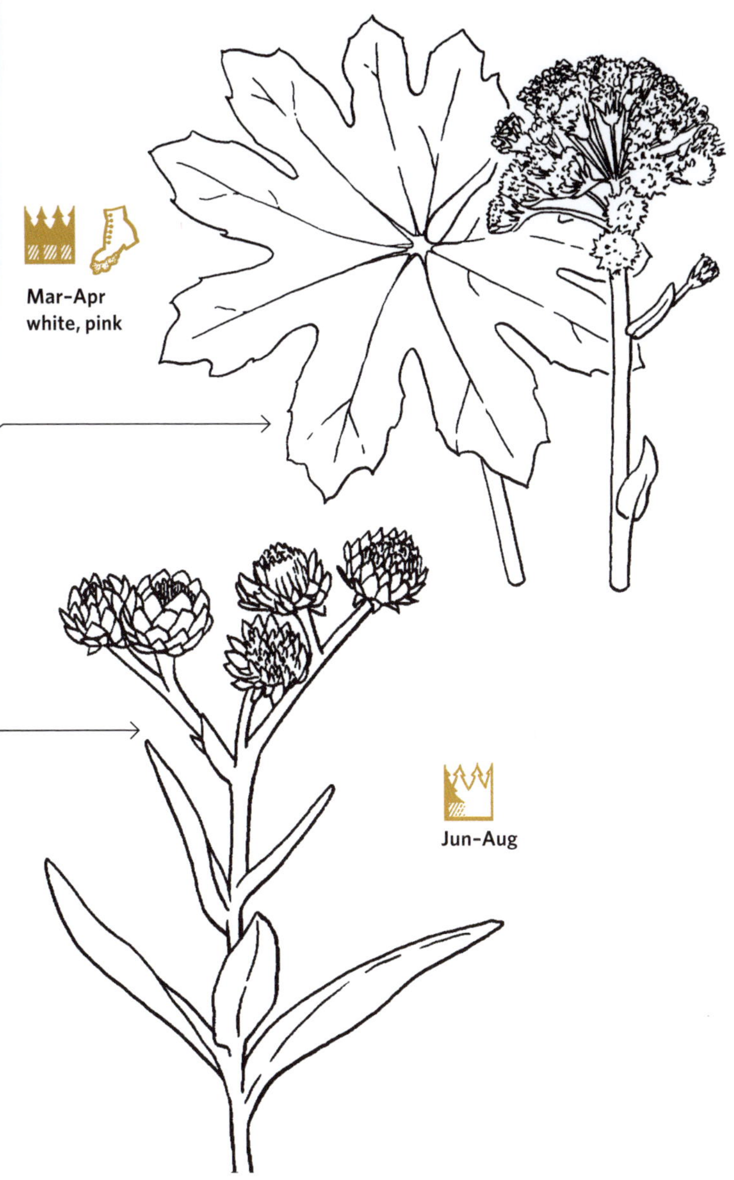

Mar–Apr
white, pink
Jun–Aug

If the ray flowers are yellow, **go to** $\longrightarrow$ 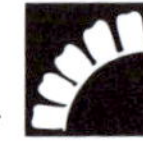

If not, **go to** $\longrightarrow$ 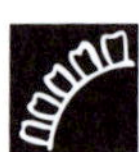

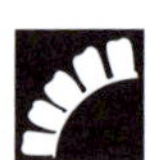

If ray flowers are yellow and turned back, it is
SNEEZEWEED *Helenium bigelovii* ——————

If they are yellow but not turned back, and the flower
heads are 1 ⅓ to 2½ inches (3.5 to 6 cm across),
leaves are 12 to 20 inches (30 to 50 cm), and it is
ear shaped, it is **MULE-EARS** *Wyethia glabra* ——————

If leaves and flower heads are smaller, simple, and
more or less entire, and the flowers are yellow, it is
WOODLAND TARWEED *Madia madioides* ——————

If the leaves are lobed or divided, **go to** $\longrightarrow$

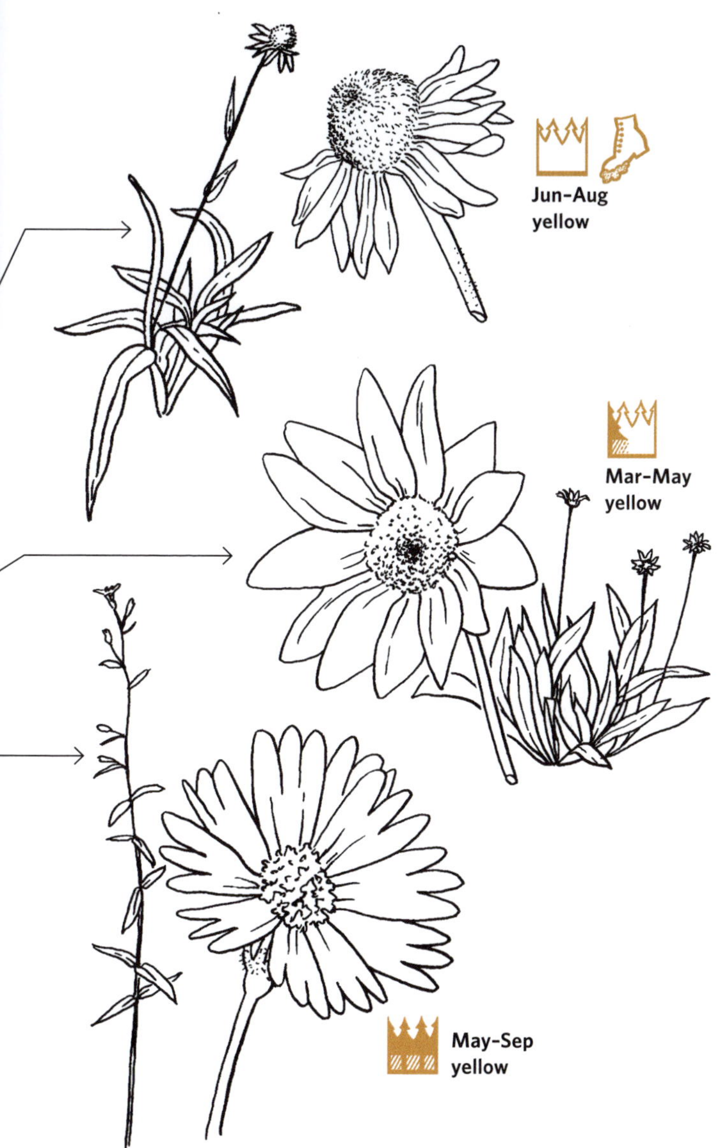

Jun–Aug
yellow

Mar–May
yellow

May–Sep
yellow

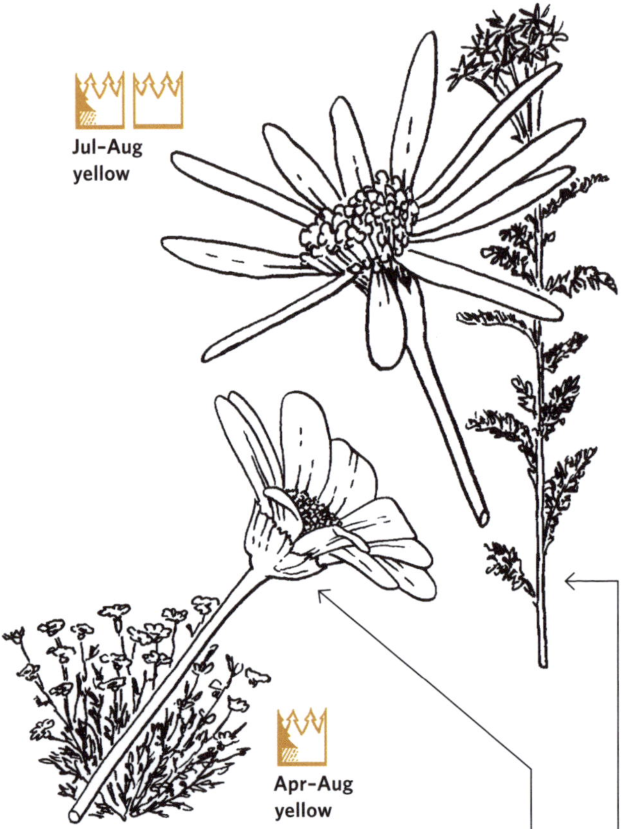

**Jul–Aug
yellow**

**Apr–Aug
yellow**

If the pappus is scaly or absent; older leaves
are woolly; and flowers are yellow, it is **OREGON
SUNSHINE** *Eriophyllum lanatum*

If the pappus is hairy; older leaves lack wool underneath; and flowers are yellow, it is **GROUNDSEL**
Senecio jacobaea

If ray flowers are blue to pink to purple, **go to** ⟶ 

If ray flowers are white, and leaves are

finely divided, **go to** ⟶

If leaves are *not* finely divided, and

in basal rosettes; flower heads under 1¼ inches (3 cm) across, with thirty to eighty rays, it is **ENGLISH DAISY** *Bellis perennis*

on stems; flower heads larger, with fifteen to thirty rays, it is **OXEYE DAISY** *Leucanthemum vulgare*

Apr–Sep
white

Jun–Aug
white

If tiny flower heads are in flat-topped umbels, it is **YARROW** *Achillea millefolium*

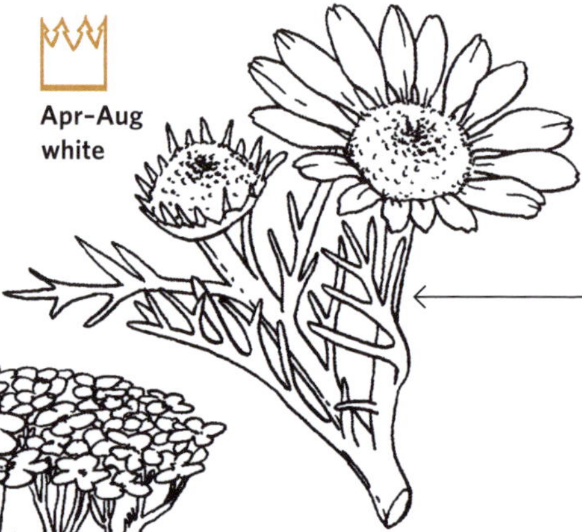

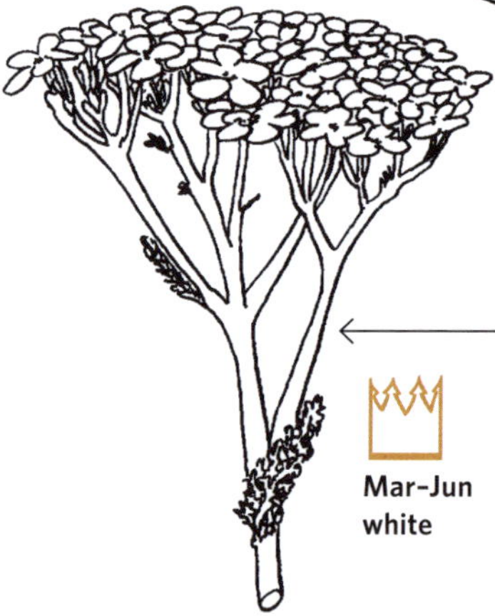

If the flower heads are not in umbels; leaves are foul smelling, it is **MAYWEEK** *Anthemis cotula*

If your white-rayed flower doesn't match the above, you may have a pale specimen of a different flower. **Go to**

If there are three rows of overlapping bracts and the flowers are purple, it is **COAST ASTER** *Aster chilensis*

If there are one or two rows of bracts that overlap only slightly, and the flowers are pink, it is **WILD DAISY** *Erigeron philadelphicus*

Jun–Aug purple

Jun–Aug pink

These two plants are over 16 inches (40 cm) tall. If rays are pink and the plant is much smaller, see English Daisy above.

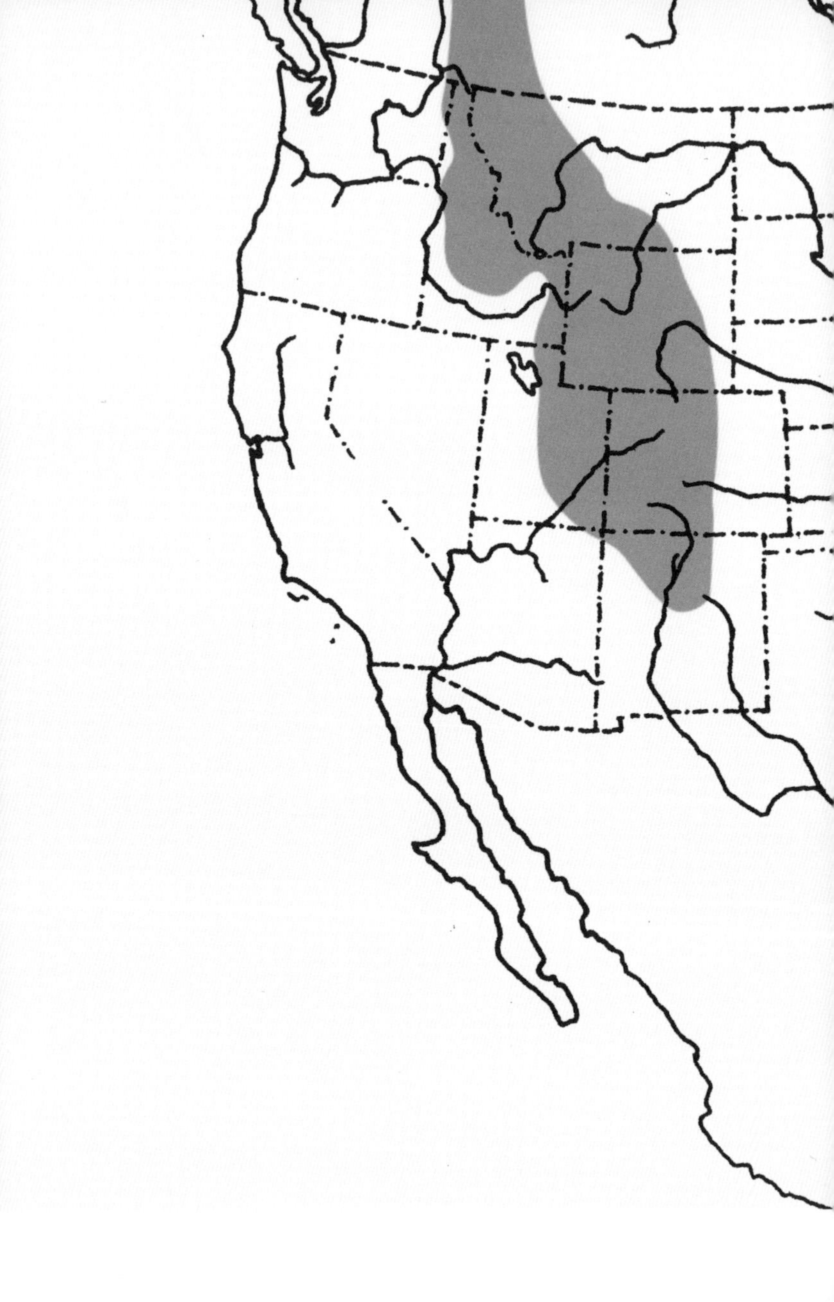

Flowers Below the Tree Line in the Rocky Mountains

Habitats

Here are the habitats of the Rocky Mountains:

- wet meadows, aspen groves

- dry, rocky slopes

- open gravelly meadows

- moist woods, thickets

- roadsides, disturbed soil

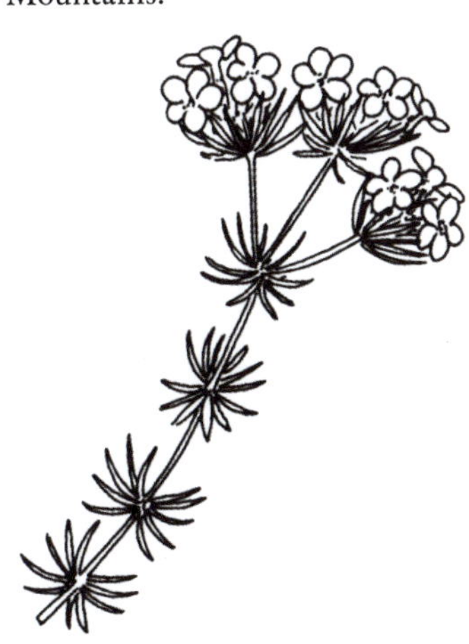

Life Zones

The Rocky Mountain life zones:

A **Alpine Zone**—The highest-elevation zone is above tree line.

S **Subalpine Zone**—Spring comes late to lush meadows and bogs, with Engelmann spruce, subalpine fir, and often aspen. Limber, whitebark, or bristlecone pine may occupy ridges.

M **Montane Zone**—Meadows are warmer and drier, with Douglas fir, aspen, and lodgepole pine; ponderosa pine on warm slopes; white fir or blue spruce in canyons, along streams.

F **Foothill Zone**—Spring comes earliest, meadows may be dry by July. Ponderosa pine gives way to grass, sagebrush, and shrubs at lower elevations.

P **Plains Zone**—Low-elevation non-mountainous areas.

The life zones are at lower altitudes to the north, higher in the south. The subalpine zone begins at about 10,000 feet (3,050 meters) in southern Colorado and 8,000 feet (2,450 meters) in Montana. And the zones are higher on sunny south-facing slopes than on cooler north-facing slopes. But the trees will tell you what zone you're in.

Identifying Tiny Flowers of the Mustard and Mint Families

 **Note:** *The whole flower should be small enough to fit into a circle ¼ inch (6 mm) in diameter*

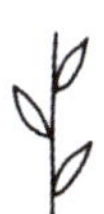 If the leaves are alternate and simple, with four flower petals and ovary superior; or leaves alternate and compound, with four flower petals and fruit below flowers, it is from the **MUSTARD FAMILY** —————————————————————→ 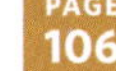

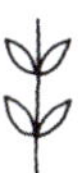 If the leaves are opposite and simple, with square stems, it is from the **MINT FAMILY** —————————————→ 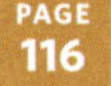

Mustard Family

(Brassicaceae or Cruciferae)

Regular flowers with four sepals, four petals, six stamens, superior ovary. The fruits (seed pods) necessary for identification are found below the flowers.

Flowers not yellow, **go to** ⟶ **PAGE 112**

Flowers yellow; fruit is

short, not more than three times as long as wide, **go to** ⟶ **PAGE 110**

longer; flowers are

in long showy raceme, **PRINCE'S PLUME**
Stanleya pinnata

not as above, flowers are

less than ½ inch (12 mm) wide, **go to** ⟶ **PAGE 108**

larger, **WESTERN WALLFLOWER**
Erysimum asperum

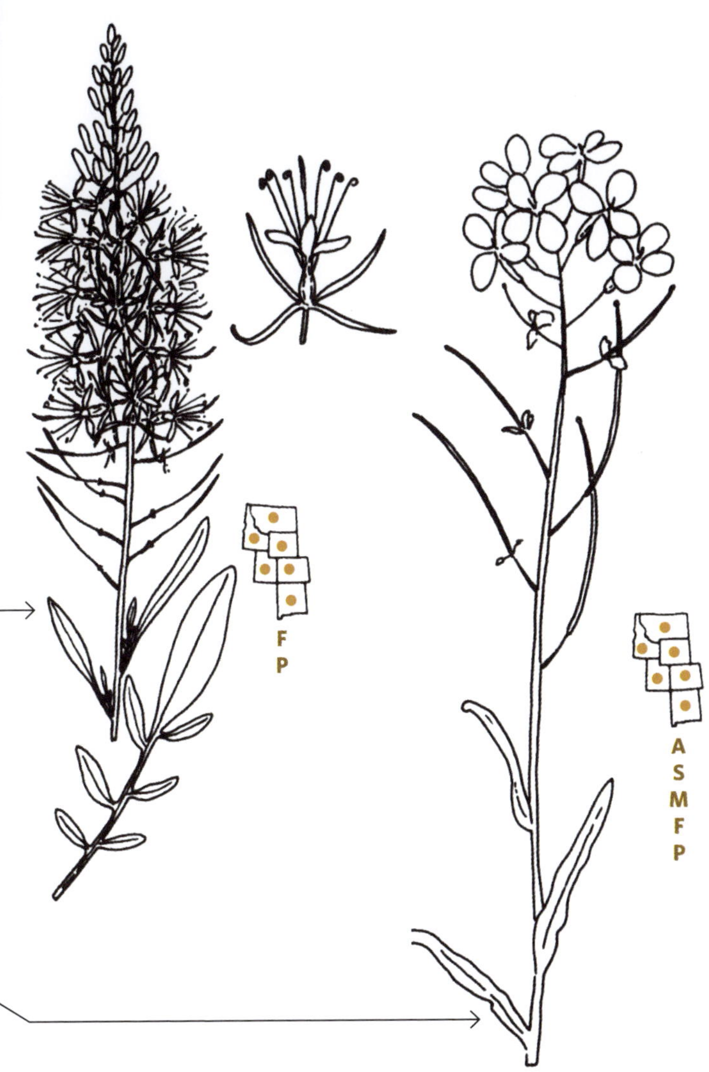

F
P
A
S
M
F
P

All leaves simple, entire or with small teeth, DRABA *Draba* sp. **Go to** ⟶

PAGE 110

Leaves (at least lower ones) compound or deeply lobed, with segments of upper leaves

narrow, nearly threadlike, **JIM HILL MUSTARD** *Sisymbrium altissimum* ——

not as above; leaves have large terminal lobes, and are

glaucous (white-coated), **FIELD MUSTARD** *Brassica campestris* ——

not glaucous, **WINTERCRESS** *Barbarea orthoceras* ——

not as above, **TANSY MUSTARD** *Descurainia* sp. *D. richardsonii, D. pinnata, D. sophia*, illus. ⟶

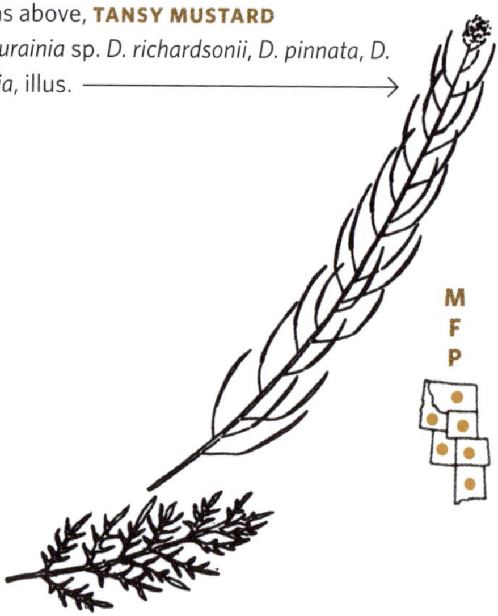

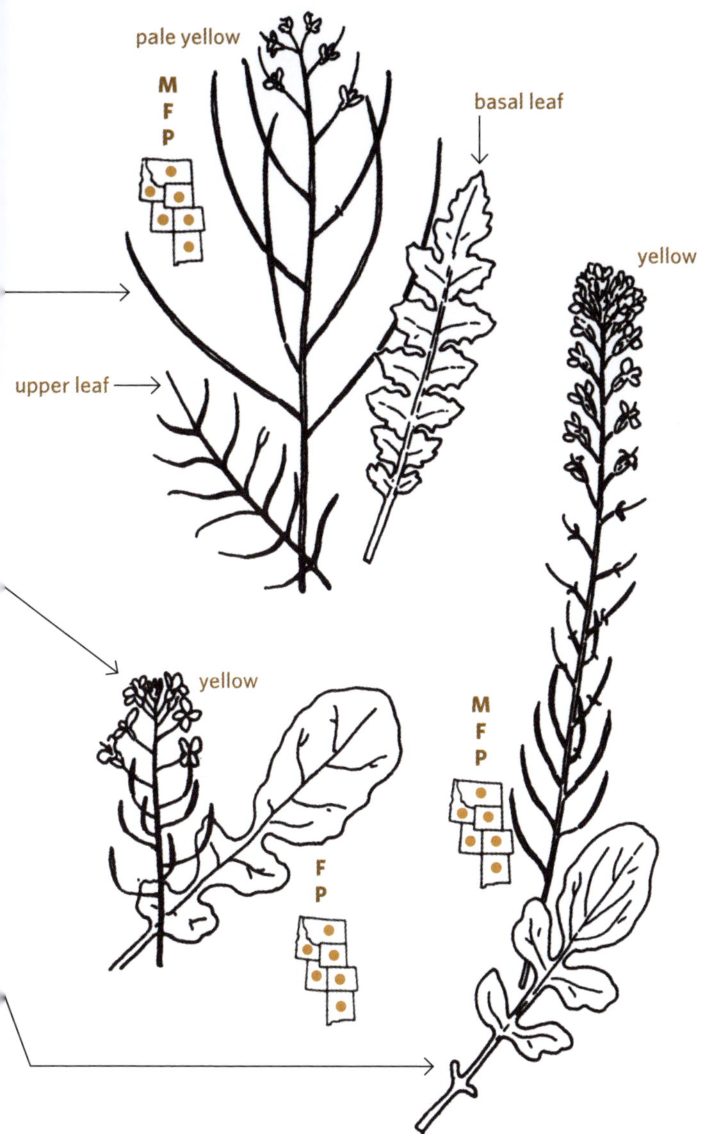

pale yellow
M
F
P
basal leaf
yellow
upper leaf
yellow
M
F
P
F
P

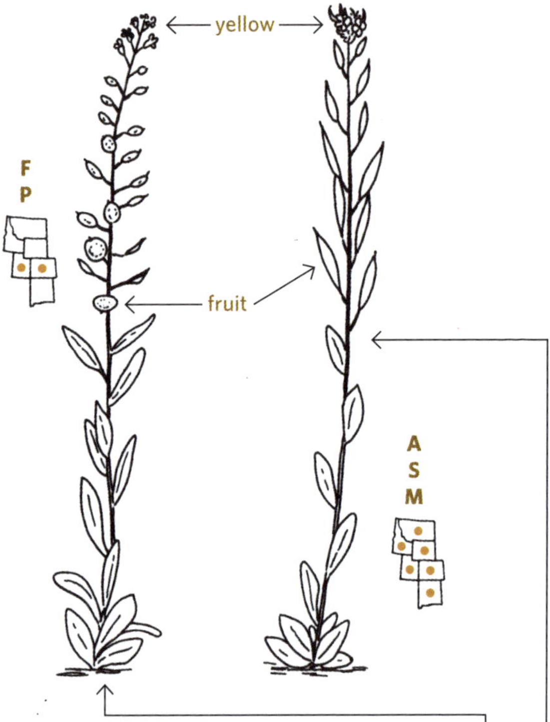

Fruit flattened; and

> *round in face view,* **WILD ALYSSUM** *Alyssum* sp.
> *A. minus,* illus.; *A. alyssoides,* sepals remain on
> mature fruit

> *not as above,* **DRABA** *Draba* sp. *D. aurea,* illus.;
> *D. crassifolia; D. streptocarpa* (Twisted-pod
> Draba); others

Fruit not flattened or only slightly so, often
inflated, **go to**

Plant of moist areas, **CRESS,** *Rorippa* sp. *R. teres*
illus.; *R. curvipes* var. *alpina*; *R. palustris* ssp. *hispida*;
R. sinuata; others

Plant of dry areas; fruit is

    double; some basal leaves fiddle-shaped,
    DOUBLE BLADDERPOD *Physaria vitulifera*

    single, **MOUNTAIN BLADDERPOD** *Lesquerella*
    montana

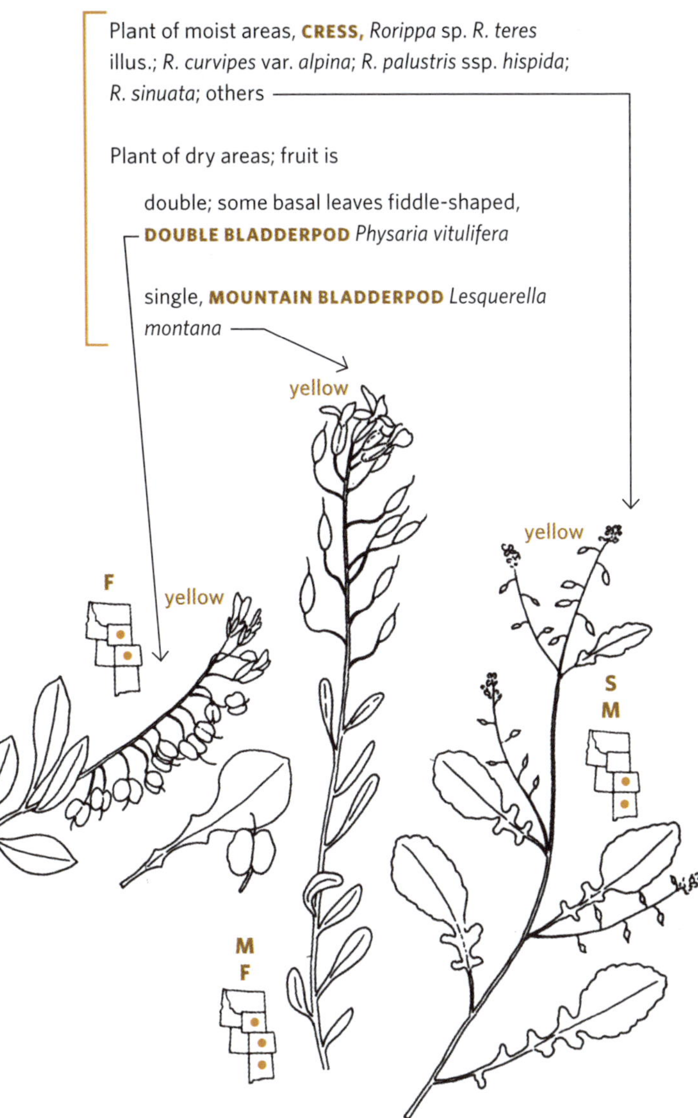

Fruits short, less than three times as long as wide, **go to** ⟶ PAGE 114

Fruits long; base of leaf is

clasping stem, and flowers are pink **ROCKCRESS** *Arabis* sp. *A. divaricarpa, A. fendleri,* illus., *A. hirsuta,* others ⟶

not as above; flowers are

pink, lavender, or purple, **go to** ⟶ PAGE 113

white; leaves are

simple, **BROOKCRESS** *Cardamine cordifolia*

compound, **WATERCRESS** *Nasturtium officinale* ⟵

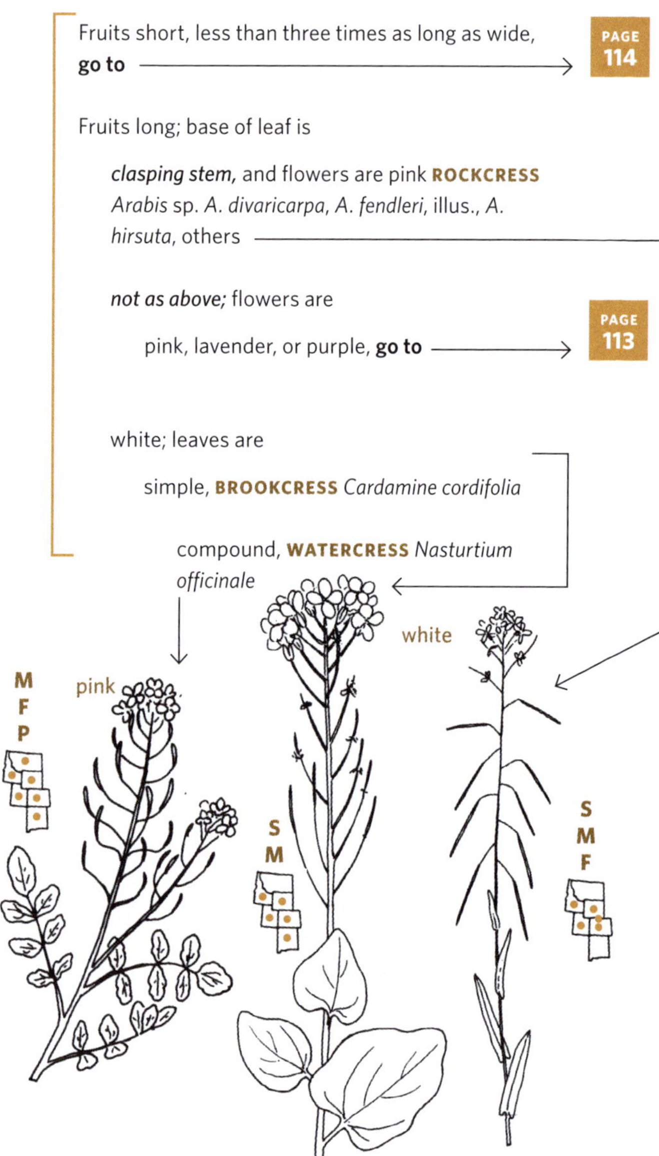

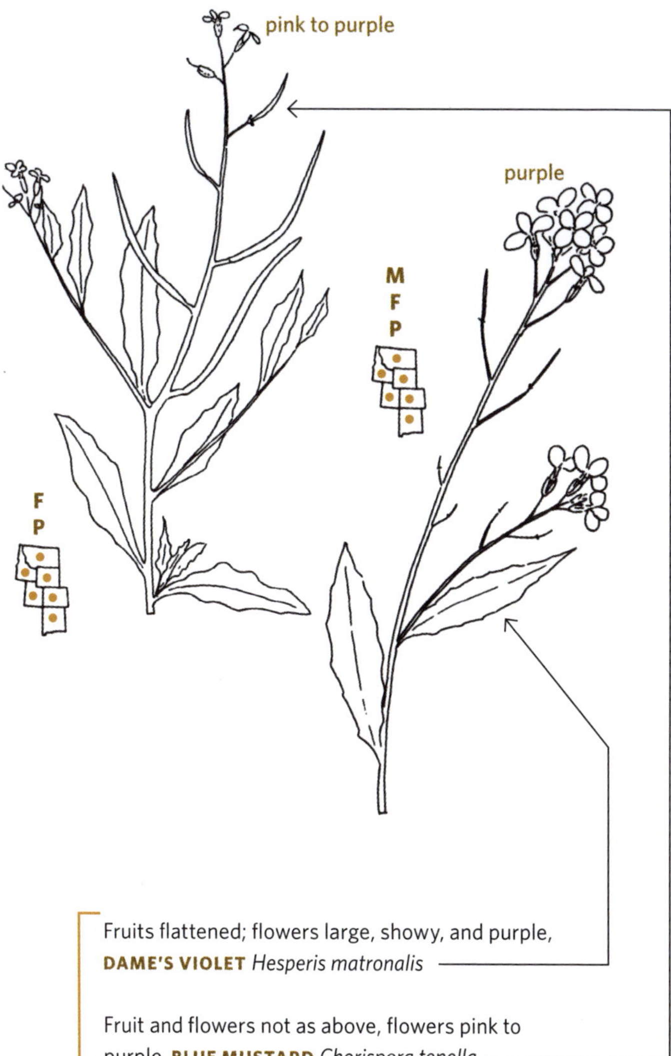

Fruits flattened; flowers large, showy, and purple, **DAME'S VIOLET** *Hesperis matronalis*

Fruit and flowers not as above, flowers pink to purple, **BLUE MUSTARD** *Chorispora tenella*

Flowers white and fruit shaped like this: 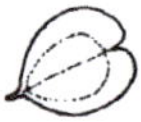
PENNYCRESS *Thlaspi arvense* ───

Flowers white and fruit shaped like this: 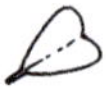
SHEPHERD'S PURSE *Capsella bursa-pastoris* ───

Fruit not as above, petals are

    white and deeply lobed **FALSE ALYSSUM**
    Berteroa incana ───

    white but not deeply lobed; stem leaves are

        clasping stem; plant is a

            small wildflower, **WILD CANDYTUFT**
            Thlaspi montanum ───

            weed, **WHITEWEED** *Cardaria draba* ───

        not clasping, **PEPPERWEED** *Lepidium*
        montanum ───

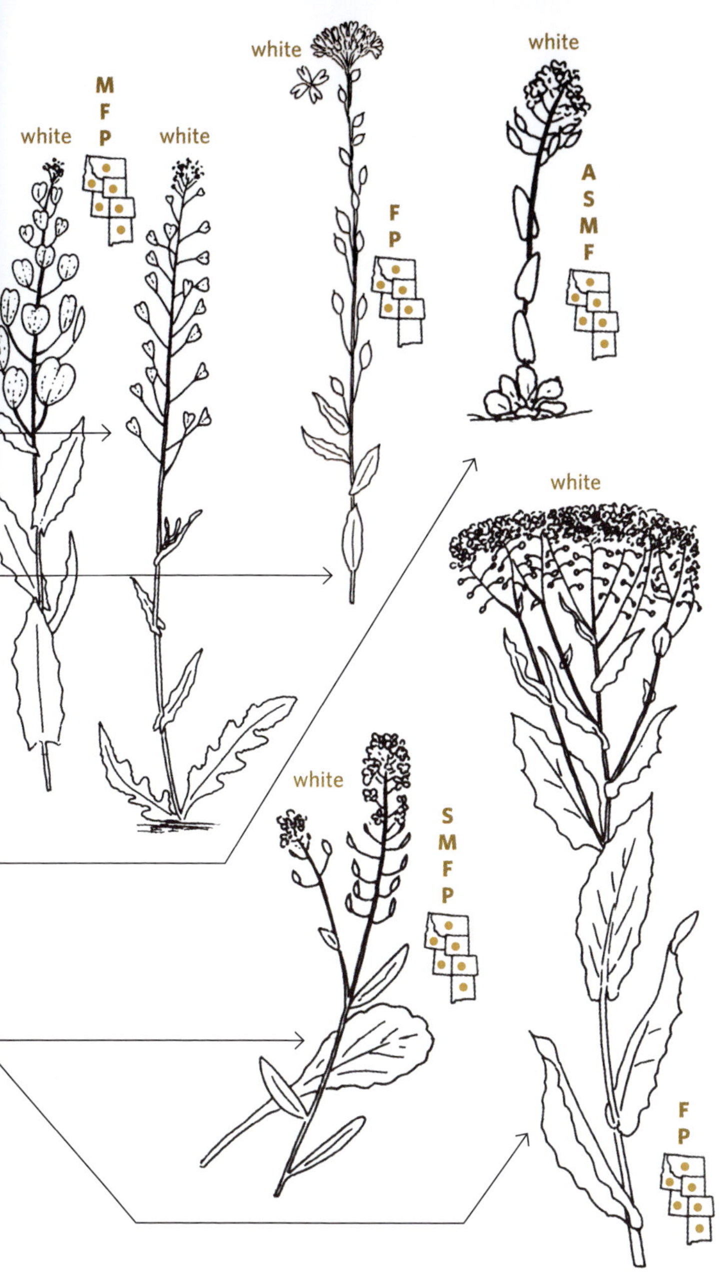

white
white
M
F
P
white
white
F
P
white
A
S
M
F
white
white
S
M
F
P
F
P

Mint Family

(Lamiaceae or Labitatae)

Mostly aromatic herbs with irregular two-lipped flowers with five united petals. Stems square with simple, opposite leaves. Stamens two or four, ovary superior with four lobes maturing into four nutlets.

Flowers in dense clusters in leaf axils, **go to** ⟶ **PAGE 118**

One to few flowers in leaf axils, **go to** ⟶ **PAGE 119**

Flowers in heads and purplish pink, **HORSEMINT** *Monarda fistulosa* ⸻

Flowers in spikes that are

spiny, and flowers are pink or blue, **DRAGON-HEAD** *Moldavica parviflora* ⸻

not spiny; stamens

extend beyond petals, and flowers are white to violet, **GIANT HYSSOP** *Agastache urticifolia* ⸻

not as above, flowers lavender-purple, **PRUNELLA** *Prunella vulgaris* ⸻

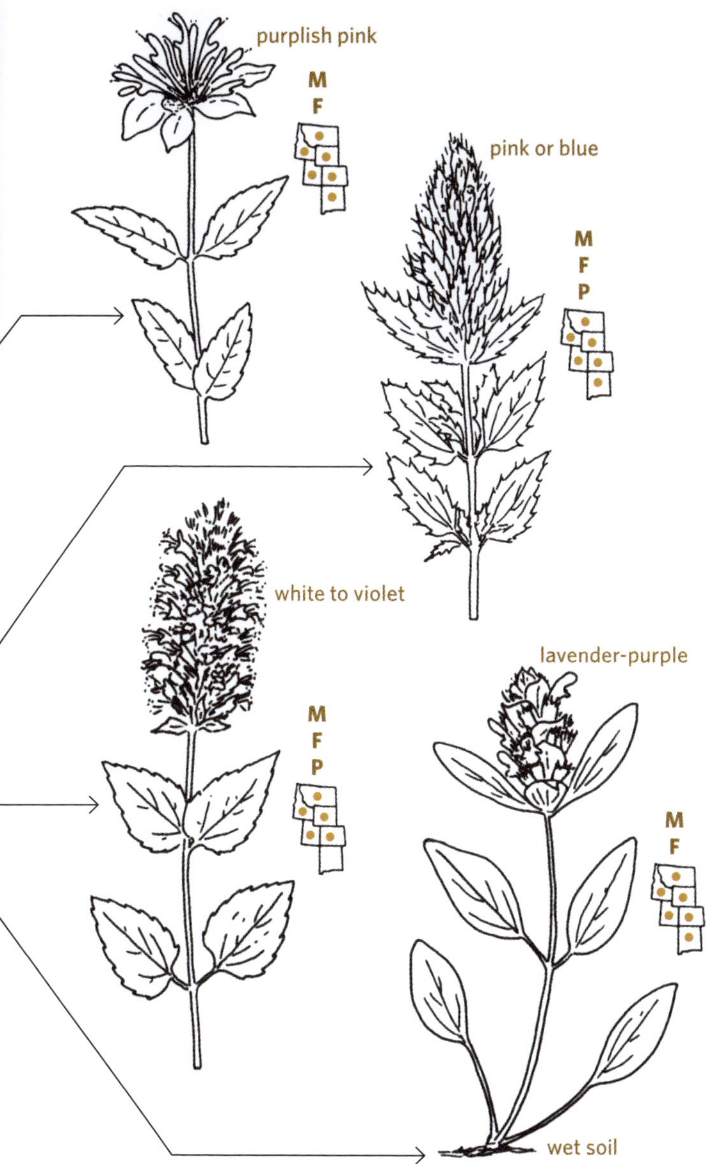

purplish pink
M
F
pink or blue
M
F
P
white to violet
M
F
P
lavender-purple
M
F
wet soil

Plant woolly, hairy, flowers white, **COMMON HOREHOUND** *Marrubium vulgare*

Plant not as above; calyx teeth are

bristlelike, flowers large and white, **WHORLED MONARDA** *Monarda pectinate*

not as above, flowers tiny and lavender to pink, **WILD MINT** *Mentha arvensis*

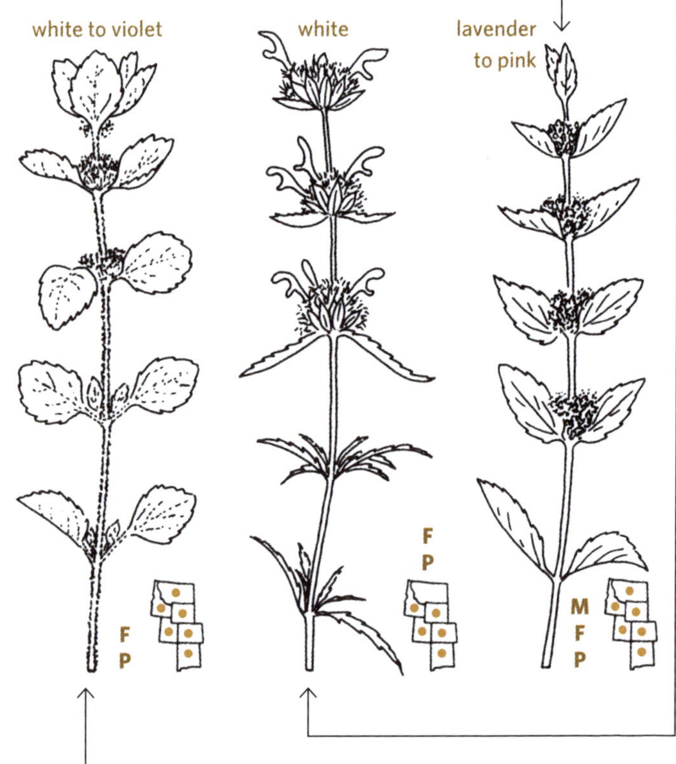

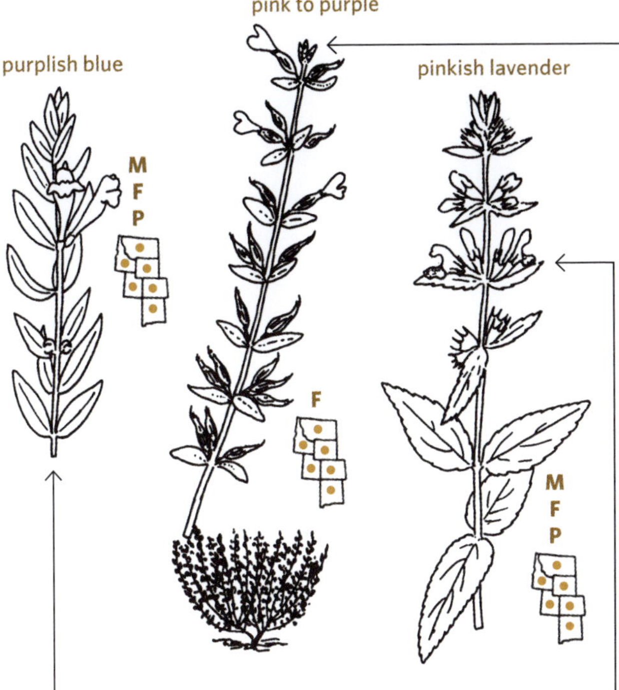

Calyx with ridge on top, purplish-blue flowers, **BRITTON'S SKULLCAP** *Scutellaria brittonii*, *S. galericulata* (Marsh Skullcap), leaves toothed, in wet areas ————

Calyx without ridge; plant of

moist areas, pinkish-lavender flowers, **HEDGE NETTLE** *Stachys* palustris ————

dry areas, pink to purple flowers, **PENNYROYAL** *Hedeoma drummondii* ————

Flower Families East of the Rockies and North of the Smokies

Identifying the Composite Family (Compositae)

Flowers packed into heads, surrounded by leafy involucre; anthers united to form a tube around style; style two-cleft at tip.

If the flowers are of two kinds: tubular at center, closed-fan shape on outside, **go to** ⟶

If the flowers are all closed-fan shape, **go to** ⟶

If the flowers are all tubular, and the plant is white-woolly, **go to** ⟶

If the flowers are not all yellow, **go to** ⟶

If the flowers are all yellow, it is **GOLDEN RAGWORT** *Senecio aureus* ⟶

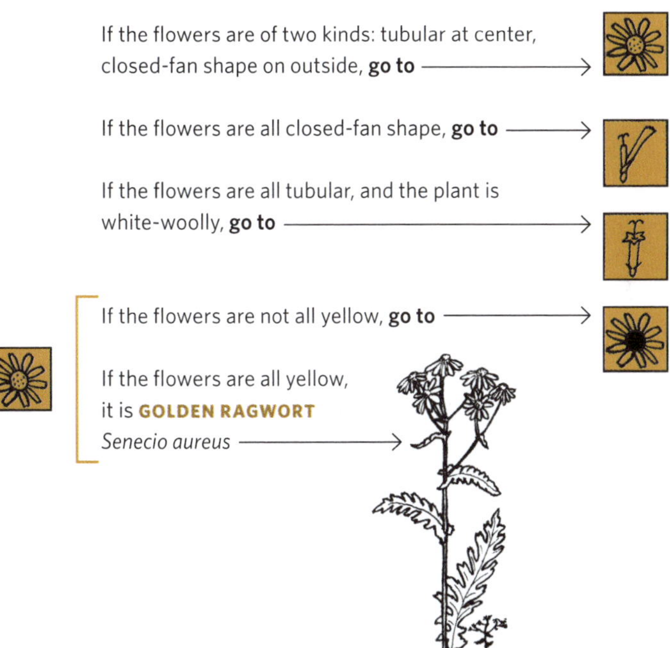

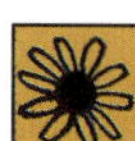

If the ray flowers are yellow, disk flowers purplish brown, it is **BROWN-EYED SUSAN** *Rudbeckia hirta*

If the flowers are not yellow, choose from the following:

ray flowers pinkish to violet, numerous, crowded

flowers blue violet, it is **ROBIN'S PLANTAIN** *Erigeron pulchellus*

flowers rose violet, it is **COMMON FLEA-BANE** *Erigeron philadelphicus*

ray flowers white, not numerous or crowded, it is **OX-EYE DAISY** *Chrysanthemum leucanthemum*

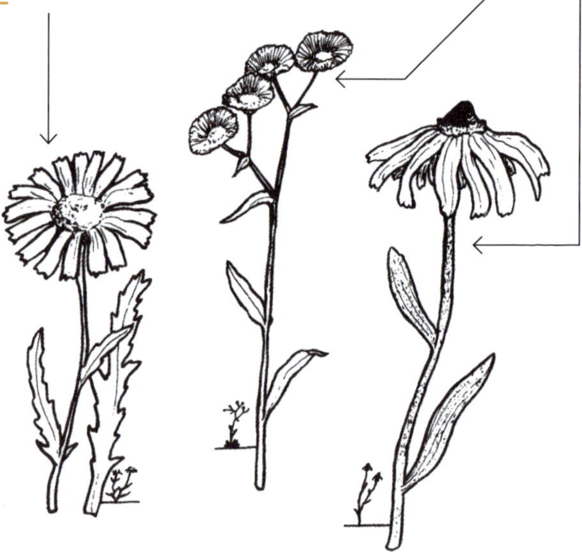

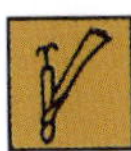

If there is only one head on a stalk, **go to** ⟶ 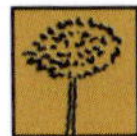

If there are several heads on a stalk, **go to** ⟶ 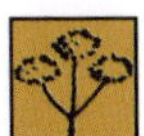

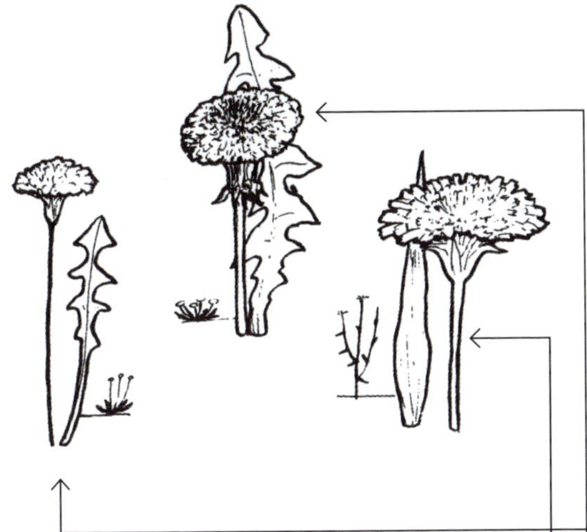

If the plants have leafy stems, the leaves long and grasslike, it is **GOATSBEARD** *Tragopogon pratensis*

If the plants have basal leaves only, choose from the following:

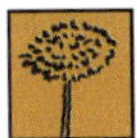

flowers crowded in head; stem hollow, it is **DANDELION** *Taraxacum officinale*

flowers fewer in head; head ¾ inch (2 cm) broad; stem slender, it is **DWARF DANDELION** *Krigia virginica*

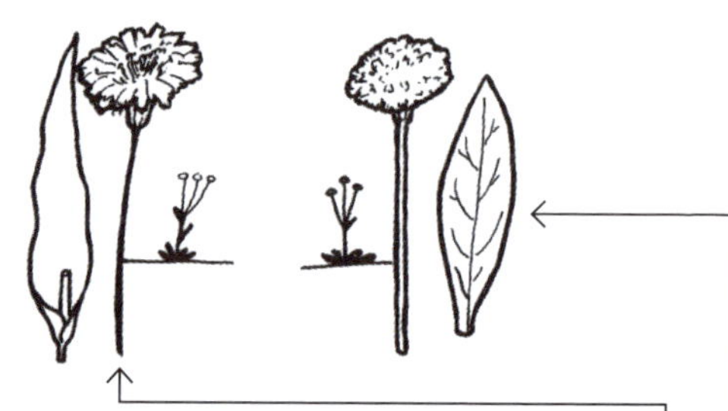

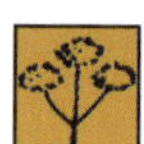

If the stem leaves have clasping bases, it is **CYNTHIA** *Krigia biflora*

If the stem leaves are lacking, or if present, do not have clasping bases; leaves purple veined, it is **PURPLE-LEAVED HAWKWEED** *Hieracium venosum*

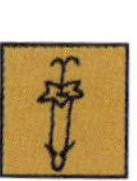

If the plant has small basal leaves, it is **EVERLASTING** *Antennaria canadensis*

If the plant has broad basal leaves, it is **PLANTAIN-LEAVED EVERLASTING** *Antennaria plantaginifolia*

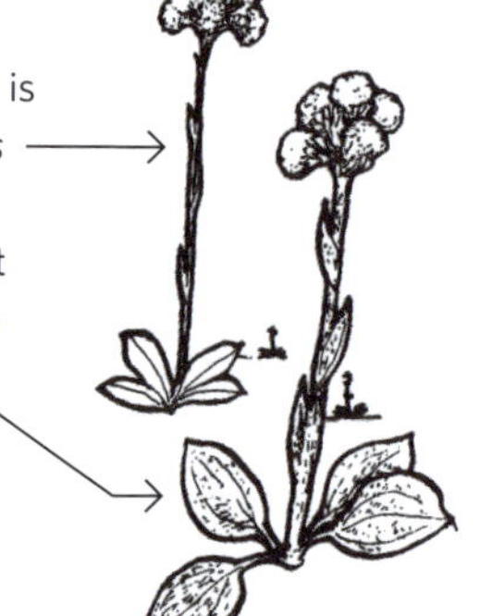

Identifying the Legume or Pulse Family (Leguminosae)

Flowers are neither on a club-like spike nor in a compact head. They have five petals, and two petals form a keel enclosing pistil and ten stamens. The flowers are polypetalous (one petal may be pulled off without tearing any others), five parted, and irregular (having petals of different shapes). The leaves are net veined, usually compound with stipules. The fruit is a pod.

If the leaves have three leaflets, choose from the following:

with end leaflet stalked, **go to** ⟶

with end leaflet not stalked, **go to** ⟶

If the leaves have more than three leaflets, **go to** ⟶

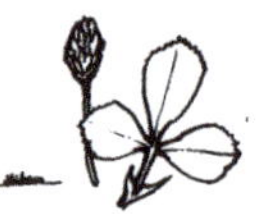

If the plant is procumbent, 6 inches (15 cm) high; flowers yellow, in a head, it is **BLACK MEDIC** *Medicago lupulina*

If the plant is upright, 1 to 4 feet (30 cm to 1.2 meters) high, choose from the following:

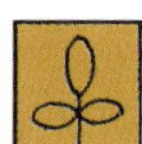

with purple flowers, it is **ALFALFA** *Medicago sativa*

with white flowers; 5 feet (1.5 meters) tall or more, it is **WHITE MELILOT** *Melilotus alba*

with yellow flowers; shorter (3 feet/1 meter), it is **YELLOW MELILOT (YELLOW SWEET CLOVER)** *Melilotus officinalis*

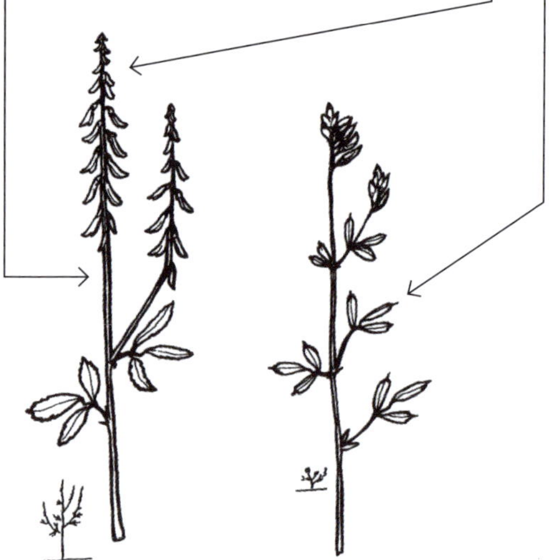

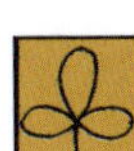

If the flowers in the head are stalked, leaves smooth, choose from the following:

white to pinkish flowers; creeping plant; leaflets indented at tip, marked with indistinct triangle, it is **WHITE CLOVER** *Trifolium repens*

pinkish flowers; erect plant; leaflets with rounded tips, no triangle, it is **ALSIKE CLOVER** *Trifolium hybridum*

If the flowers in the head are sessile, roseate; leaves hairy; distinct triangle, it is **RED CLOVER** *Trifolium pratense*

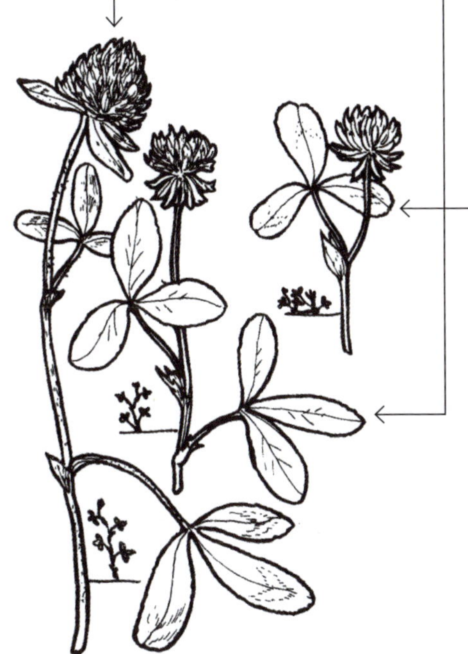

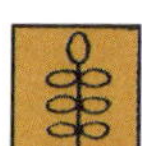

If the leaves are palmately compound, it is **LUPINE** ———————
Lupinus perennis

If the leaves are pinnately compound, tipped with tendrils; plants trailing or climbing, **go to** ————→ 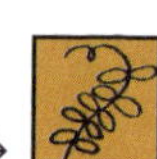

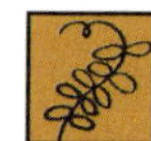

If leaves are thick, prominently veined, back and side petals curved upward; flowers purple, it is **WILD PEA** *Lathyrus venosus* ———————

If leaves are thin; side petals joined to keel, choose from the following:

    plants smooth

        four to eight flowers, purple, ¾ inch (2 cm) long, it is **AMERICAN VETCH** *Vicia americana* ———————

        ten or more flowers, pale tip of keel bluish, it is **WOOD VETCH** *Vicia caroliniana* ———————

    plant hairy, in mats; flowers many, it is **HAIRY VETCH** *Vicia villosa* ———————

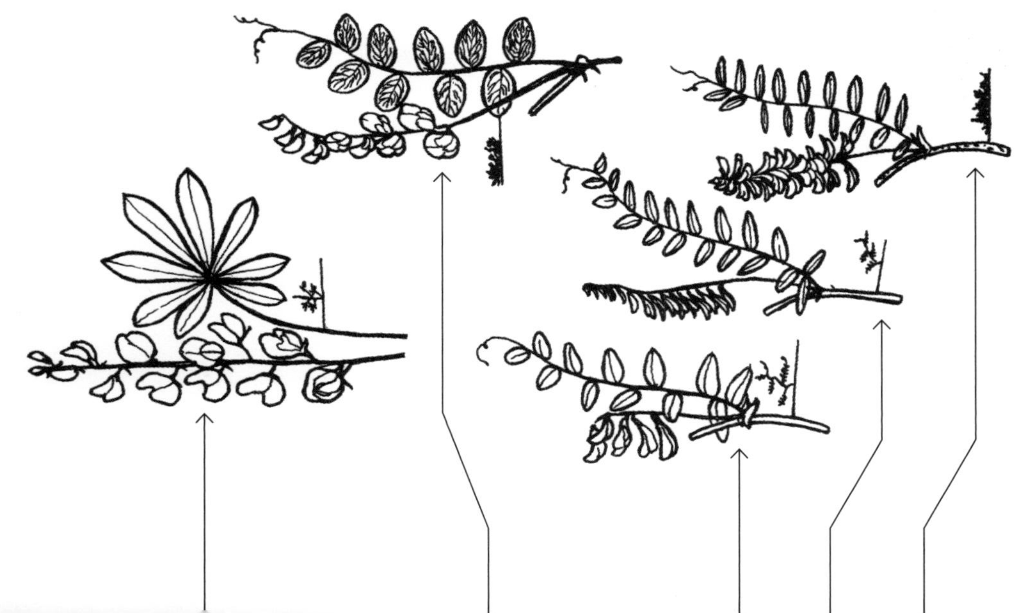

Plants of the Northeastern United States and Southeastern Canada in Winter

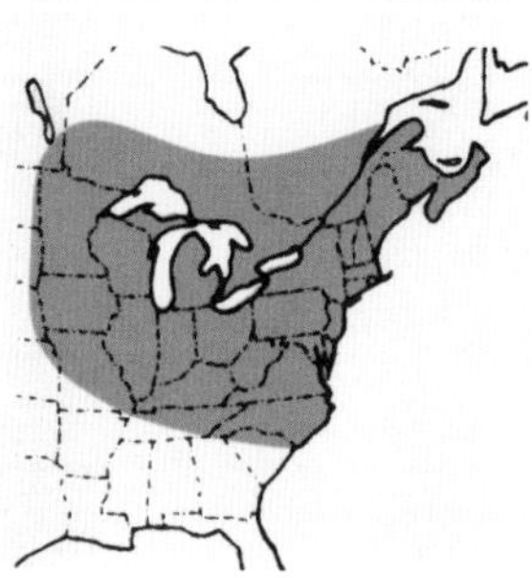

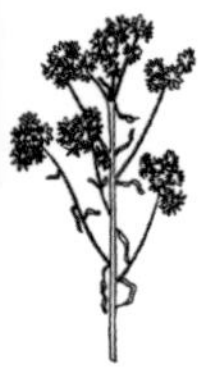

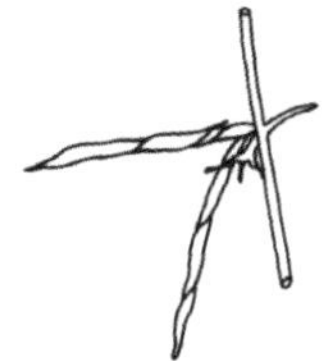

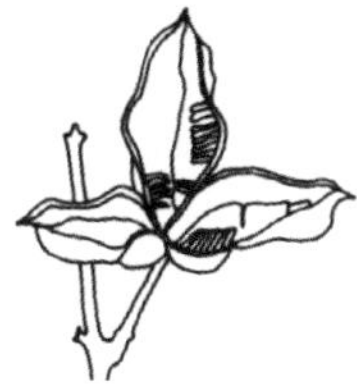

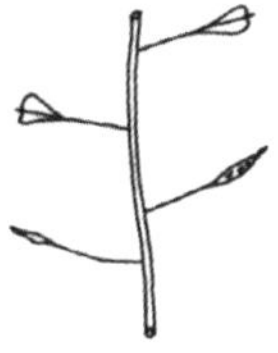

ried flowers

a pod (dry fruit that opens along its entire length along one or two seams)

a capsule (dry fruit that splits open, usually into two or more sections)

a silique (two-parted dry fruit with a translucent membrane separating the sections)

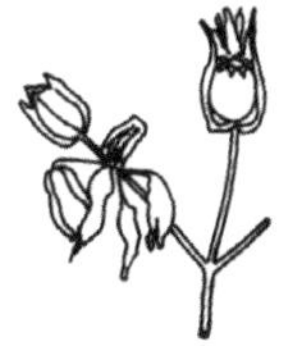

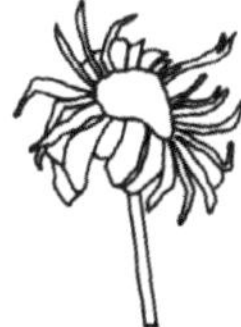

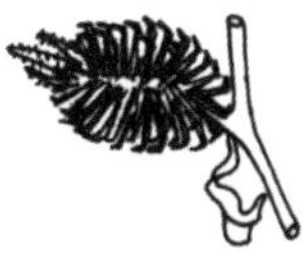

a calyx (outer circle of flower parts, sometimes a papery covering around the capsule)

bracts (modified leaves, usually at the base of a flower or flower head)

a burr (dry fruit or collection of fruits— or the covering of a fruit—that has hairs or hooks and attaches easily to clothing)

Some Shortcut Clues to Identification

If you find stalkless structures crowded on the upper portion of the stem, try

- **MULLEIN** or **TURTLEHEAD**

- **VERVAIN**

If the structure is on a vine, and it has wings, try **WILD YAM** or **DOCK**

If the structure has silky or woolly hairs floating off it, and

- it has two parts, try **MILKWEED, SWALLOW-WORT,** or **DOGBANE**

- it has spines, try **THISTLE**

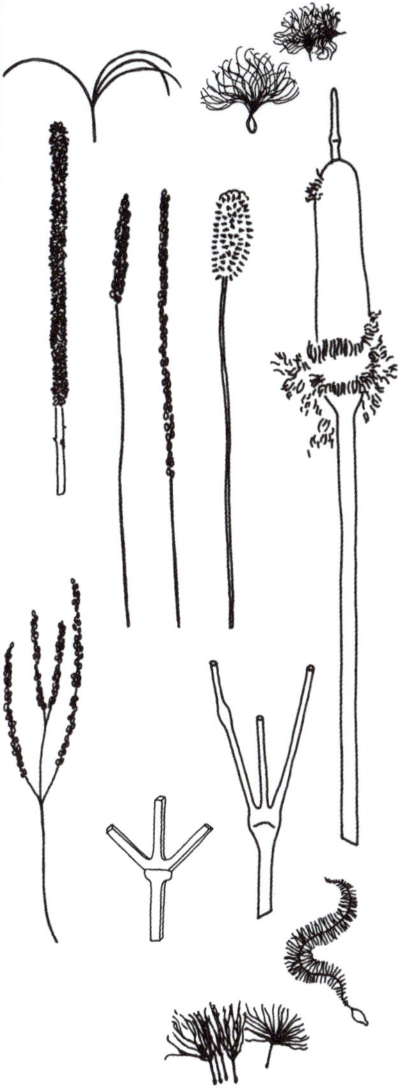

Some terms used in this section:

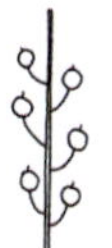

ALTERNATE BRANCHING

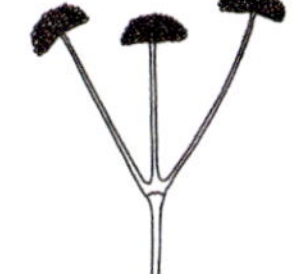

OPPOSITE BRANCHING

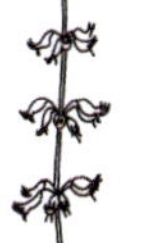

WHORLED BRANCHING

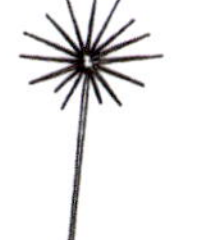

SIMPLE UMBEL

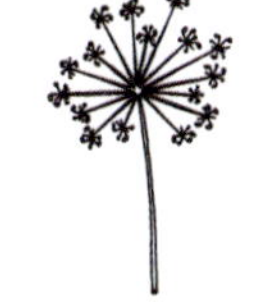

COMPOUND UMBEL

Identifying plants with barbs or needles or a structure (pod, capsule, dried flower, calyx, bract) that sticks to clothing

If the structure has needles or hairs that are not hooked, **go to** ⟶ PAGE 137

If structure has curved, hooked barbs and is less than ½ inch (12 mm) long, and it's

teardrop shaped and entirely covered with barbs, it is **ENCHANTER'S NIGHTSHADE** *Circaea quadrisulcata*

cone shaped with barbs at one end, it is **AGRIMONY** *Agrimonia* sp.

more than ½ inch (12 mm) long, **go to** ⟶ PAGE 136

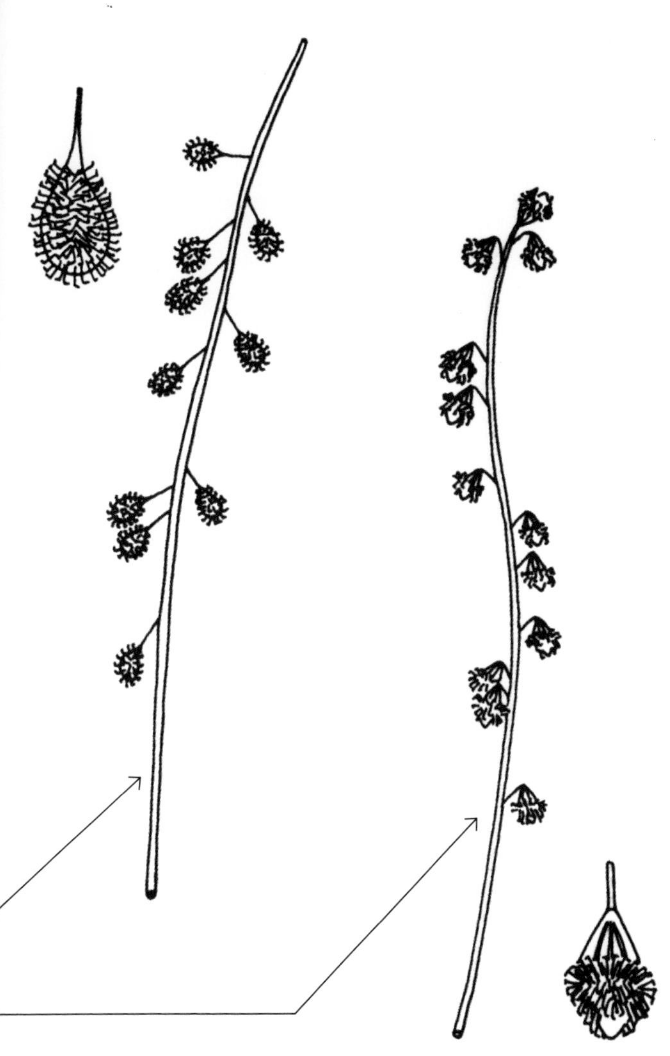

If structure (burr) is round, and

barbs pull out of structure easily, it is **AVENS** *Geum* sp.

barbs don't pull out easily, it is **BURDOCK** *Arctium* sp.

If burr is not round, it is **COCKLEBUR** *Xanthium* sp.

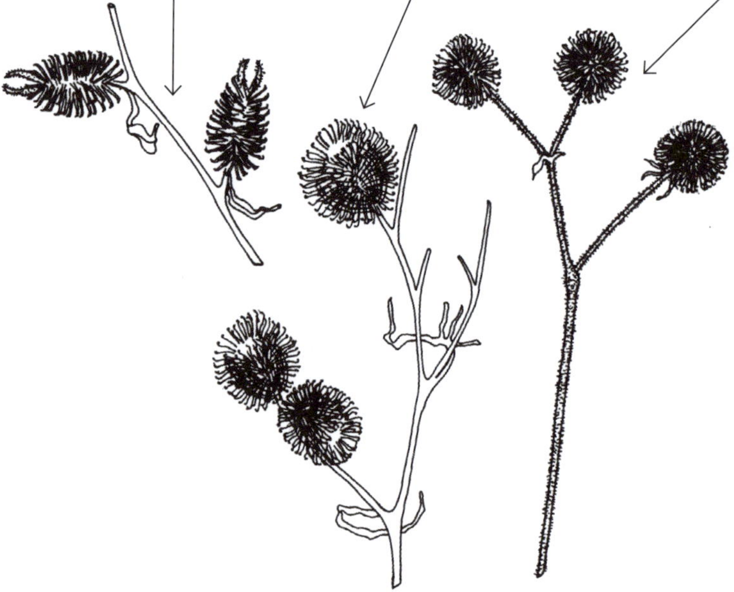

If the plant is entirely covered with sharp white hairs, it is **VIPER'S BUGLOSS** *Echium vulgare*

If structure is a dense cylinder that breaks easily into woolly seeds, it is **THIMBLEWEED** *Anemone* sp.

If the plant is not as above and its structures have needles but are not funnel shaped, **see below**

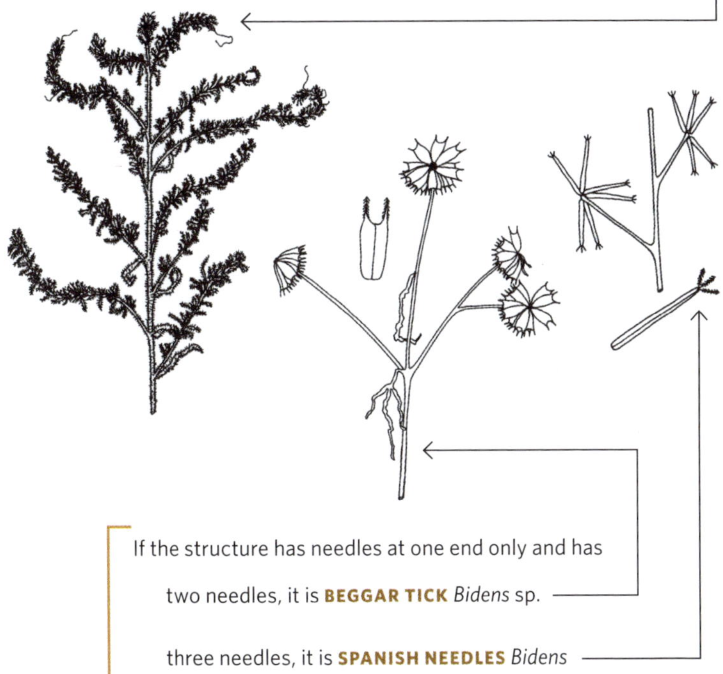

If the structure has needles at one end only and has

two needles, it is **BEGGAR TICK** *Bidens* sp.

three needles, it is **SPANISH NEEDLES** *Bidens bipinnata*

If not as above, **go to** → **PAGE 138**

If structure has two to four parts, and is

not an inflated, papery sack, it is **JIMSON WEED** *Datura stramonium* ————

an inflated, papery sack, it is **WILD CUCUMBER** *Echinocystis lobata* ————

If structure does not have two to four parts and is

egg shaped, it is **TEASEL** *Dipsacus sylvestris* ————

bell shaped, perhaps with long, fluffy white hairs, it is **THISTLE** *Cirsium* sp. ————

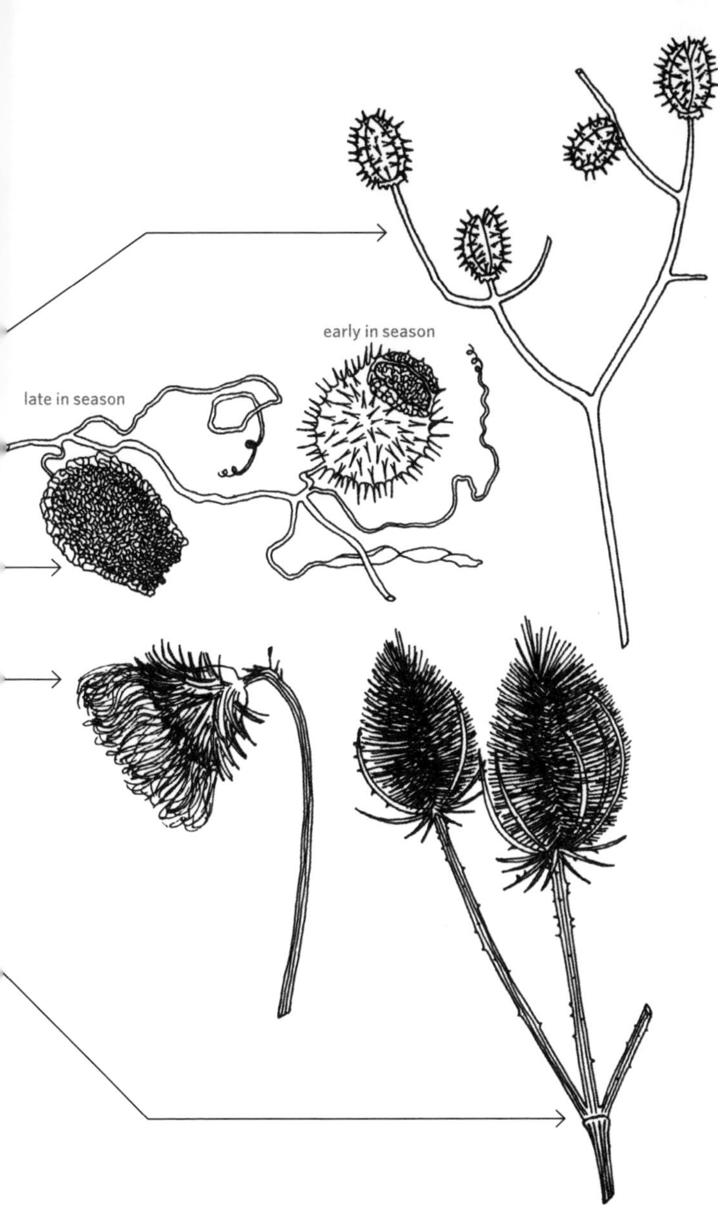
early in season
late in season

Identifying Plants with Umbels

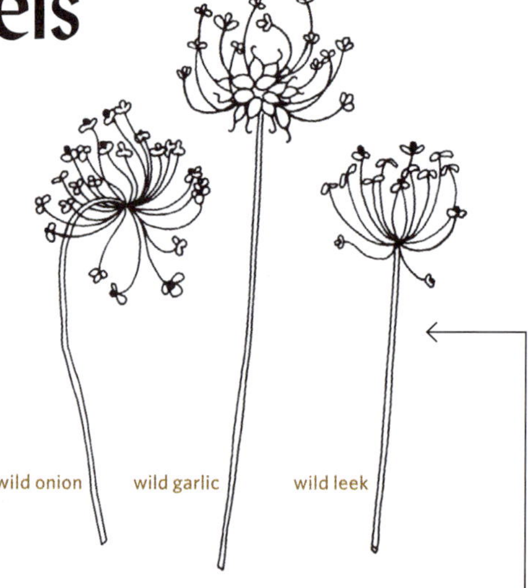

If plant has a simple umbel with three-parted capsules (which may contain small, round, black seeds), it is **WILD ONION, WILD GARLIC,** or **WILD LEEK** *Allium* sp.

If it has compound umbels, it's in the Parsley Family. **Go to** ⟶ PAGE 141

If plant appears to have a compound umbel, but the outer structure is actually a seed with three needles at one end, **see** Spanish Needles, page 137.

The Parsley Family is a large and complex group, so it's difficult to identify individual species without the flower. Common species include:

WILD PARSNIP *Pastinaca sativa*

SWEET CICELY *Osmorhiza claytonii*

QUEEN ANNE'S LACE *Daucus carota*

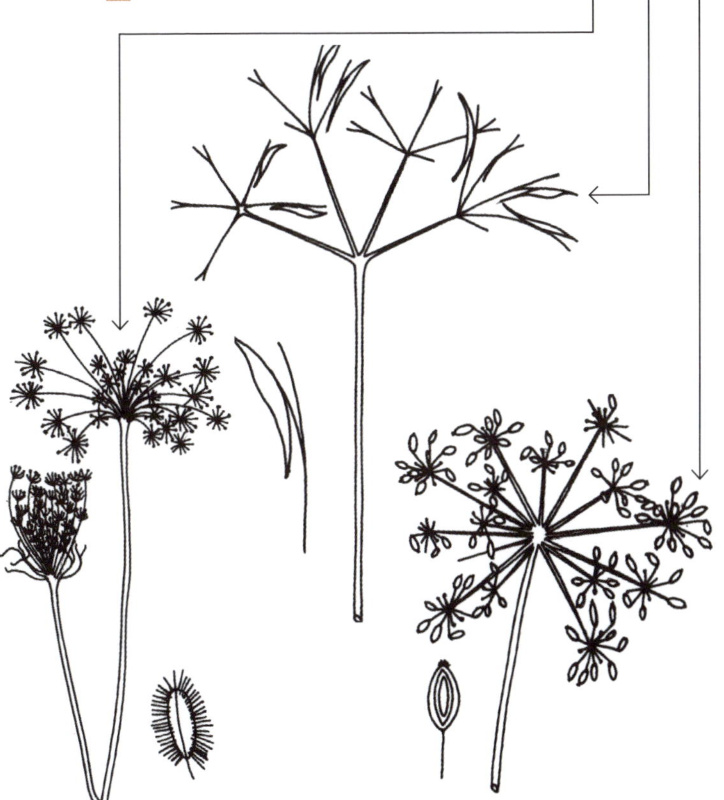

Identifying Plants with Flat, Papery Structures (pod, capsule, or silique)

If structure is heart shaped, with stalk that is not as long as capsule, it is **COMMON SPEEDWELL** *Veronica officinalis.* ⟶

If structure looks like a pea pod (lacks an inner partition, splits along entire length into two parts), it's in the **PEA FAMILY. Go to** ⟶ **PAGE 144**

If it's not as above, it's in the **MUSTARD FAMILY.** ⟶

Members of the Mustard Family bear seeds in papery pods called siliques. After time, the outer part of the pod falls off, leaving a translucent membrane. Eventually the membrane also falls off, leaving a skeleton. By winter, many of the outer coverings have fallen off.

The Mustard Family is a large one. Without flowers and leaves, it's hard to identify individual species. Common members of the family are:

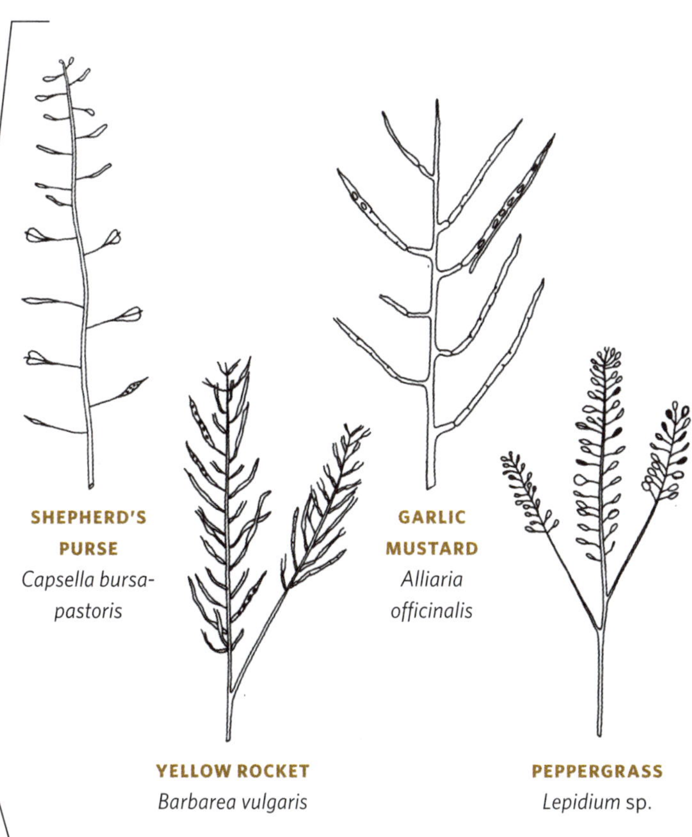

SHEPHERD'S PURSE
Capsella bursa-pastoris

GARLIC MUSTARD
Alliaria officinalis

YELLOW ROCKET
Barbarea vulgaris

PEPPERGRASS
Lepidium sp.

Identifying Certain Plants with Two-Parted Structures (pods, dried flower, calyx, or bract)

If structure looks like a pea pod, of if pod has opened and the two parts are tightly spiraled, it's in the Pea Family. Three common species are:

WILD PEA or **VETCH** *Vicia sp. Lathyrus* sp. (pod smooth) ——————————————————————

TICK TREFOIL *Desmodium* sp. ——————————

LUPINE *Lupinus perennis* (pod fuzzy) ——————

If structure is not as above, **go to** ————————→ **PAGE 146**

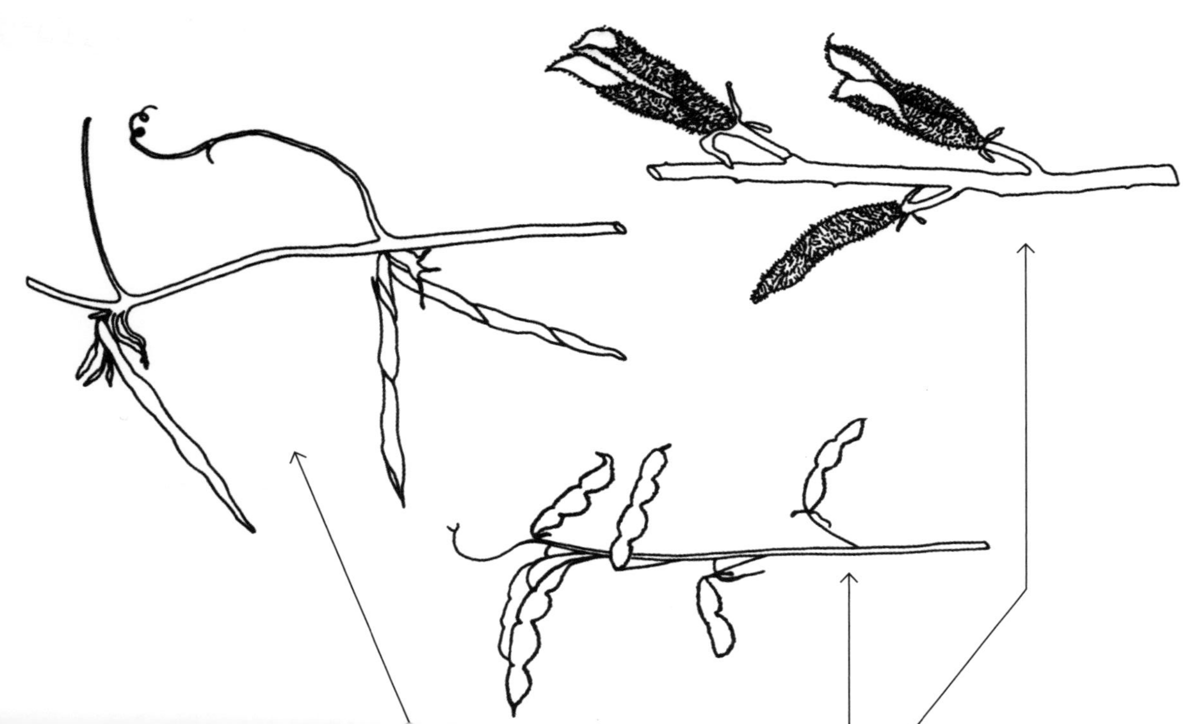

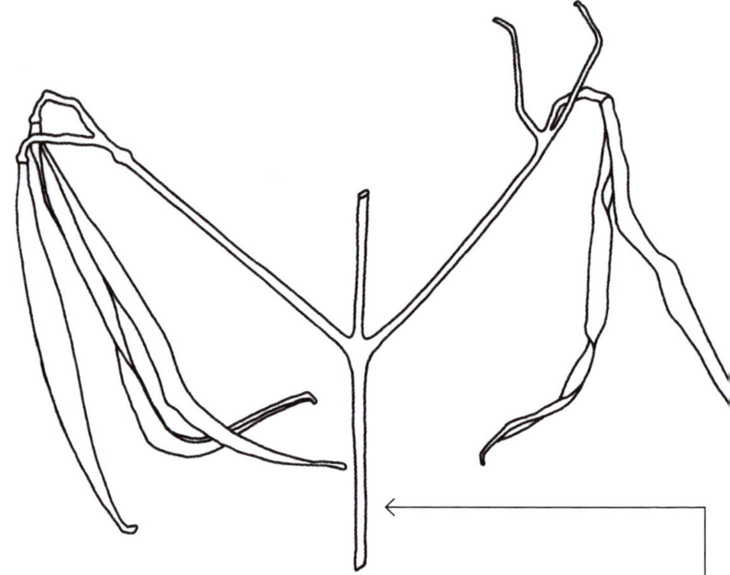

If structure is 1 inch (2.5 cm) long or longer, and

is long and thin or if it has split into two papery strips, it is **DOGBANE** *Apocynum* sp.

is a fibrous, netlike cylinder, see **WILD CUCUMBER** *Echinocystis lobata*, see page 138.

is not as above, and plant is:

not a vine, it is **MILKWEED** *Asclepias* sp.

a vine, it is **SWALLOW-WORT** *Cynanchum* sp. or *Matelea* sp.

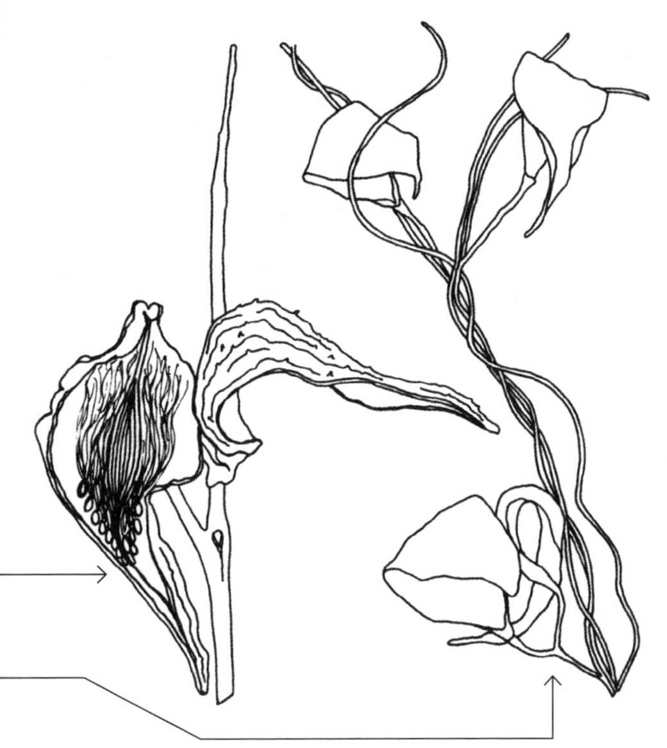

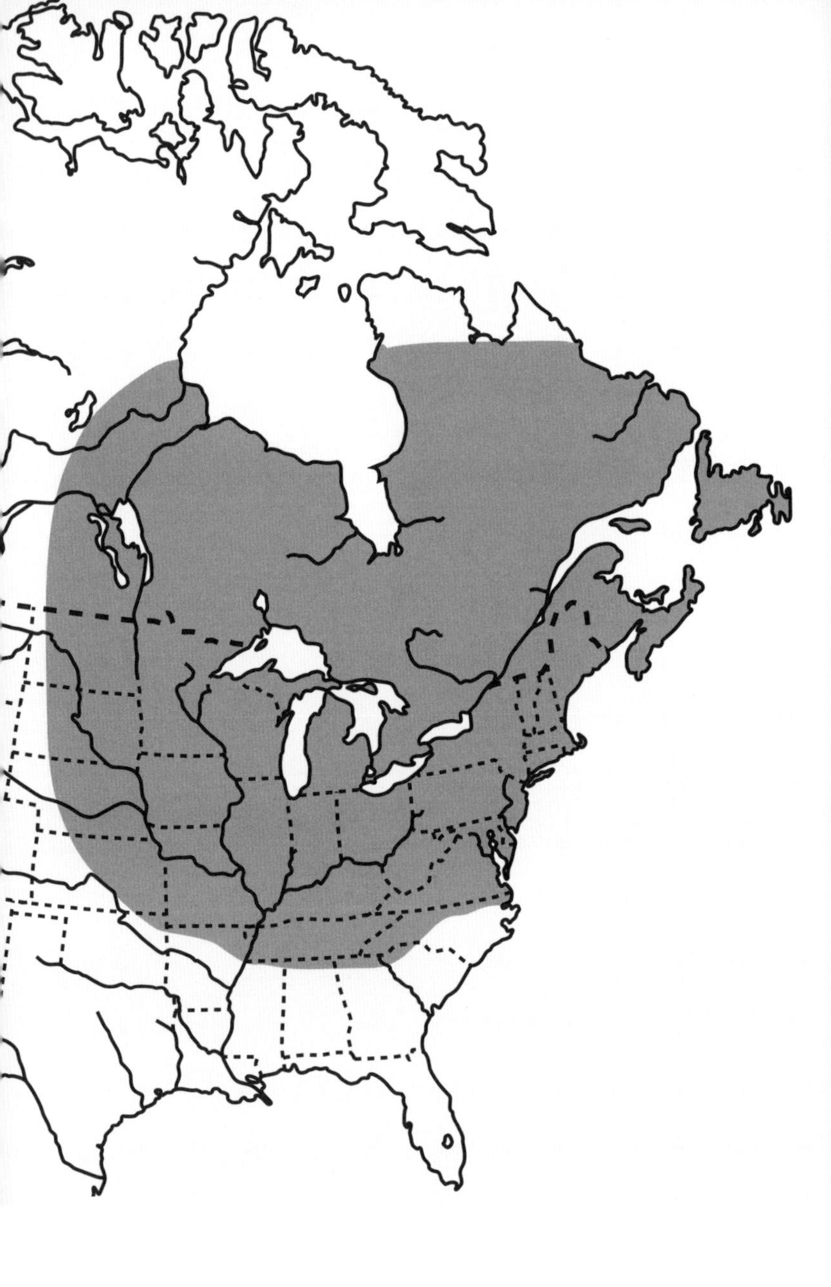

Berries of the Eastern United States and Canada

THIS CHAPTER WILL HELP YOU IDENTIFY plants with fleshy fruit 1 inch (2.5 cm) in diameter or smaller. It describes ripe fruits. Most unripe fruits are green or white and may turn several colors before maturing.

Read this first:

Some berries illustrated are marked with a danger sign.

 **Known to be poisonous**

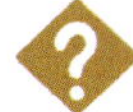 **Danger suspected but unproved**

DON'T TASTE THESE BERRIES EVER. NEVER TASTE A BERRY UNLESS YOU'RE SURE YOU KNOW WHAT IT IS.

Why this caution?

Some berries are poisonous even in small quantities. Eating them can cause mild to severe discomfort, vomiting, diarrhea, convulsions, even death. You can't tell if berries are safe by tasting them. Be warned that birds and mammals often eat berries that are deadly to humans.

Should you eat berries *not* marked with a danger sign in this book?

Few wild fruits have been systematically tested for edibility. Most species, the rarer ones especially, are seldom gathered or eaten in significant amounts. In some cases there may not be enough reliable evidence to declare a species either "safe" or "poisonous." Roadside plants may have been sprayed with toxic herbicides. Proceed with caution.

This is a plant identification guide, not a guide to edible plants.

Some terms used in this chapter:

Leaves may be

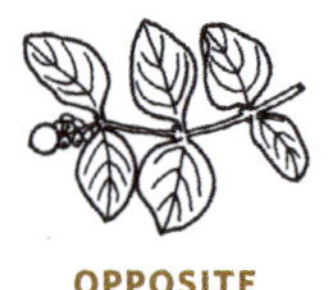

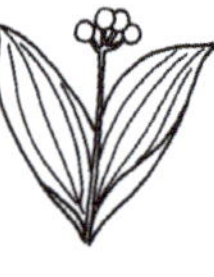
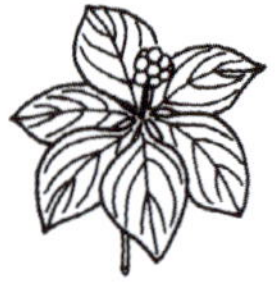

OPPOSITE   **ALTERNATE**   **BASAL**   or **WHORLED**

Leaf or leaflet edges may be

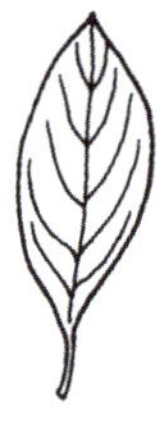
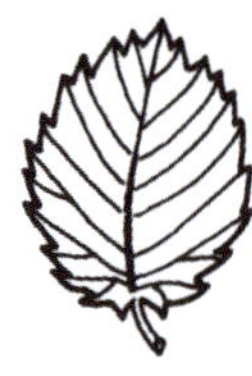
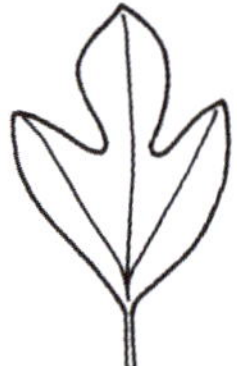

ENTIRE   **TOOTHED**   **LOBED**

Leaves may be

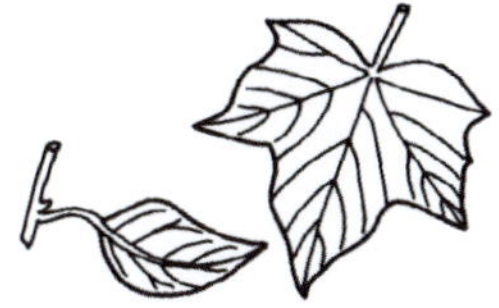 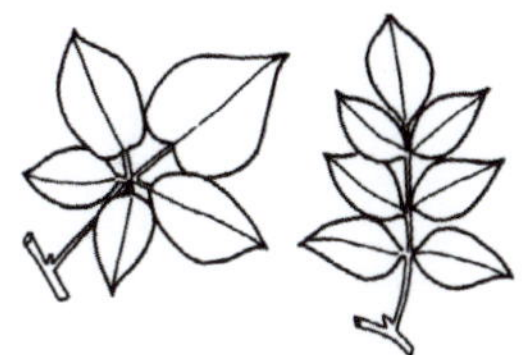

SIMPLE (a single blade with a bud at the base)

or **DIVIDED** into leaflets, with a bud at the base.

The position of berries on a plant may be

TERMINAL (at the end of a branch)

or **AXILLARY** (from the angle between the leaf and stem, or leaf axil)

Berries may grow

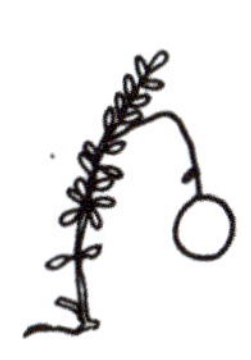  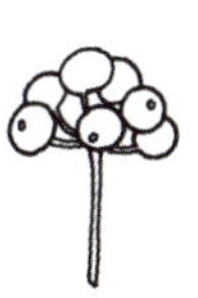 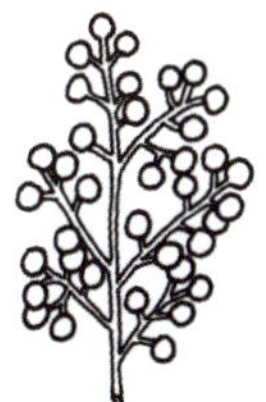

SINGLY

in **RACEME** (on separate stalks)

in an **UMBEL**

in a **PANICLE** (a compound raceme or umbel)

or in various other kinds of clusters.

This book calls all fleshy fruits "berries," whether they are drupes, pomes, accessory fruits, aggregates, or true berries. It includes native species as well as some cultivated species that have escaped to the wild.

BERRY: soft fruit with seeds embedded in the flesh, as in Grape.

DRUPE: fleshy part of the fruit encloses a hard "stone" that contains a seed, as in Wild Plum or Cherry.

HIP: a hollow, leathery receptable containing several to many seedlike fruits. Only in Rose.

POME: fleshy part of the fruit encloses a papery inner wall around the seeds, as in Apple or Hawthorn.

ACCESSORY: soft part of the fruit is covered with numerous small, dry, hard seeds. Only Strawberry and False Strawberry.

AGGREGATE: formed of many tiny berries growing clumped together, as in Raspberry or Blackberry.

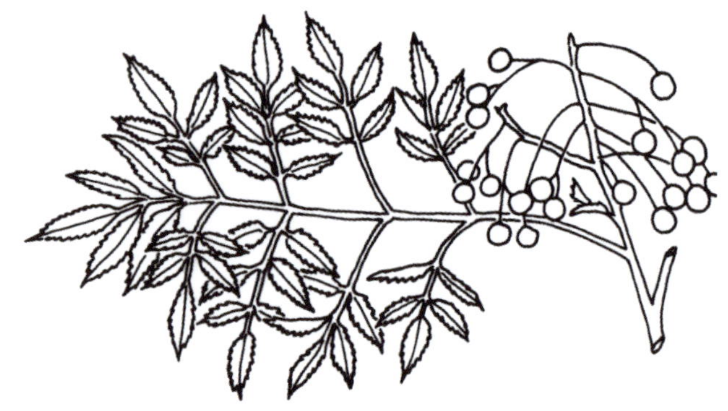

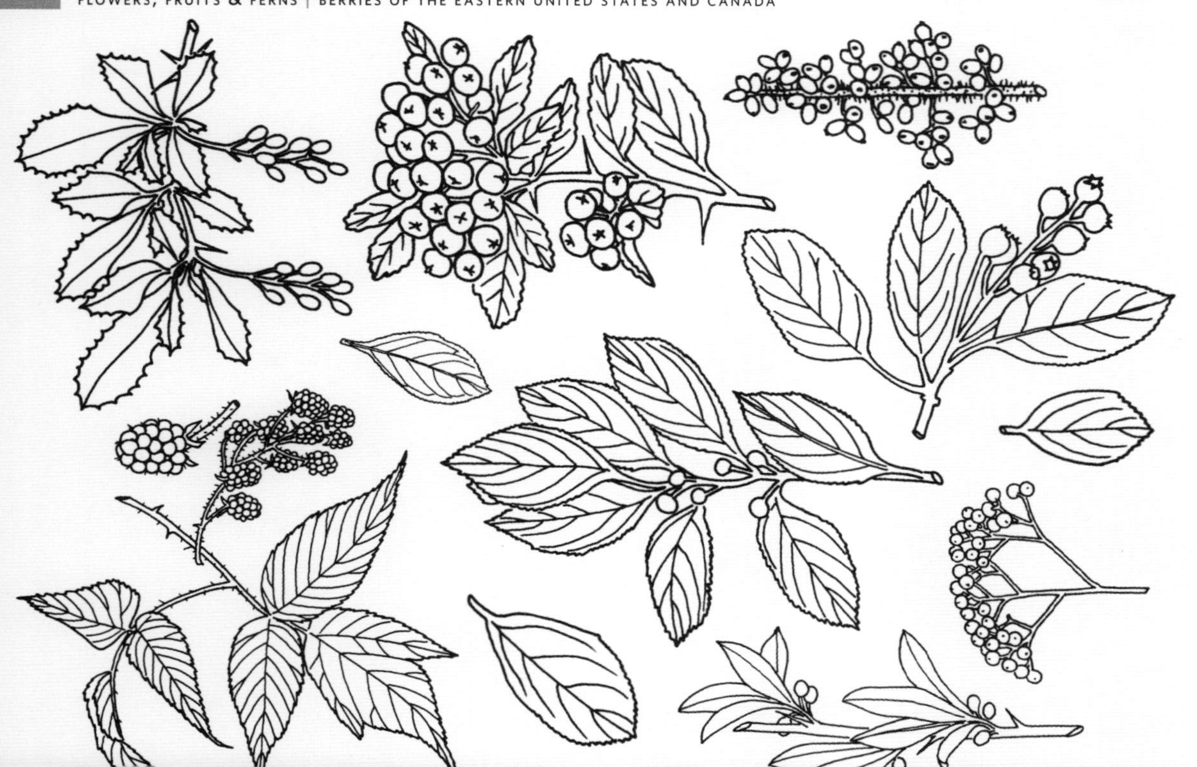

Identifying the Berries of Herbs (nonwoody plants)

Select the leaf arrangement, leaf shape, and other clues that match your plant, and go to the section indicated.

Basal or whorled leaves

simple leaves with entire edges

red berries, **go to** ⟶ PAGE **158**

blue, purple, or black berries, **go to** ⟶ PAGE **159**

divided leaves

red berries, **go to** ⟶ PAGE **160**

blue berries, it is **BLUE COHOSH** *Caulophyllum thalictroides*

purple-black berries, it is **SARSAPARILLA** *Aralia nudicaulis*

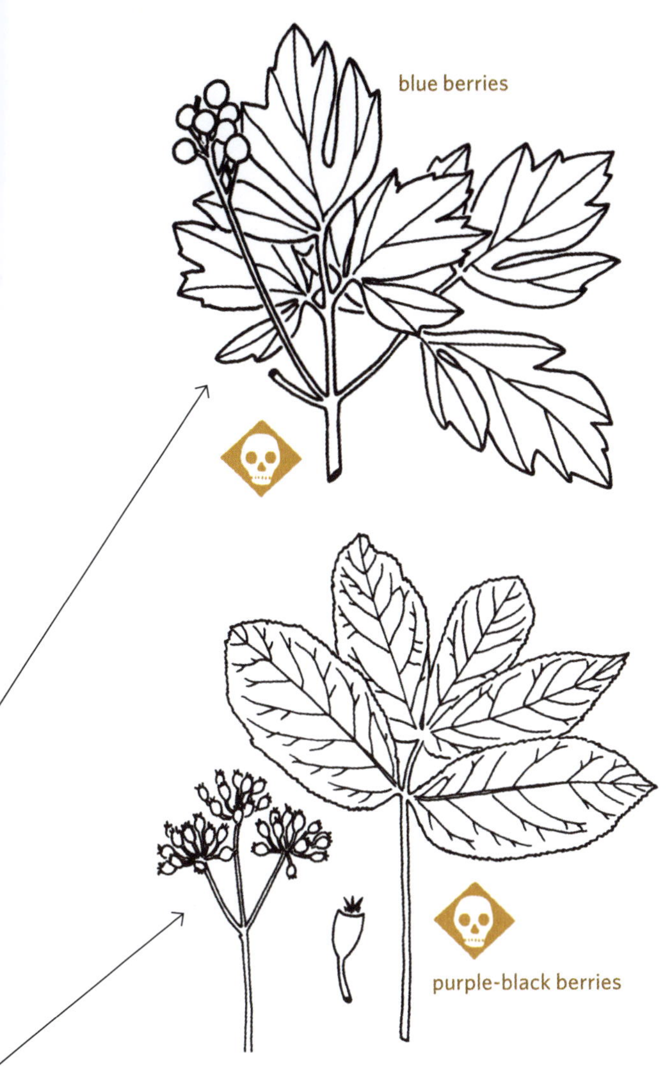
blue berries
purple-black berries

OPPOSITE LEAVES

simple leaves with entire edges

red-orange or yellow-orange berries, in
leaf axils, it is **FEVERWORT** or **WILD COFFEE**
Triosteum spp.

red berries, at ends of stems, it is
PARTRIDGEBERRY *Mitchella repens*

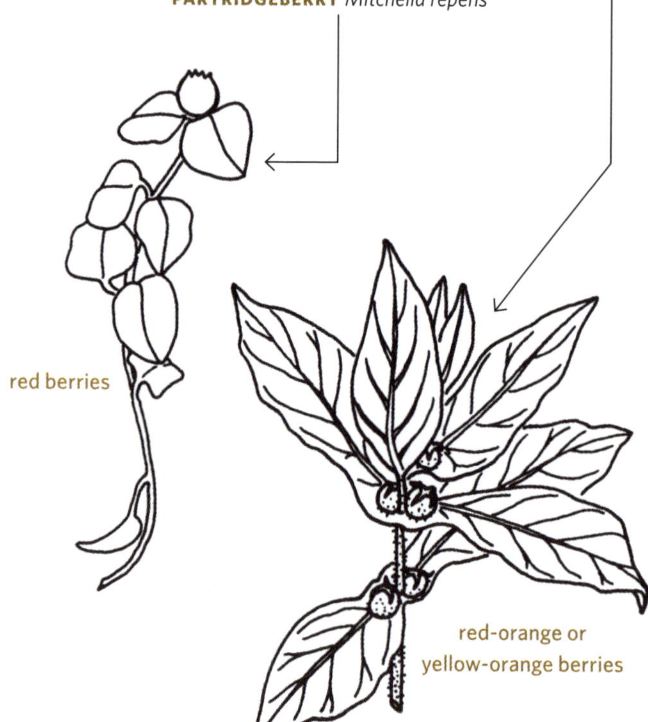

ALTERNATE LEAVES

simple leaves with entire edges

    red berries, **go to** ⟶ PAGE **161**

    yellow berries, **go to** ⟶ PAGE **165**

    green berries turning red, **go to** ⟶ PAGE **162**

    black berries, **go to** ⟶ PAGE **163**

simple leaves with toothed edges

    red, yellow, or black berries, **go to** ⟶ PAGE **165**

simple leaves with lobed edges (toothed or not)

    berry yellow and not an aggregate, it is
    HORSE NETTLE *Solanum carolinense*

    berry and an aggregate, **go to** ⟶ PAGE **166**

divided leaves

    white, red, or dark purple/black berries,
    go to ⟶ PAGE **167**

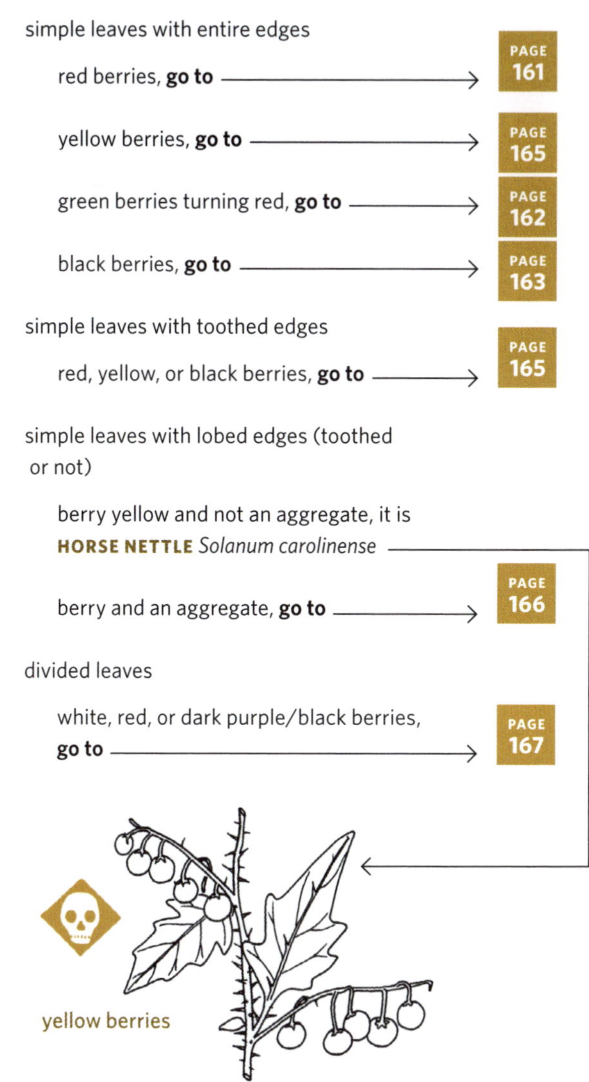

yellow berries

157

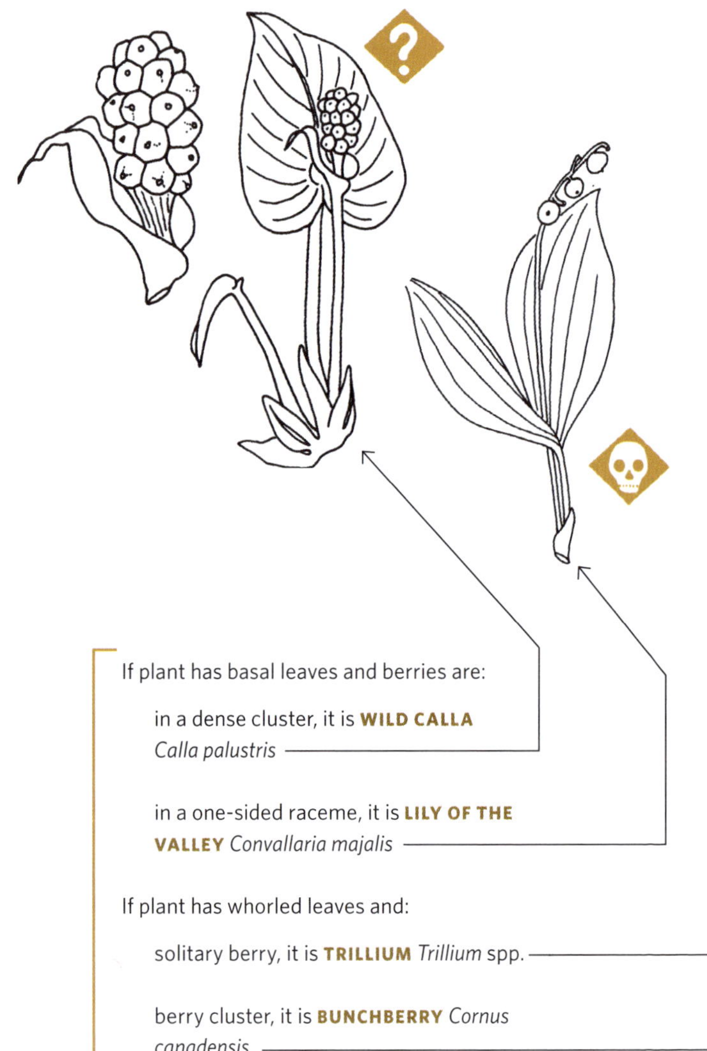

If plant has basal leaves and berries are:

in a dense cluster, it is **WILD CALLA** *Calla palustris*

in a one-sided raceme, it is **LILY OF THE VALLEY** *Convallaria majalis*

If plant has whorled leaves and:

solitary berry, it is **TRILLIUM** *Trillium* spp.

berry cluster, it is **BUNCHBERRY** *Cornus canadensis*

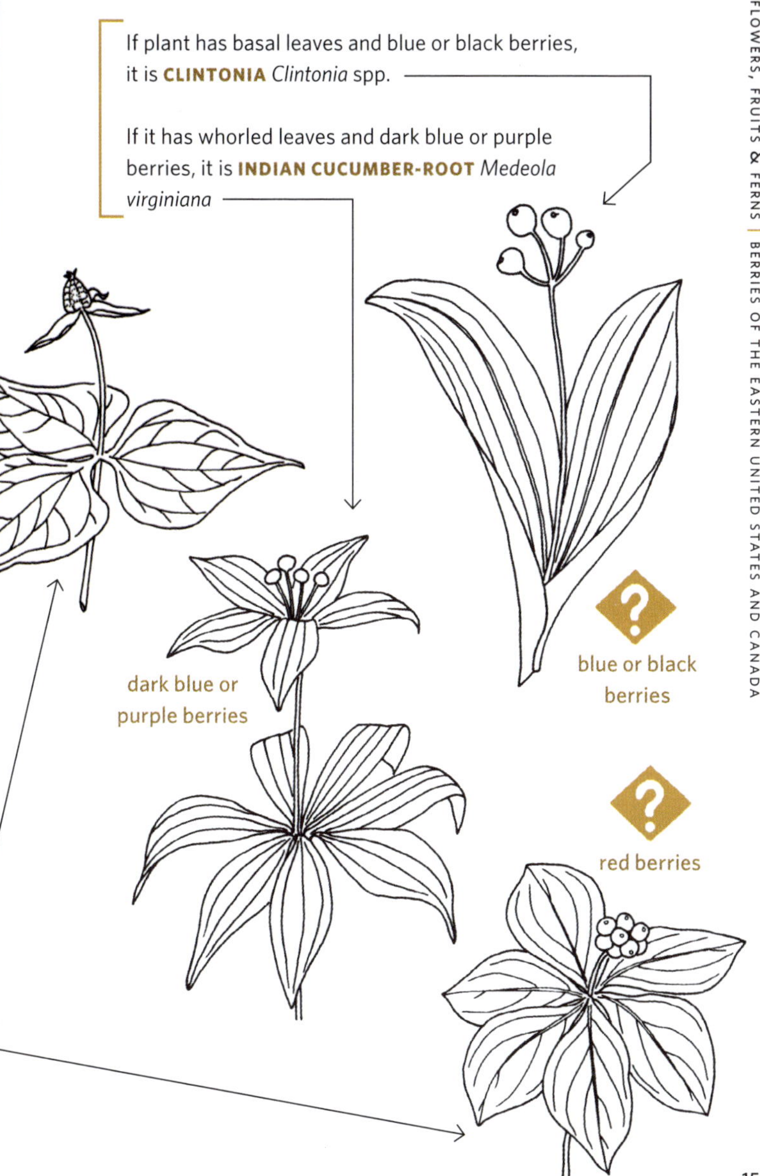

If plant has basal leaves and blue or black berries, it is **CLINTONIA** *Clintonia* spp.

If it has whorled leaves and dark blue or purple berries, it is **INDIAN CUCUMBER-ROOT** *Medeola virginiana*

dark blue or purple berries

blue or black berries

red berries

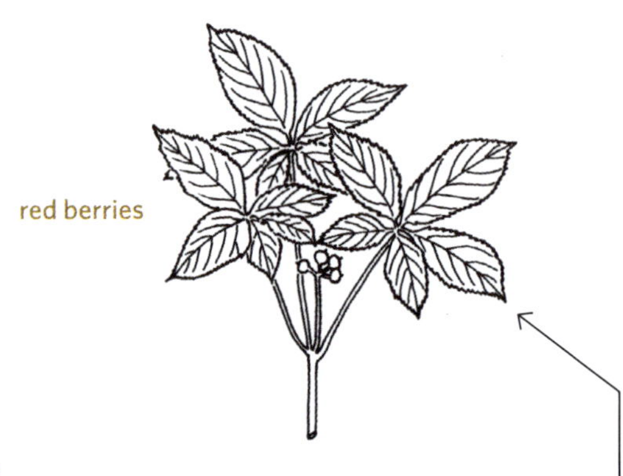

If plant has whorled leaves and red berries, it is
GINSENG *Panax quinquefolius*

If it has basal leaves and:

five to fifteen leaflets, berries on a club-like
structure, it is **GREEN DRAGON** *Arisaema
dracontium*

three leaflets, berries on club-like structure, it
is **JACK-IN-THE-PULPIT** *Arisaema* spp.

three leaflets, berries borne singly, it is
WILD STRAWBERRY *Fragaria* spp.

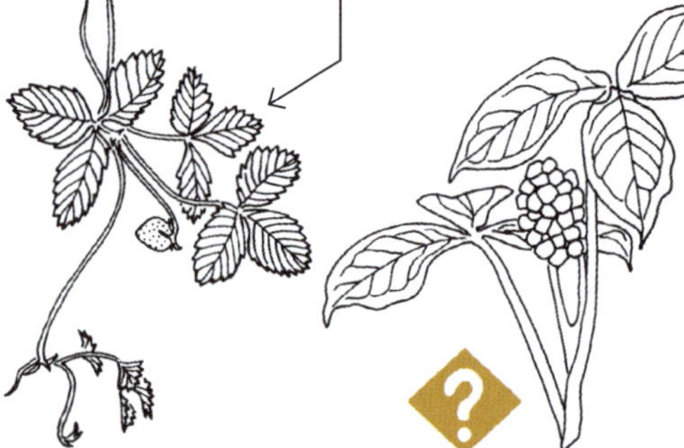

If berries are in leaf axils and are:

in a pod, **go to, GROUND CHERRY** ⟶ PAGE 165

not in a pod and

plant is feathery, it is **ASPARAGUS** *Asparagus officinalis*

plant has broad leaves, it is **TWISTED STALK** *Streptopus* spp.

If berries are borne terminally, **go to** ⟶ PAGE 162

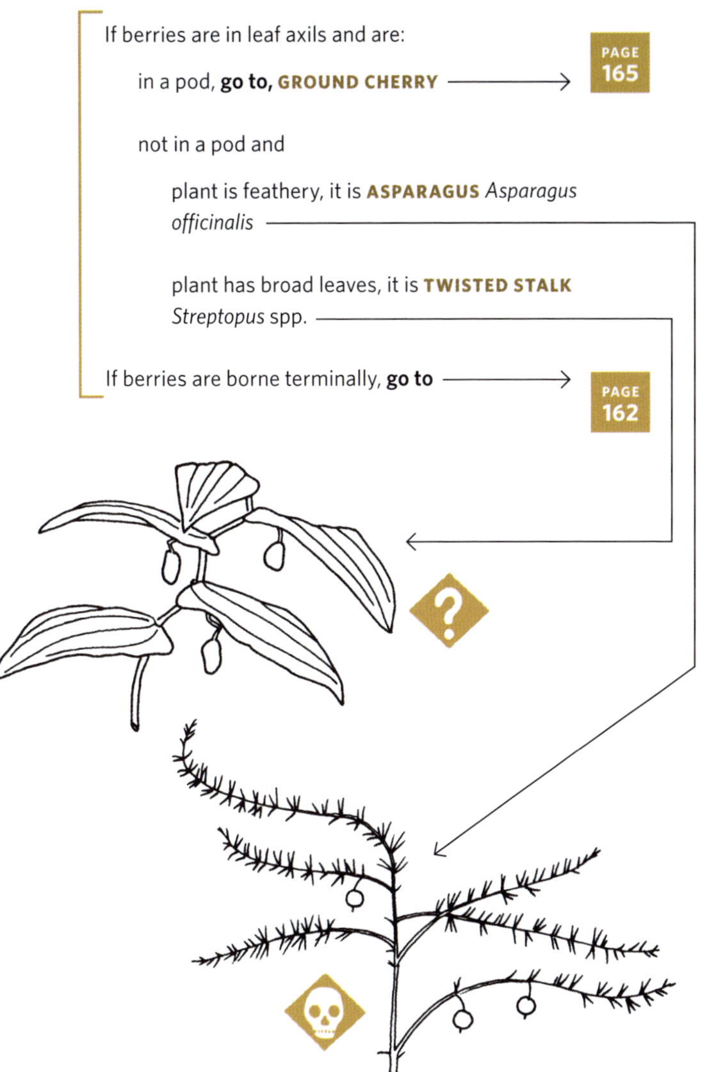

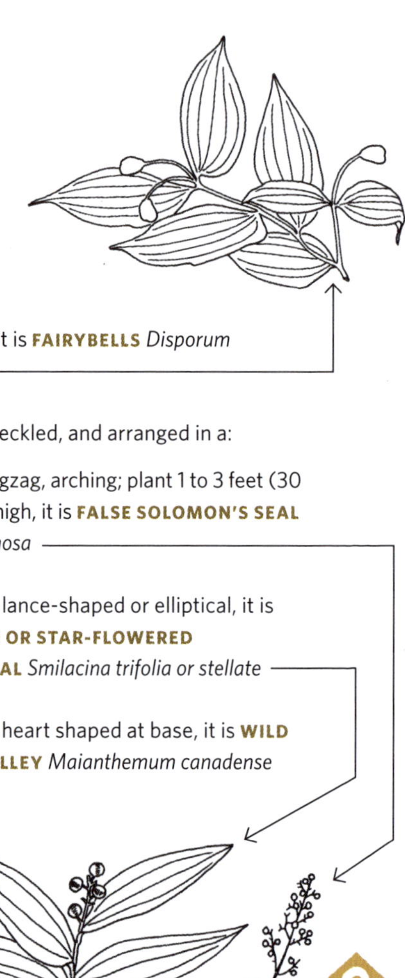

If berries are oval, it is **FAIRYBELLS** *Disporum lanuginosum*

If they're round, speckled, and arranged in a:

panicle, stem zigzag, arching; plant 1 to 3 feet (30 cm to 1 meter) high, it is **FALSE SOLOMON'S SEAL** *Smilacina racemosa*

raceme; leaves lance-shaped or elliptical, it is **THREE-LEAVED OR STAR-FLOWERED SOLOMON'S SEAL** *Smilacina trifolia or stellate*

raceme; leaves heart shaped at base, it is **WILD LILY OF THE VALLEY** *Maianthemum canadense*

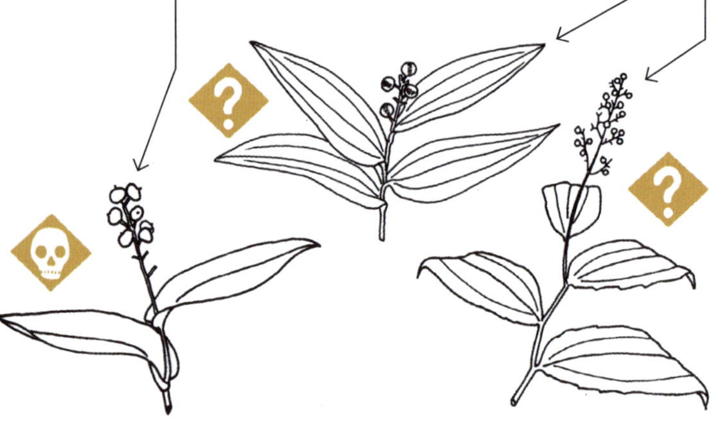

If berries grow:

from leaf axils, **go to** → PAGE 164

in umbels, it is **BLACK NIGHTSHADE, see page 165**

in a pod, it is **BLACKBERRY LILY** *Belamcanda chinensis*

in racemes (button-shaped berries), it is **POKEWEED** *Phytolacca americana*

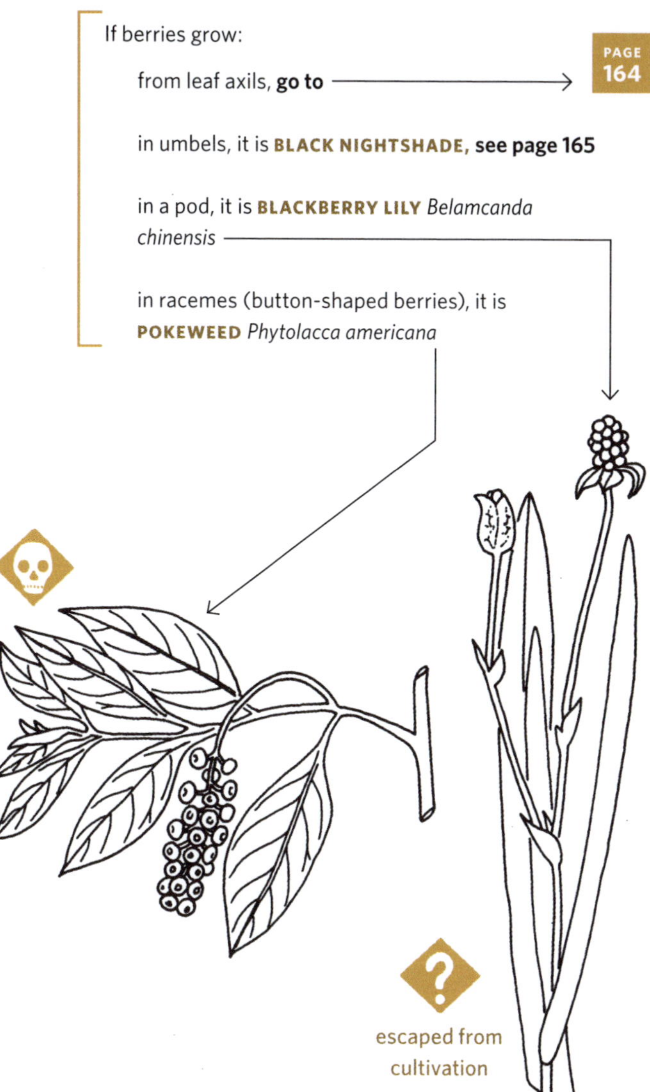

escaped from cultivation

If berries are blue black and in long-stalked umbels, it is **CARRION FLOWER** *Smilax* spp.

If they are long stalked, often in pairs, it is **SOLOMON'S SEAL** *Polygonatum* spp.

If they are small, stalkless bulblets, it is **TIGER LILY** *Lilium tigrinum*

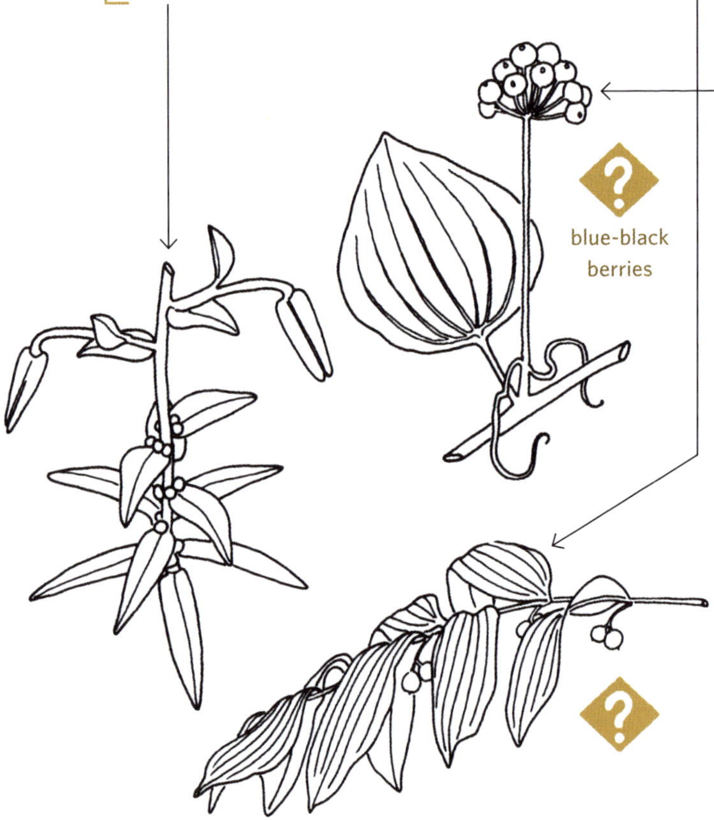

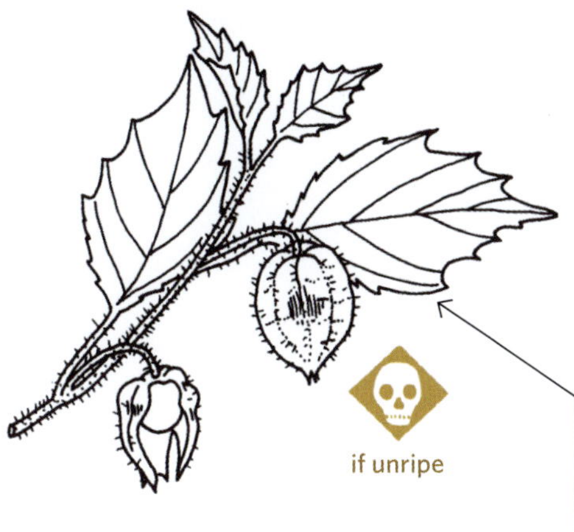

If berries are red or yellow, in pods, it is
GROUND CHERRY *Physalis* spp.

If berries are black; some leaves may be broadly
toothed or entire, it is **BLACK NIGHTSHADE**
Solanum nigrum

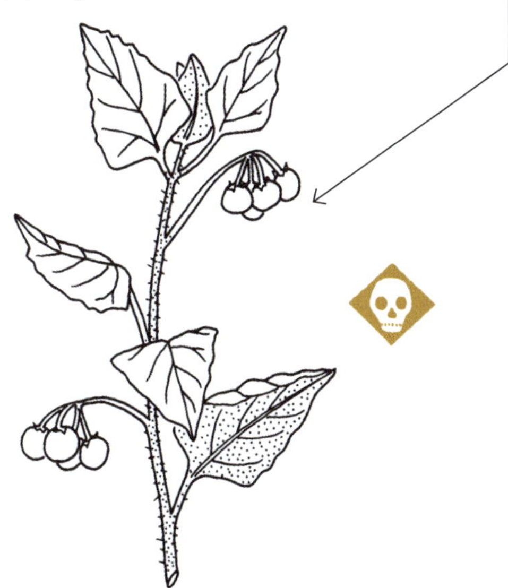

If berries are:

orange red; plant has a hairy stem, grows in the woods, it is **GOLDENSEAL** *Hydrastis canadensis*

gold or pale red; plant grows in a bog, it is **CLOUDBERRY** *Rubus chamaemorus*

bright red; plant has triangular leaves, it is **STRAWBERRY BLITE** *Chenopodium capitatum*

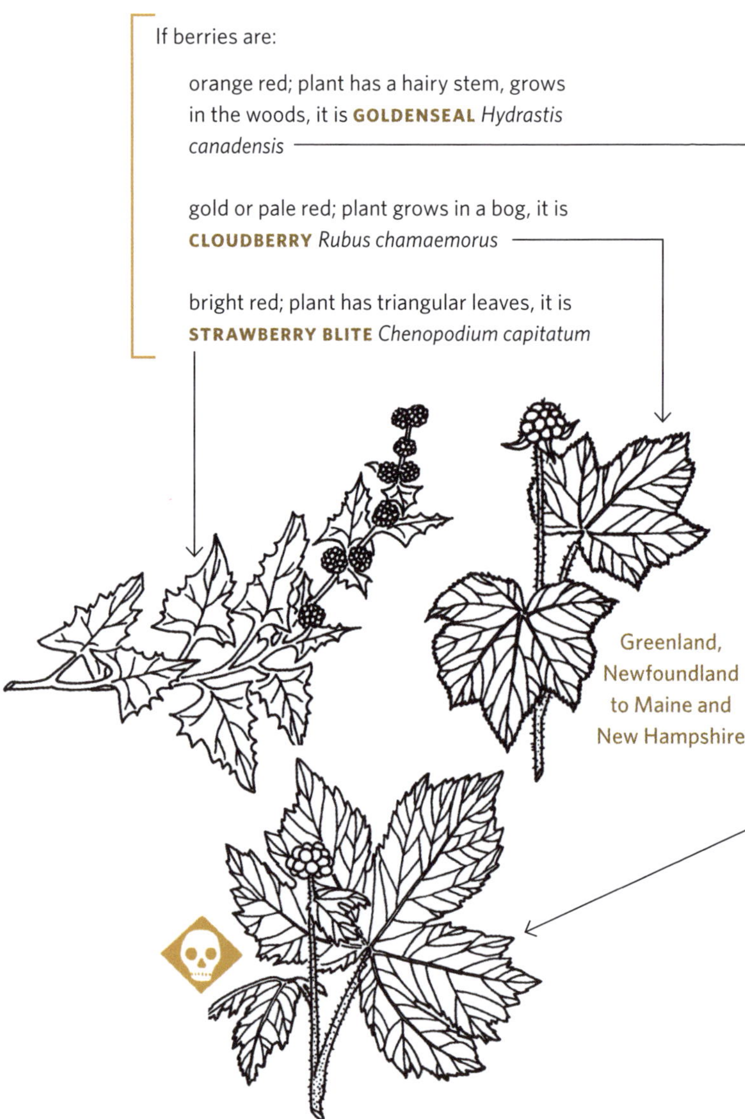

If plant has red or white berries that:

look like strawberries; three leaflets per leaf, it is **INDIAN STRAWBERRY** *Duchesnea indica*

are in a 2- to 4-inch (5- to 10-cm) showy cluster; many leaflets per leaf, it is **BANEBERRY** *Actaea* spp.

If plant has dark purple-black berries that are in numerous umbels; plant 3 to 5 feet (1 to 1½ meters) high; stem smooth, it is **AMERICAN SPIKENARD** *Aralia racemosa*

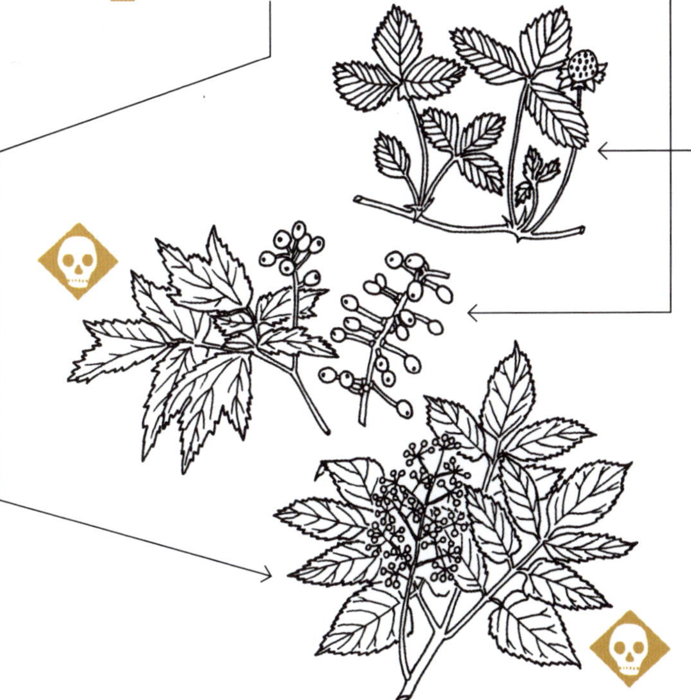

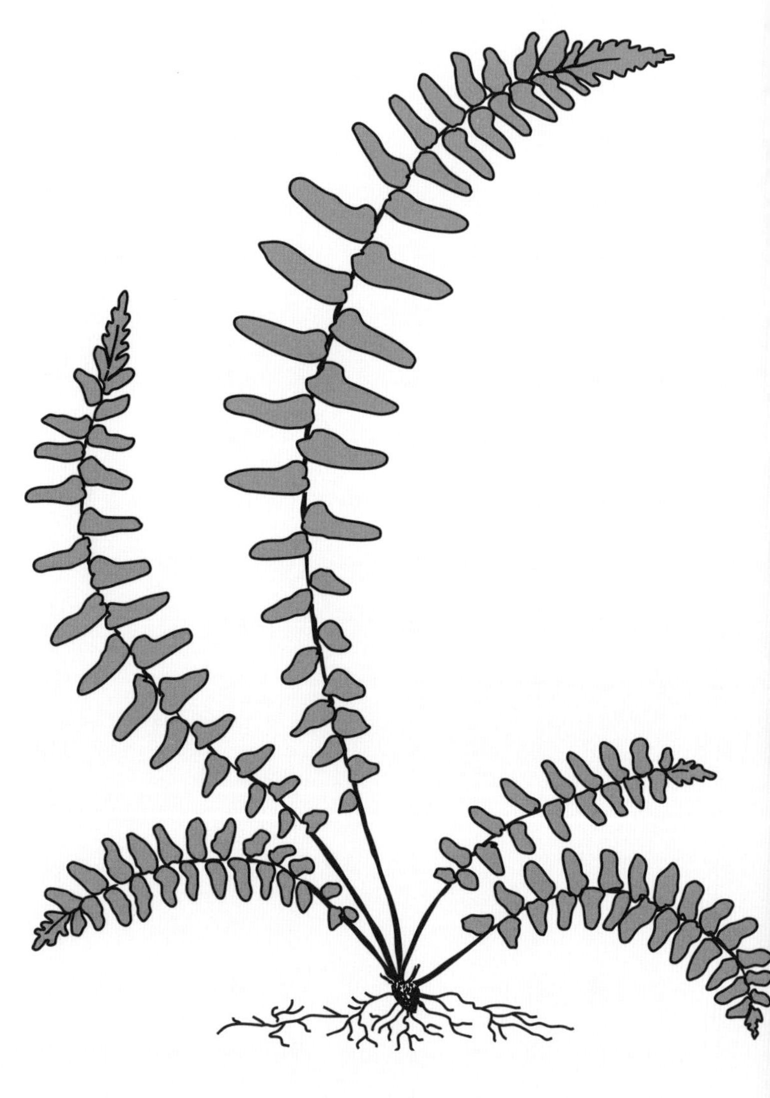

Ferns

Parts of a fern

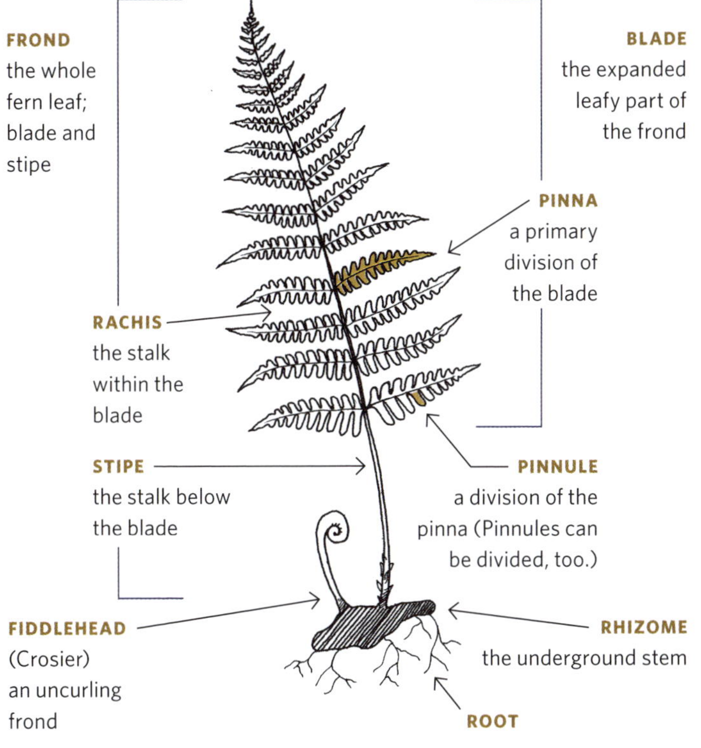

To identify a fern, look at:

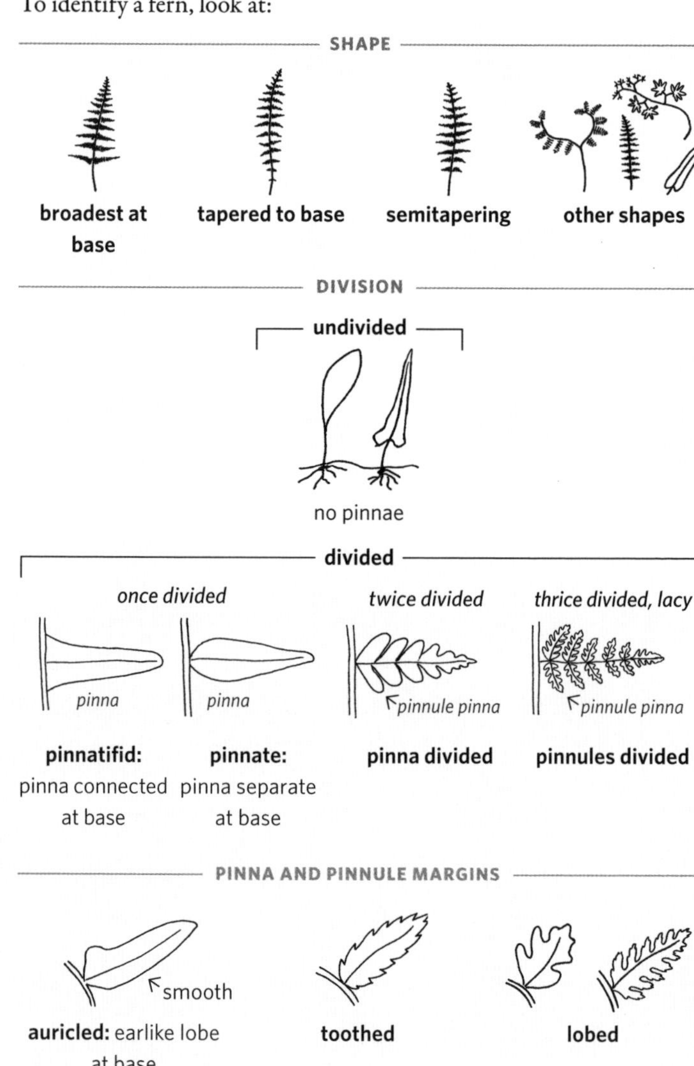

GROWTH FORM

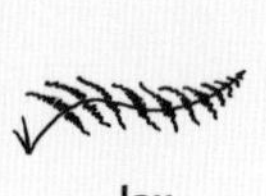

lax

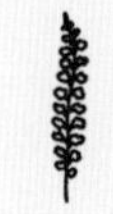

erect

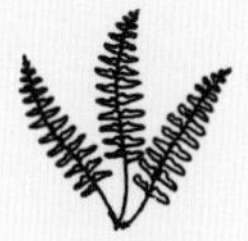

clustered

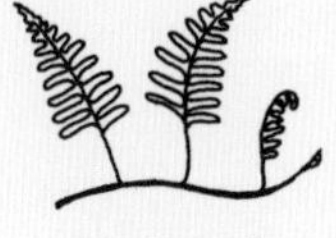

scattered

STIPE AND RACHIS FEATURES

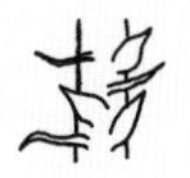

scaly

smooth

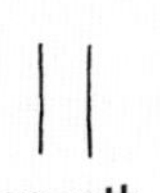

hairy

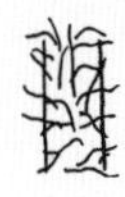

colored

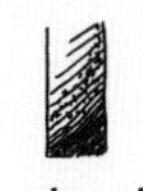

wooly

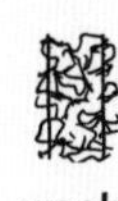

glandular

winged

OTHER FEATURES

pinnae or blade

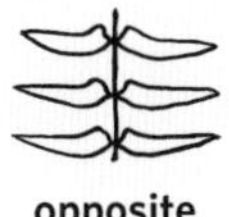

**opposite
pinnae**

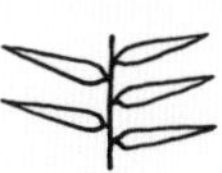

**alternate
pinnae**

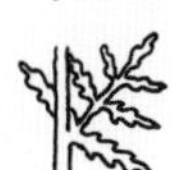

sessile

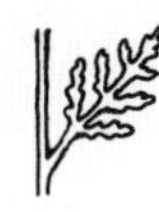

stalked

REPRODUCTIVE FEATURES

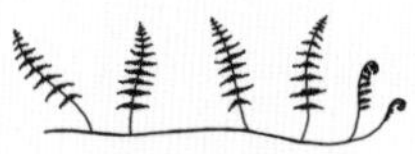

**shoots form
rhizomes,** like
hay-scented fern

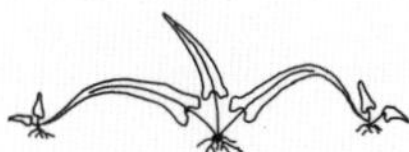

**new plants form leaf
tips,** like walking fern

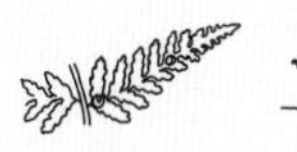

bulblets which drop off
and take root, like bulblet
fern spore-bearing
parts, see next page

171

- Ferns produce nearly microscopic reproductive cells called **SPORES.**

- Spores develop inside spore cases called **SPORANGIA.** →

- A fern frond with sporangia is a **FERTILE FROND.**

- A fern frond without sporangia is a **STERILE FROND.**

- Sporangia are usually grouped into clusters called **SORI** (singular sorus) or fruit dots on the backs of the blades.

Each fern species has a typical sorus shape and arrangement.

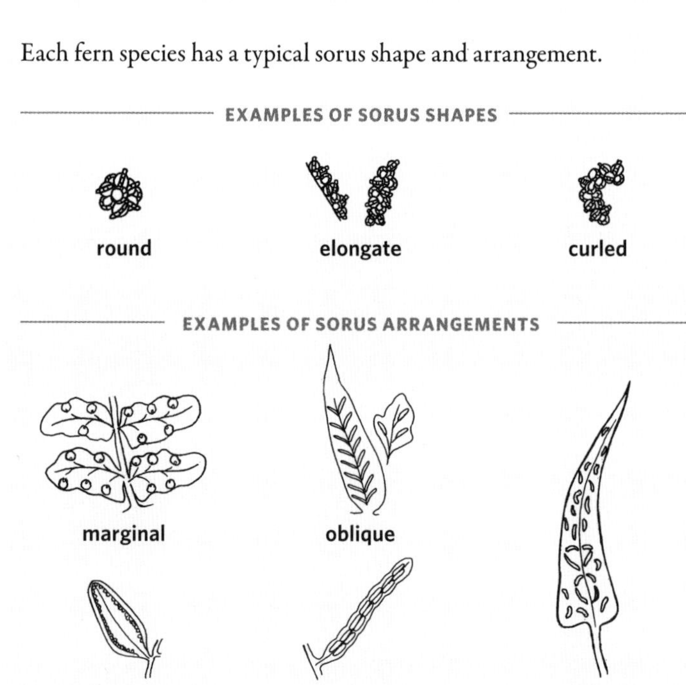

A thin layer of cells, an **indusium**, usually covers the sorus while the spores are developing inside the sporangia.

examples of indusium shapes

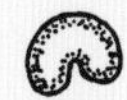 **KIDNEY-SHAPED**, attached at one side, as in wood ferns

 **ROUND, ATTACHED CENTRALLY**, as in Christmas fern

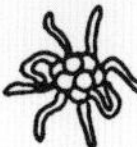 **STAR-SHAPED**, attached under sorus, as in woodsias

 **ELONGATE**, attached at one side, as in chain ferns

 **CURVED**, attached at one side, as in lady fern

 **CUP-LIKE**, as in hay-scented fern

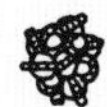 **NO INDUSIUM**, as in polypody

As spores ripen, the indusium pulls back, shrivels, and often disappears, exposing the sporangia. When spores are fully ripe, the sporangia burst open and spores spill out.

Some ferns bear spores on parts that have no green, leafy tissue.

F=*fertile (spore bearing)*
S=*sterile*

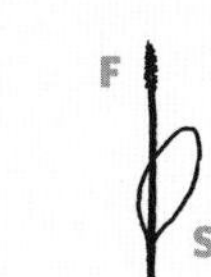

Adder's tongue

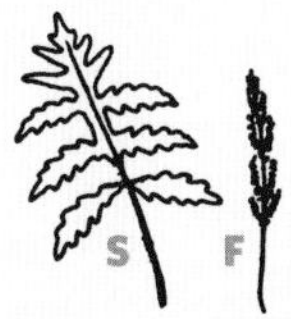

Sensitive fern

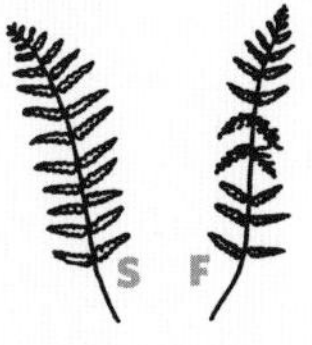

Interrupted fern

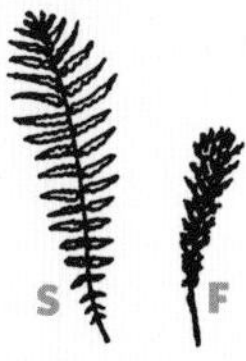

Ostrich fern

The life cycle of a fern

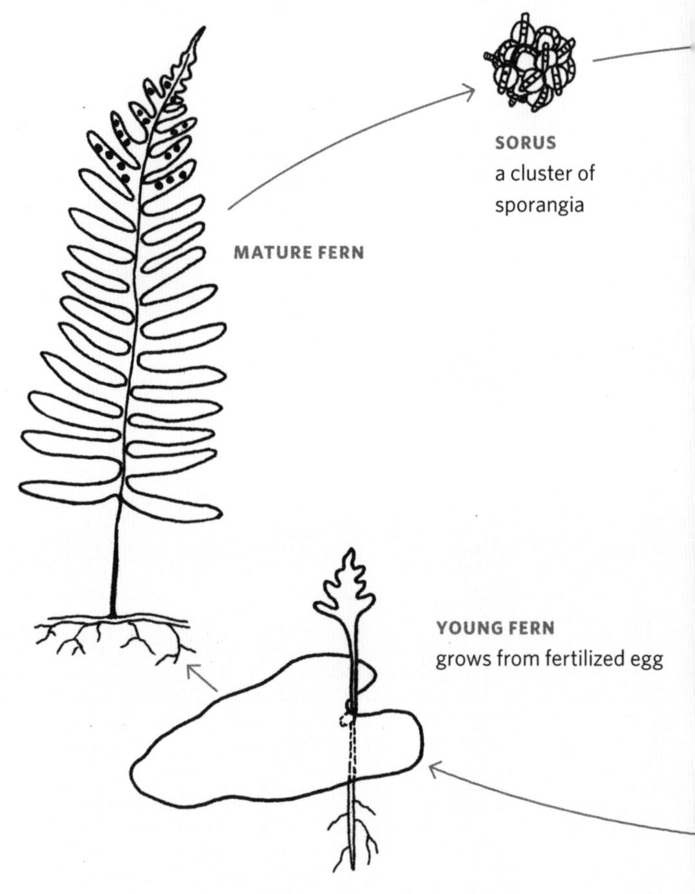

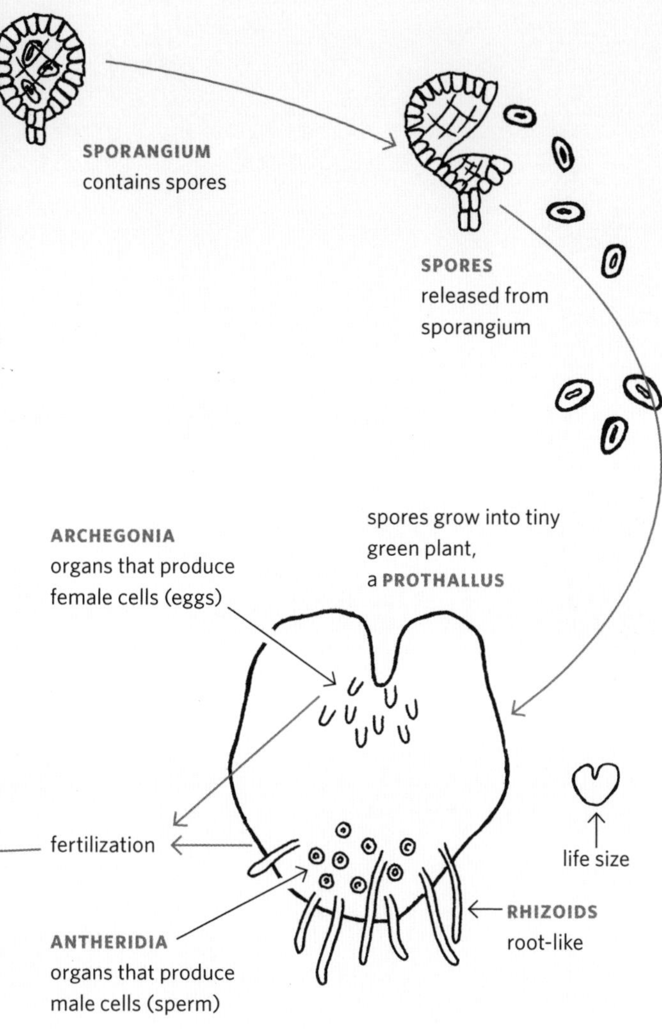
SPORANGIUM
contains spores
SPORES
released from
sporangium
ARCHEGONIA
organs that produce
female cells (eggs)
spores grow into tiny
green plant,
a PROTHALLUS
fertilization
life size
RHIZOIDS
root-like
ANTHERIDIA
organs that produce
male cells (sperm)

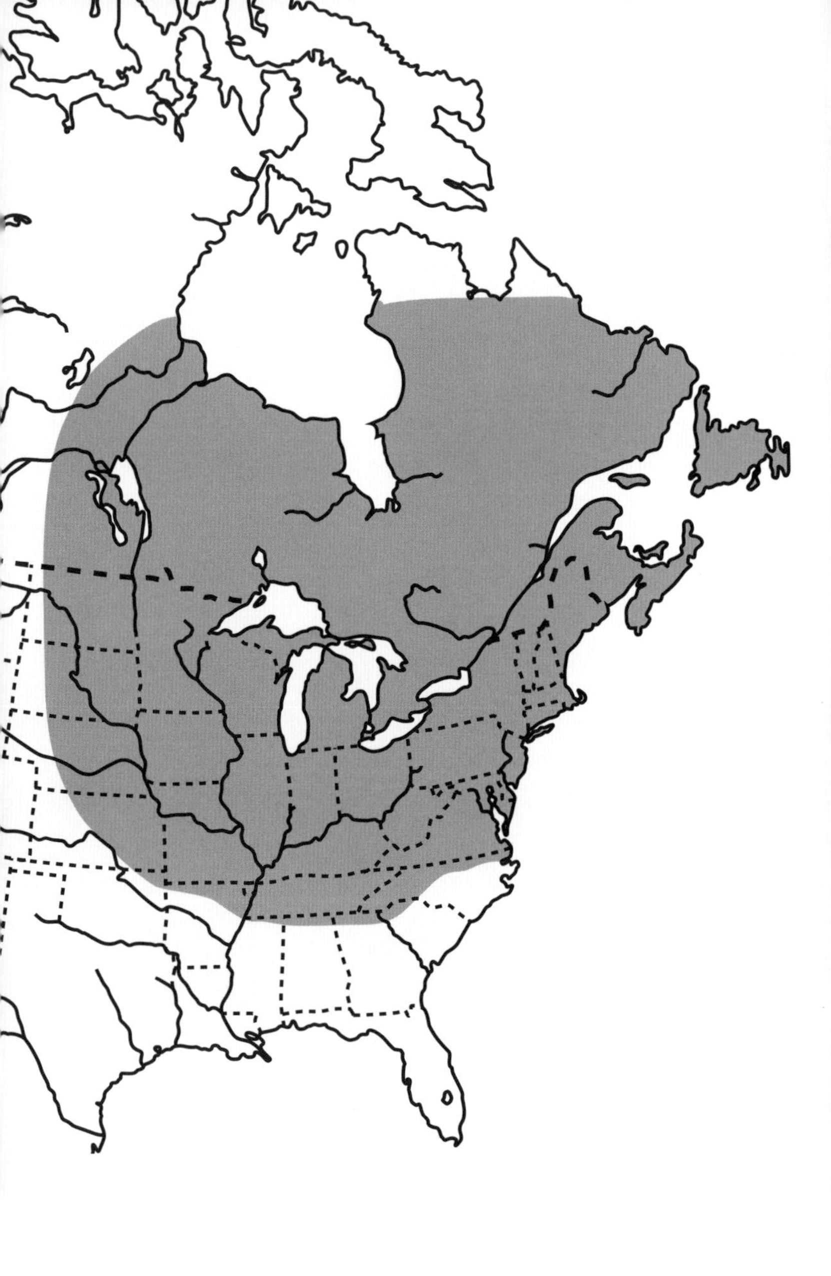

Ferns of the Northeastern United States and Eastern Canada

Habitats

 **WOODS**—in mostly shaded places

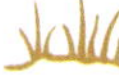 **FIELDS**—in open, weedy fields or grassy meadows

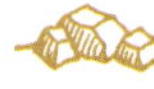 **ROCKS**—among or on rocks, cliffs, walls, or ledges

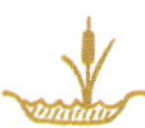 **WET AREAS**—in bogs and swamps, often "with feet wet"

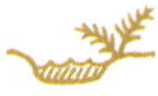 **MOIST BANKS**—along banks of streams and edges of wet areas, not "with feet wet"

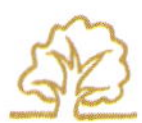 **TREES**—on tree trunks and large, spreading limbs

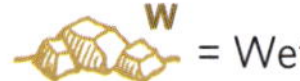 = Wet   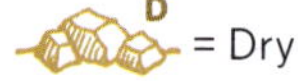 = Dry

Many ferns grow in more than one habitat. 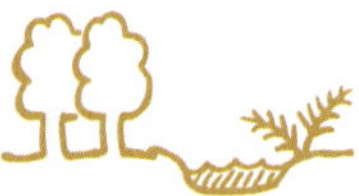

Identifying Selected Ferns with Divided and Non-Divided Blades

If blade (the green, leafy part of the frond) is divided, **go to** ⟶

If blade is not divided, **go to** ⟶

If blade is leaflike, **go to** ⟶

If blade is grasslike, wiry, and curly, it is **CURLY GRASS FERN** *Schizaea pusilla*

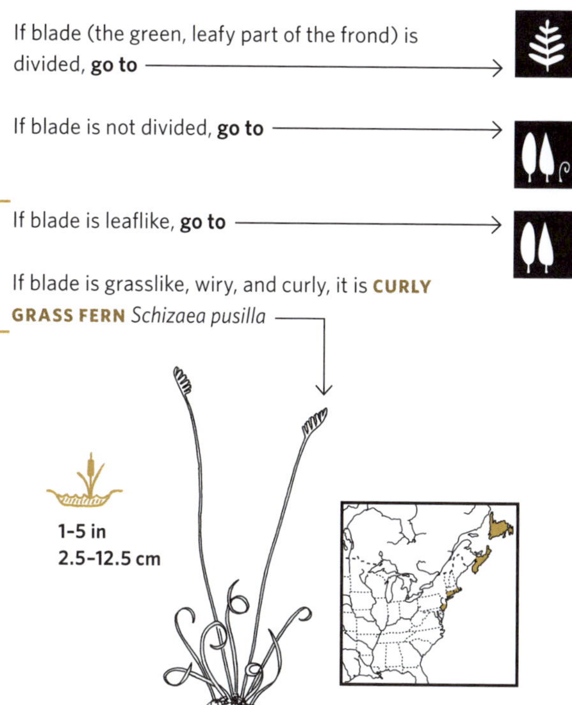

1–5 in
2.5–12.5 cm

If blade is oval or elliptical, with a fertile stalk arising from its base, **go to** ⟶ 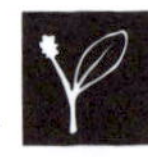

If blade has heart-shaped base and elongate sori on underside, **go to** ⟶ 

If blade tip has a tiny, sharp point, it is **LIMESTONE ADDER'S-TONGUE** *Ophioglossum engelmannii*

If tip is blunt, it is **ADDER'S-TONGUE** *O. vulgatum* (*O. pycnosticum*)

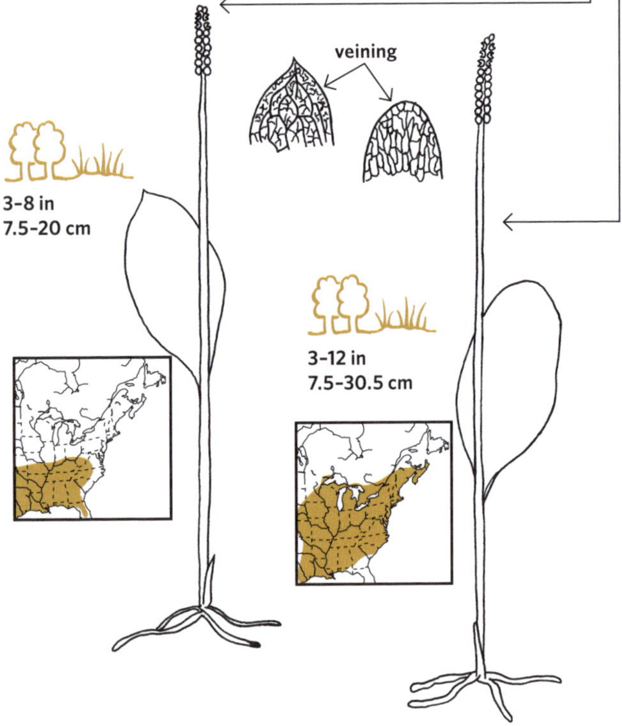

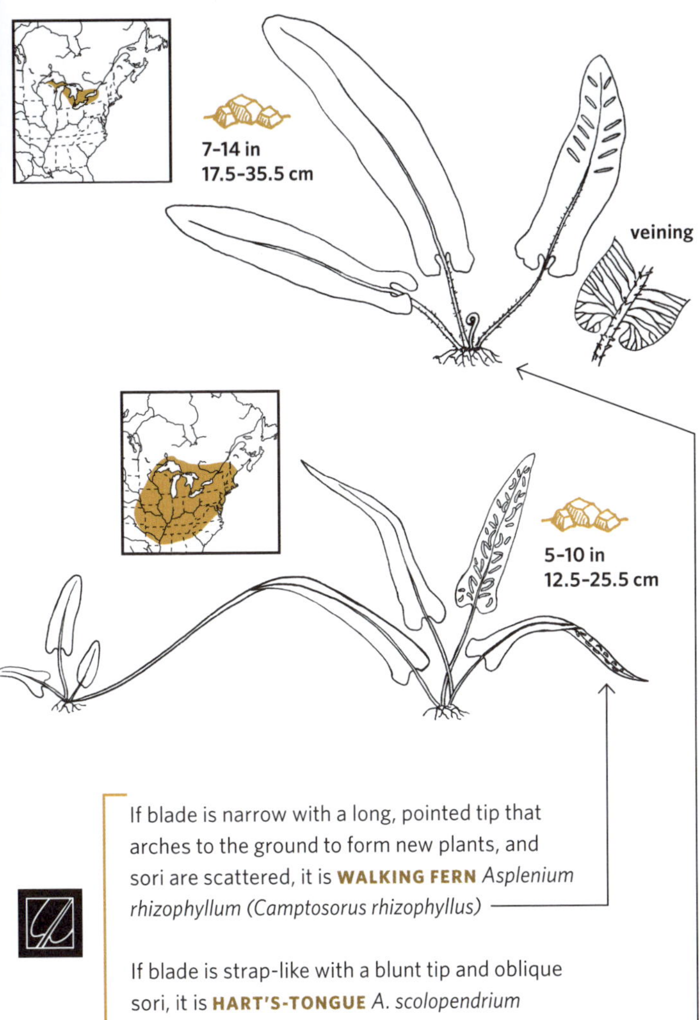

If blade is narrow with a long, pointed tip that arches to the ground to form new plants, and sori are scattered, it is **WALKING FERN** *Asplenium rhizophyllum (Camptosorus rhizophyllus)*

If blade is strap-like with a blunt tip and oblique sori, it is **HART'S-TONGUE** *A. scolopendrium (Phyllitis scolopendrium)*

If blade is vine-like, and each pinna divides into two hand-shaped parts, it is **CLIMBING FERN** *Lygodium palmatum* ————

If blade resembles a four-leaf clover, it is **WATER CLOVER** *Marsilea* spp. ————

If blades have scalelike pinnae and the plant is free-floating on water, it is **MOSQUITO FERN** *Azolla* spp. ————

If none of the above, **go to** ⟶ 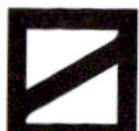

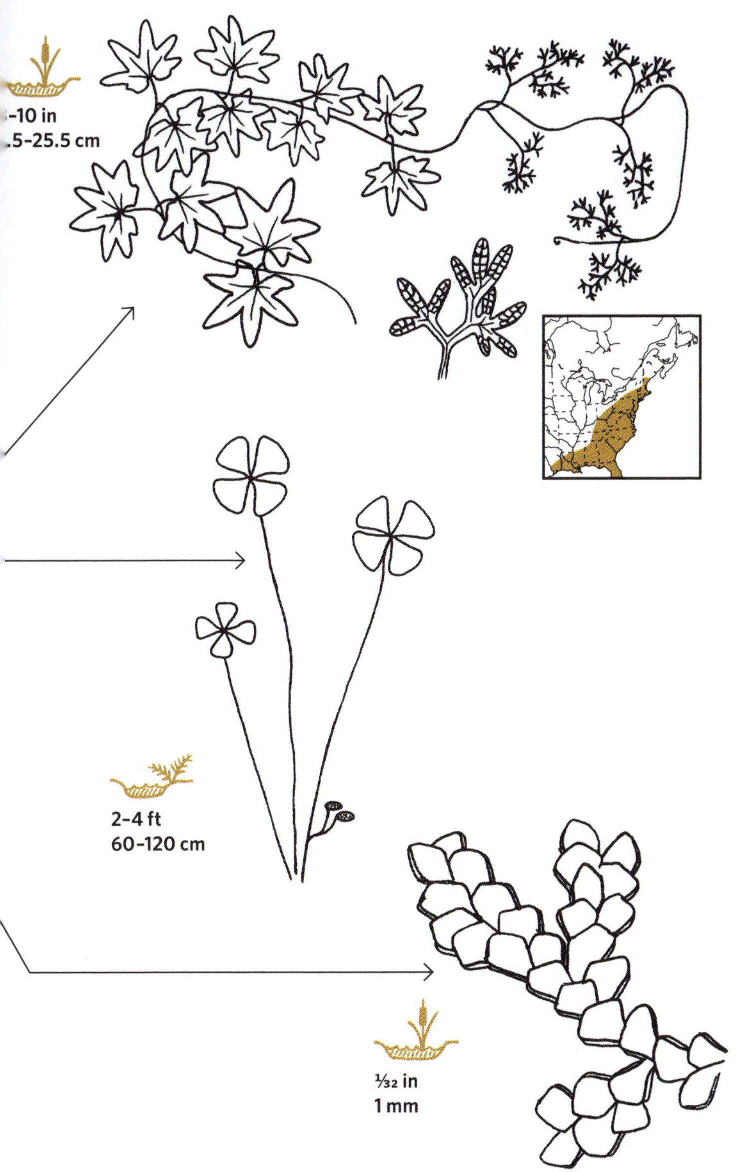

6–10 in
15.5–25.5 cm
2–4 ft
60–120 cm
1/32 in
1 mm

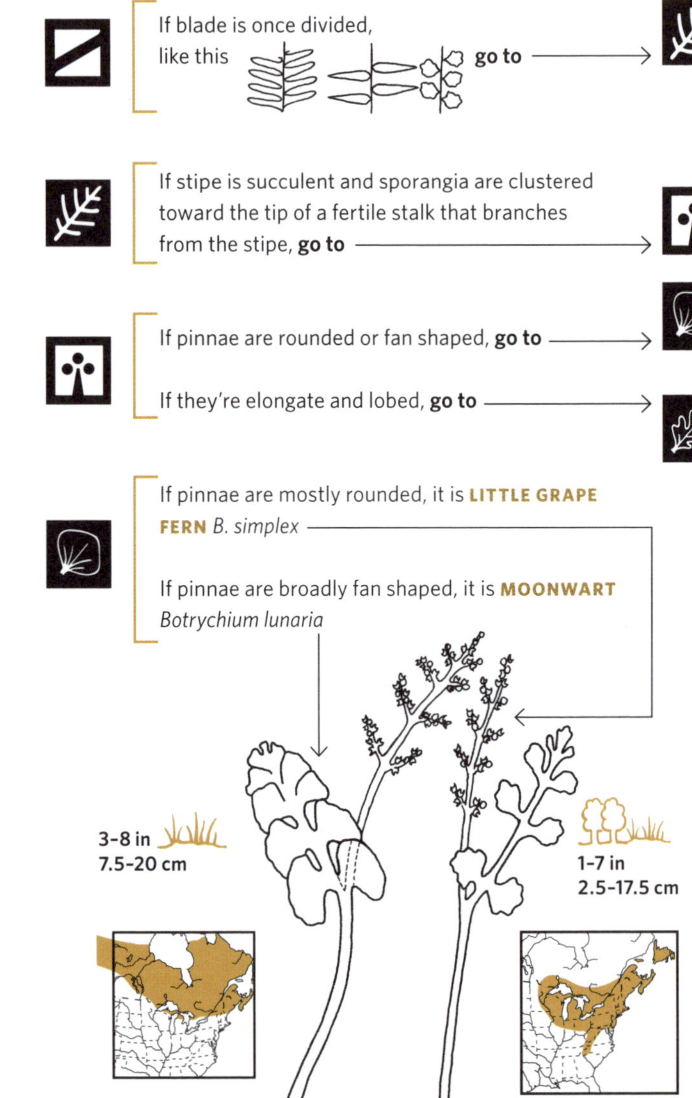

If blade is once divided, like this **go to**

If stipe is succulent and sporangia are clustered toward the tip of a fertile stalk that branches from the stipe, **go to**

If pinnae are rounded or fan shaped, **go to**

If they're elongate and lobed, **go to**

If pinnae are mostly rounded, it is **LITTLE GRAPE FERN** *B. simplex*

If pinnae are broadly fan shaped, it is **MOONWART** *Botrychium lunaria*

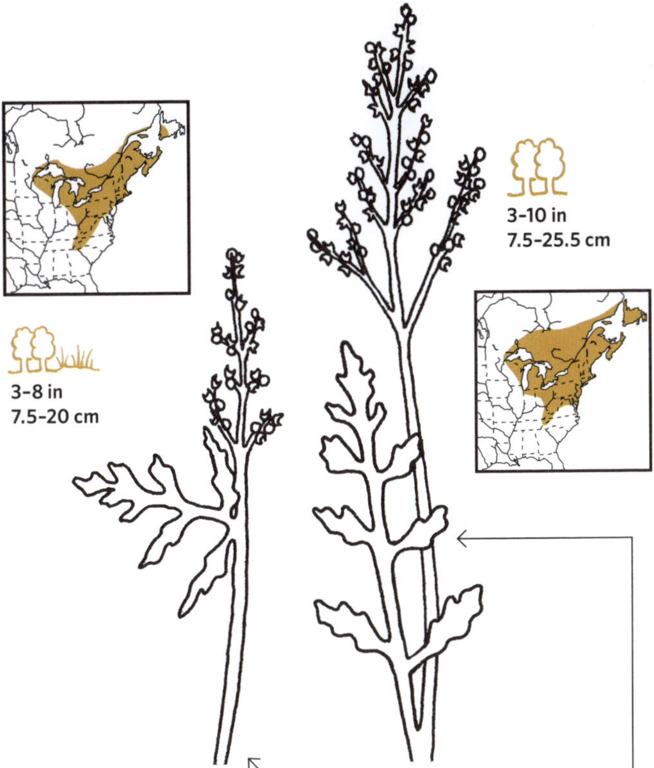

3–10 in
7.5–25.5 cm

3–8 in
7.5–20 cm

If blade is a broad triangle and is sessile, it is
LANCE-LEAVED GRAPE FERN *B. lanceolatum*

If blade is a narrow triangle with a short stalk, it is
DAISY-LEAVED GRAPE FERN *B. matricariifolium*

185

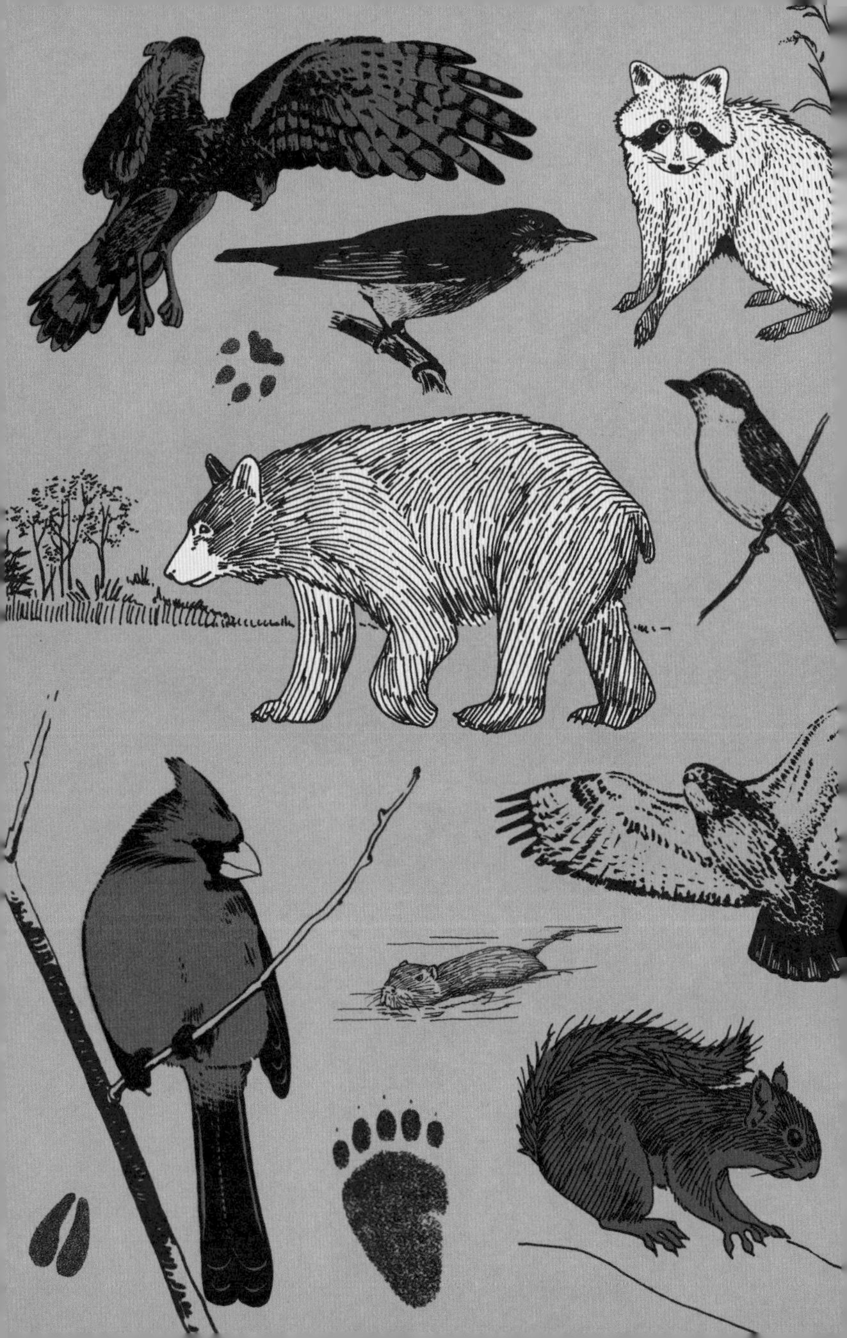

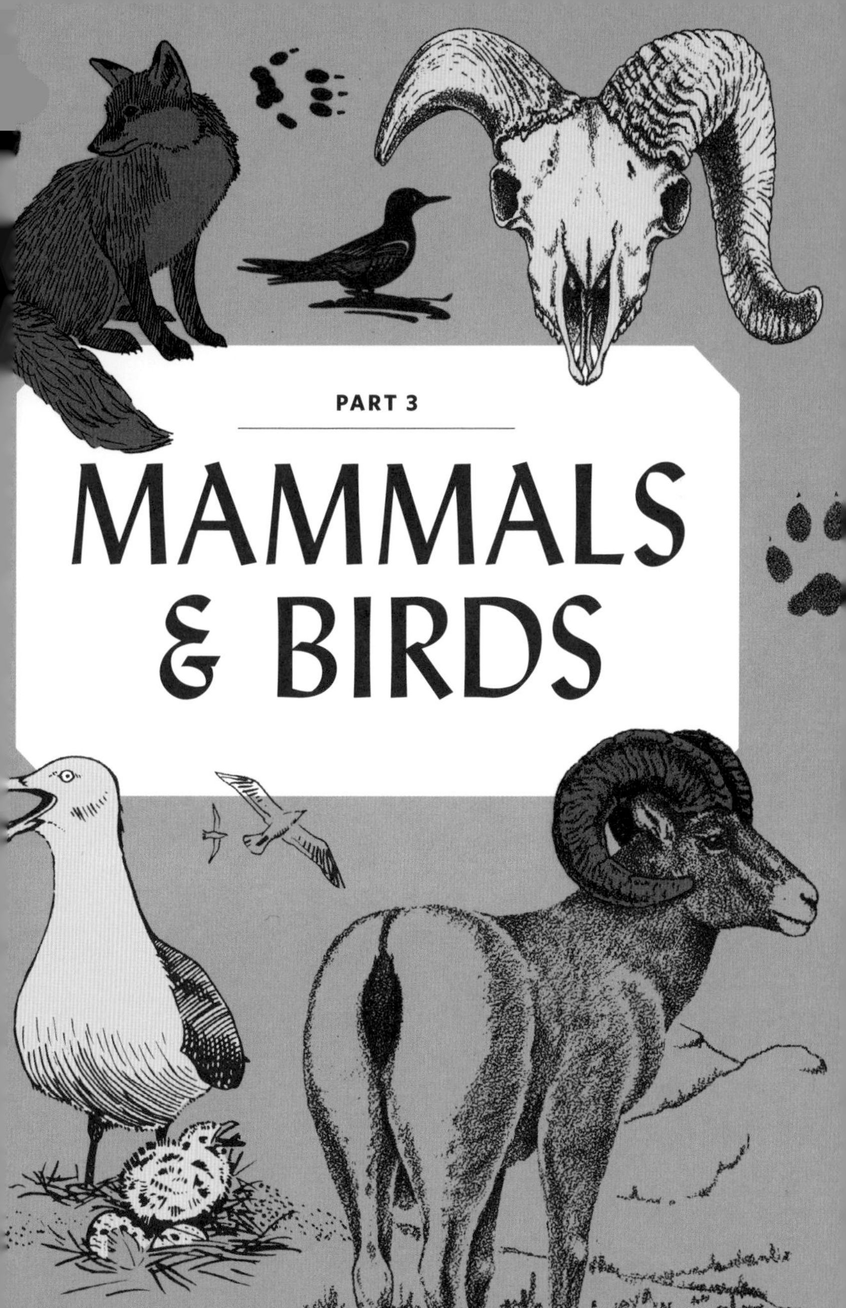PART 3

MAMMALS
& BIRDS

Mammals

MAMMALS DIFFER FROM OTHER ORGANISMS in having fur, large brains, a highly refined sense of hearing, and a variety of skin glands. Mammals also suckle their young and spend an extended time caring for them. Here are some other traits.

Body temperature

Mammals keep warm under a coat of hair or fur. To protect against winter temperatures, many mammals develop thicker, longer fur, which they shed in the spring. Deer get extra insulation from the trapped air inside their hollow hairs. Each species keeps a fairly constant body temperature by panting, sweating, and confining activity to the cooler hours of summer, or to warmer slopes in winter. Body temperature varies, however, among different groups of mammals.

Mammals living in desert regions must confine their activity to the cool of night to avoid overheating and loss of precious body water. Many desert mammals get all the water they need from the food they eat. They control sweating and concentrate their urine and droppings, retaining as much water as possible. Kangaroo rats have gone a step further in having developed nasal traps that remove moisture from their exhaling breath.

Mammals, in any environment, that are active by night and quiet by day are called **nocturnal**. Sometimes you can spot nocturnal mammals by the reflected shine when their eyes, specially equipped for night vision, are struck by the light of a car or campfire.

Camouflage

The color of mammal hair generally blends with the terrain, which helps mammals to avoid detection by prey or predators. When a predator approaches, many mammals, like hares and rabbits, remain motionless and trust their safety to camouflage and stillness. Most mammals see no color, only shades of black and white. A still, well-concealed animal may never be seen, even at a distance of a yard or so. If the predator gets too close the prey will attempt to flee, a tactic also used by quail and pheasants. In some mammals, like squirrels, all-black (melanistic) individuals are occasionally seen.

Reproduction

Mammals have one of the most successful forms of reproduction in the animal kingdom. The development of the young within an amniotic sac provides nourishment directly from the mother and protection from bacteria. The survival rate of fetuses tends to be high as is the survival of newborns because of extended postbirth care. Although many are born naked and blind, others are ready to go. Antelope fawns are ready to stand and run within a few minutes after birth.

Antlers and horns

Some mammals have antlers or horns growing from their skulls. Male members of the deer family and both sexes of caribou produce antlers. Antlers grow externally from calcium deposited by blood-filled capillaries underneath furred skin. By late summer, growth stops and the furry skin, called **velvet**, is rubbed off against tree trunks and small saplings. Antlers are sharpened for use and display during the fall mating season. They are shed in winter, regrown in spring.

In contrast, horns grow from an inner core of calcium-rich blood tissue. Antelope horns are shed annually, but those of bighorn sheep are permanent and grow to spectacular sizes and shapes. Males competing for females judge each other's status by the size and shape of horns, antlers, and physique. In pronghorn antelope and bighorn sheep, females have smaller horns than males.

Ever-growing teeth

Rabbits and rodents have incisors that grow continuously, an adaptation to grit in their diet that would otherwise wear their teeth to the gumline. Ever-growing teeth are kept under control normally by grit, or by gnawing action. If these animals don't have enough grit in their food, or are unable to gnaw regularly, their incisors will grow outside of their skulls and protrude from their mouths. If unchecked, these teeth will continue to grow until they turn backward toward the skull, preventing the animal from eating. Starvation and death result.

Territories

Mammals often restrict their activities to a definite area, called a **home range**. In order to get enough to eat, attract mates, and survive, mammals often defend all or a part of their home range as their **territory**, and routinely drive off intruders of the same species. A defended territory usually has enough food, shelter, and nesting material to support a male and female of the same species. Nonbreeding animals usually exist in unclaimed zones or strips found between two or more territories of their species. Territory owners usually patrol their boundaries, marking them in prescribed locations with scent, urine, or droppings. The size of territories varies from year to year as the availability of food and other essential resources changes. If food is plentiful, territories are smaller than in a lean year. Territories expand in response to scarcity. The territories of noncompeting species usually overlap.

Parasites

Wild animals support a wide variety of hitchhiking parasites like fleas, ticks, blood-sucking flies, and lice. Parasites can weaken their hosts, and occasionally kill them. Because some of these parasites can transmit diseases like plague and tularemia to people, it is a good idea to **avoid handling dead animals**, especially rodents and rabbits.

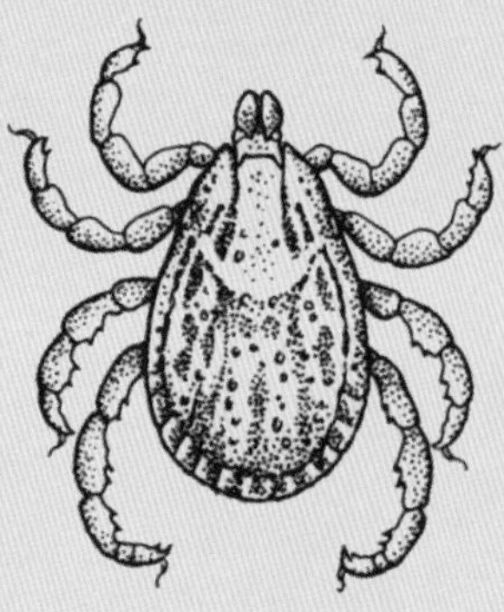

Rocky Mountain Wood Tick
Dermacentor andersoni

Skin glands

Mammals have five different kinds of skin glands. Mammary glands in females produce milk to nurse young. Sweat glands help to cool mammals and get rid of waste products through pores of the skin. Oil glands lubricate skin and hair. Scent and musk glands produce chemicals to mark territories and communicate.

Skulls

Dental Formulas

The dental formula for a skull or jaw is made by noting the numbers of each kind of tooth, from the front center to the rear of one side. Separate counts are made for the teeth attached to the skull and those in the jawbone.

A formula of:

$$\frac{1\ 0\ 1\ 3}{1\ 0\ 1\ 3}$$

indicates that one incisor, no canine, one premolar, and three molars should normally be present on each side of both the skull and the jawbone. This formula matches the squirrel teeth illustrated on page 193.

Health and age may cause teeth to be missing. So a specimen you find may not precisely match the stated formula.

If you find a skull or jawbone, try to match it with the ones shown in this chapter.

Carnivores

COYOTE

$$\frac{3\ 1\ 4\ 2}{3\ 1\ 4\ 3}$$

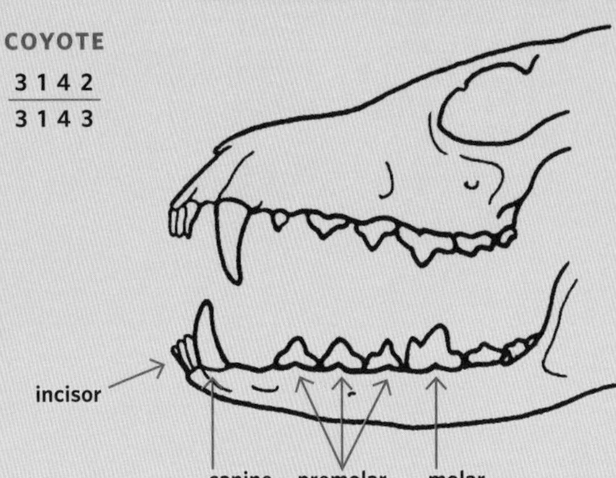

Herbivores

SQUIRREL

$$\frac{1\ 0\ 1\ 3}{1\ 0\ 1\ 3}$$

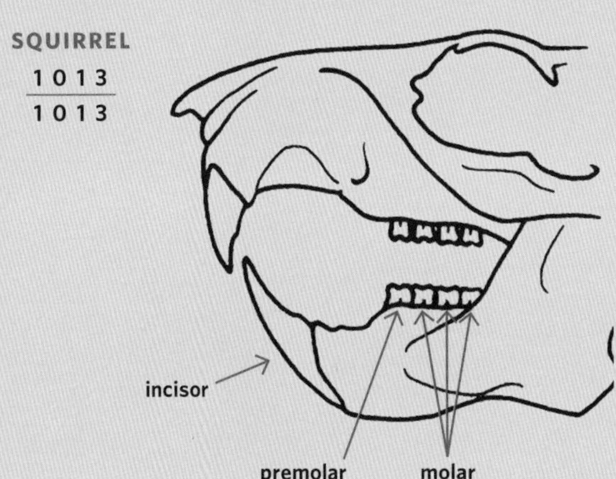

Identifying Select Mammals of the Pacific Coast

Nests, Burrows, Mounds, Depressions

If you've found:

- a linear ridge of earth in lawn or soil, which is either pushed up or caved in, and there is no obvious plugged tunnel entrance, **see MOLE, page 197**

- a nest or house of twigs made of stacked twigs in mounded piles 6 to 10 feet (2 to 3 meters) high/wide with mud packed on top, near twig dam, near or in water, **see BEAVER, page 198**

- a depression or matted-down area, **go to** ⟶ 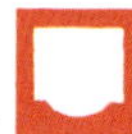

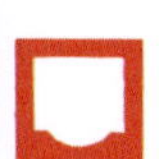

If the depression or matted area is:

- composed of tules or streamside grasses, matted down, with tufts of twisted grass, droppings, musky odor, **see river otter, page 206**

- in meadow, forest, chaparral area, with or without strong urine scent and droppings, over 3 feet (1 meter) long, **see elk, page 202**

Chew, Scratch, Browse, or Other Signs

If you've found:

- tree trunks/limbs with chew marks, gouges, or missing bark, **go to** ⟶ 

- tree trunks with vertical, obvious, long scratch or claw marks, usually associated with extensive damage to tree, **see bear, page 212**

- all of lower branches of a tree pruned evenly and horizontally, 6 to 10 feet (2 to 3 meters) above ground, **see elk, page 202**

- signs that don't fit into the above categories, **go to** ⟶ 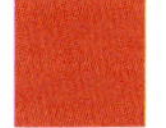

If the chew marks or gouges are:

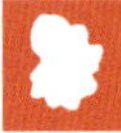

- many, on one side or all around lower part of tree trunk, deep into wood, often felling tree, near pond, stream, **see beaver, page 198**

- linear gouges that break through the bark, but not deep into wood, extending several feet above the ground, or sharp, with shredded bark, one meter or so above ground, common on saplings and low branches, **see elk, page 202**

- vertical, sharp, on underlying wood with bark torn away, 6 feet (2 meters) or more above ground, **see bear, page 212**

If you've found:

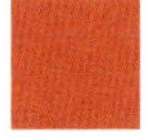

- a dam of sticks across a stream forming a pond, **see beaver, page 198**

- long, muddy slides on riverbanks, see **river otter, page 206**

A note on dimensions Unless otherwise indicated, dimensions given are for length. A single figure (1¾ inches/4.5 cm) represents a measurement made on a single specimen. A range (1½ to 2 inches/4 to 5 cm) indicates least and greatest dimensions of a number of specimens. Track measurements given are lengths of hind-foot prints left by the animal and are not the length of its whole hind foot.

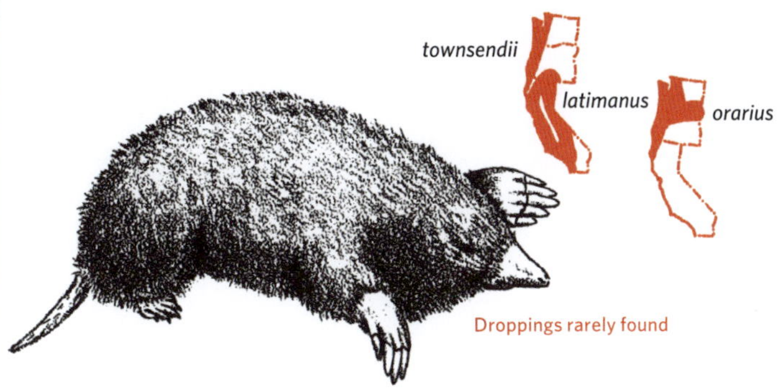

Mole

Scapanus townsendii, S. latimanus, S. orarius

Fur blackish brown or gray brown, velvety, flexible; allows forward or backward movement in tunnels with equal ease. Eyes pinhead size. Earholes concealed. Naked nose is most vital sense organ. Mole digs with broad front feet in breaststroke motion. Can tunnel 12 inches (30 cm) per minute. Mounds appear to erupt from earth. Tunnel ridges collapse in time; improve soil aeration, rain penetration, reduce erosion. Mole feeds on worms, insects, centipedes, snails, slugs, some root crops. Young born March to April, two to six per litter. Active day and night. Rarely above ground, thus low predation rate.

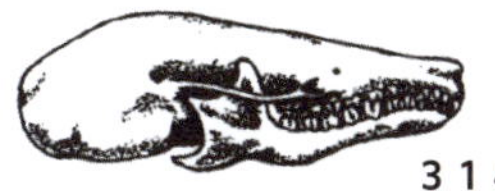

$$\frac{3\ 1\ 4\ 3}{3\ 1\ 4\ 3}$$

Skull lengths:
S. orarius 1¼ to 1½ inches (33 to 39 mm) (illus.)
S. townsendii 1½ to 1¾ inches (41 to 44 mm)
S. latimanus 1¼ to 1½ inches (29 to 37 mm)

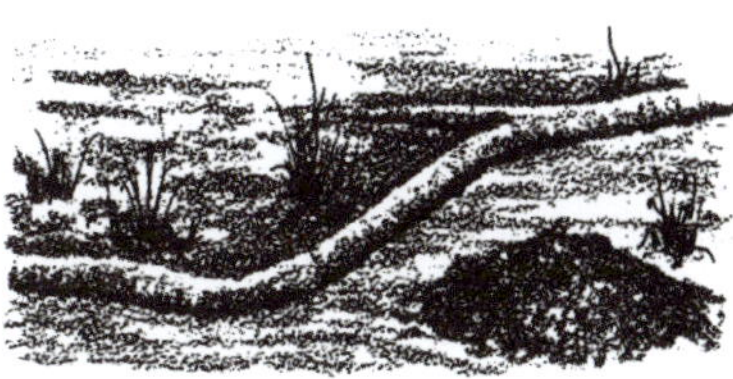

Mole runnel: ridge of pushed-up soil or grass. May collapse.
townsendii
orarius
latimanus

Beaver

Castor canadensis

Large, dark brown rodent with scaly tail. Skull with chestnut brown incisors. Lives in mountain streams, ponds, lakes. Its transparent eyelids cover eyes during dives to allow continuous vision while underwater. When alarmed, it slaps water with its tail as a warning to other beavers. Gnaws trees until they fall: for food; for building materials for bow-shaped dams; for dome-like lodges; and for scent mounds, which mark territory and attract mates. Burrow up through snow to gnaw trees at surface; as snow level changes, several gnawed bands may be left, creating a "totem pole" effect. Active throughout year. Feeds

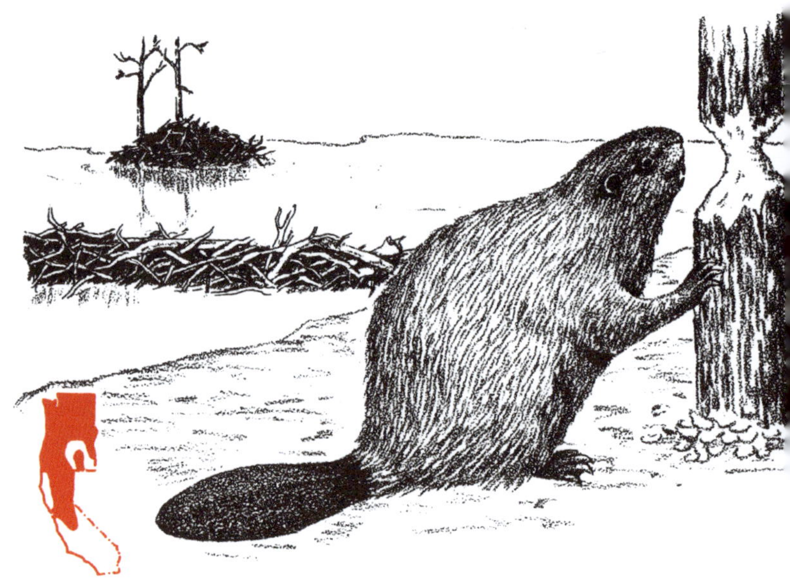

mostly on aspen, cottonwood, willow, and birch. Will also eat cattails, tules, willow roots, and pond lilies. Its droppings of coarse, sawdust-like material are rarely found because they decompose rapidly. Probably mates for life. Kits born furred, eyes open, four per litter. Stay with parents for two years. Lifespan ten to twelve years. Young killed by otters. Adults occasionally killed by coyotes, foxes, bobcats, and lions. Almost trapped out of existence in nineteenth century to supply soft fine fur for hats, robes, and coats, now reestablished over most of its former range.

4½ to 5 inches (112 to 130 mm)

1 0 1 3
1 0 1 3

hind 6½ inches (165 mm)

top view of molar

Droppings: 1⅓ to 1½ inches (35 to 40 mm), rarely found

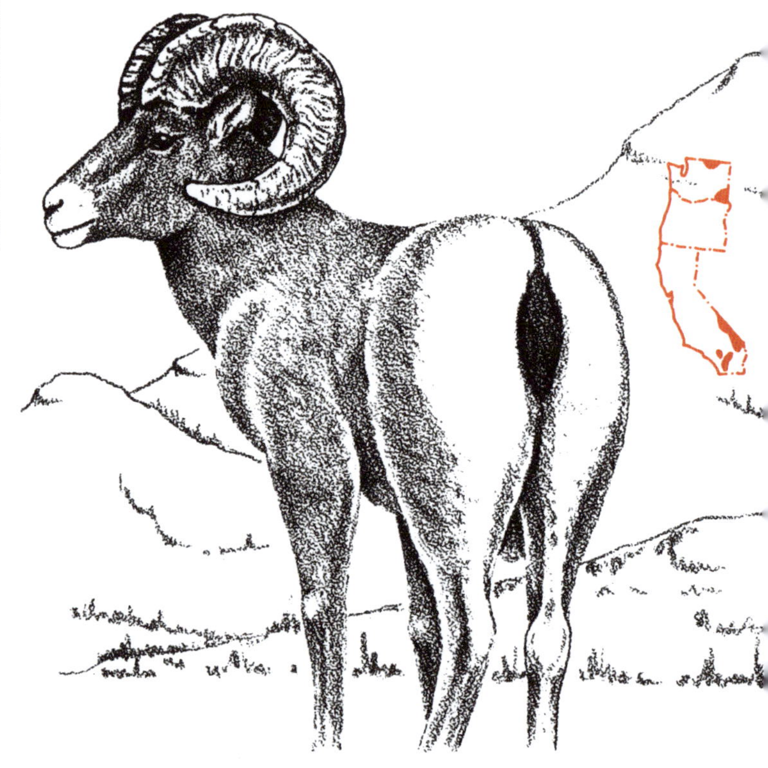

Bighorn Sheep

Ovis canadensis

Gray brown to ash gray; belly, rump white. Horns in rams thick, coiled; in ewes, not coiled, small. Sexes separate in summer, together in fall. Rams of equal size challenge each other for ewes, rear and charge at 20 mph (32 kph) to butt heads loudly. Skull double thick with struts of bone to cushion impact. Eats sedge, grass, sagebrush, and alpine plants. Bed a depression smelling of urine with droppings along edges, used for years. Single lamb born May to June, stays with herd. Lives fourteen years. Golden eagles take lambs. Threatened by weather, disease, loss of viable habitat, and intrusion. Several isolated races or populations occur.

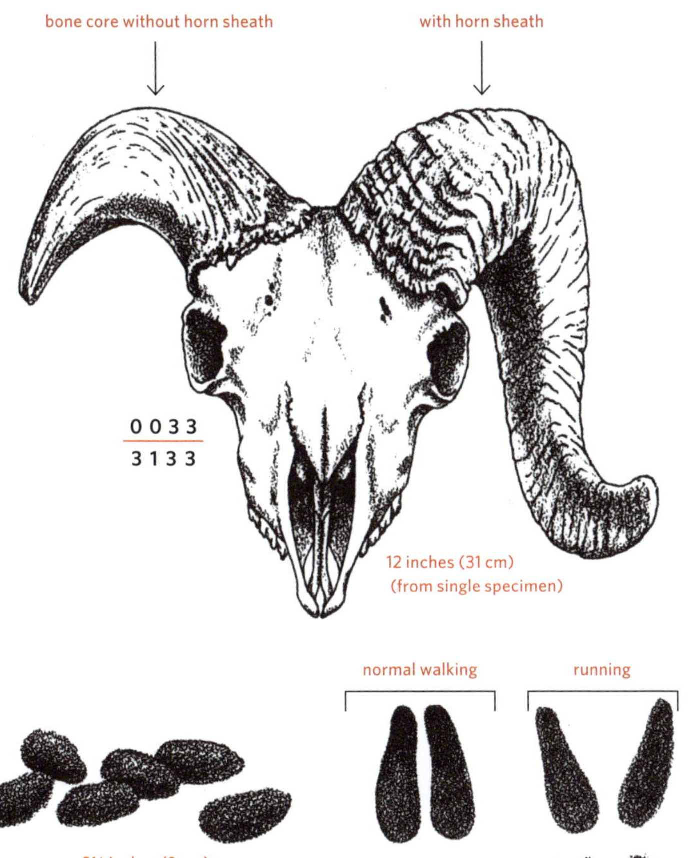

Wapiti, Canadian Elk
Cervus canadensis

Tule Elk
C. nannodes

Tail and rump patch is tawny in wapiti; white in tule elk. Males to 1,000 pounds (450 kg), females smaller. Habitat: mountain meadows in summer, lower wooded areas in winter. Tule elk prefer marshes, river bottoms, open plains. Active dusk to dawn. Can run 35 mph (56 kph). Primarily grazers. In mixed groups of twenty-five or so in winter. Older bucks separate in summer. During breeding season (rut) and August to November, bucks bugle loudly to challenge rivals and gather harems of up to sixty females. Bulls weakened after breeding go into hiding to avoid predators. Shed antlers February to March; new antlers start in April. Mountain lion is chief enemy. Bears and coyotes take calves. Wapiti were once slaughtered for their two upper canine "bugler" teeth, which were used for watch charms.

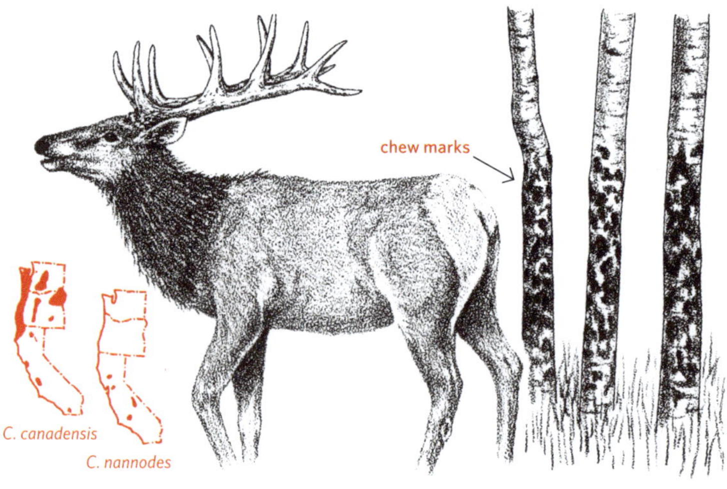

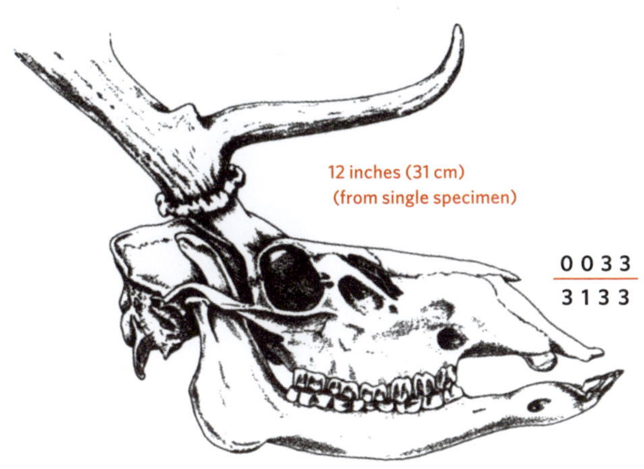

12 inches (31 cm)
(from single specimen)
0 0 3 3
3 1 3 3

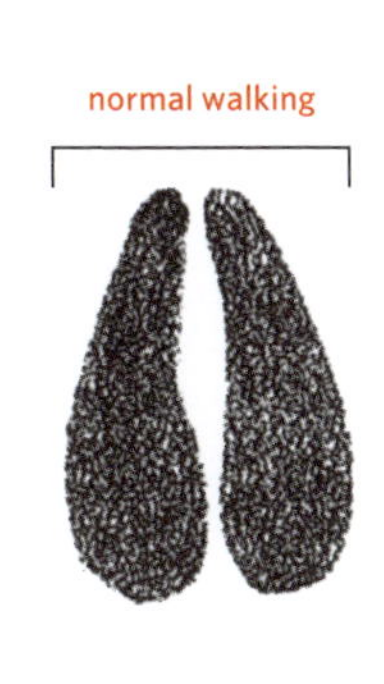

normal walking

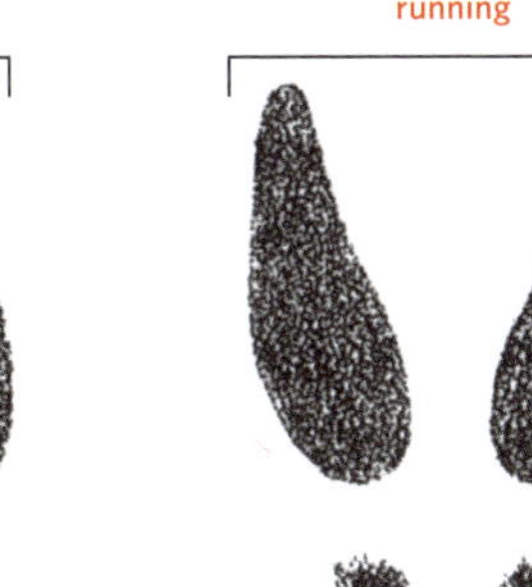

running
dew claws

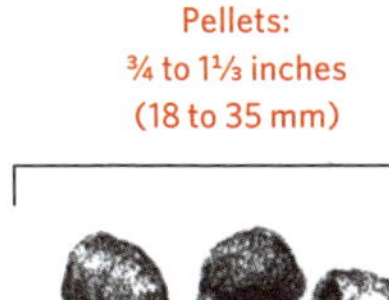

Pellets:
¾ to 1⅓ inches
(18 to 35 mm)

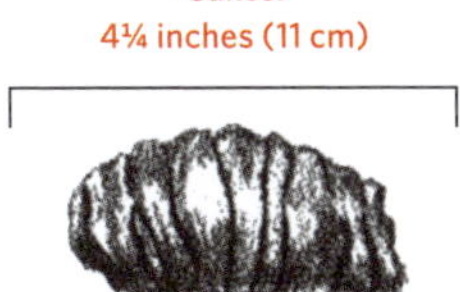

Cakes:
4¼ inches (11 cm)

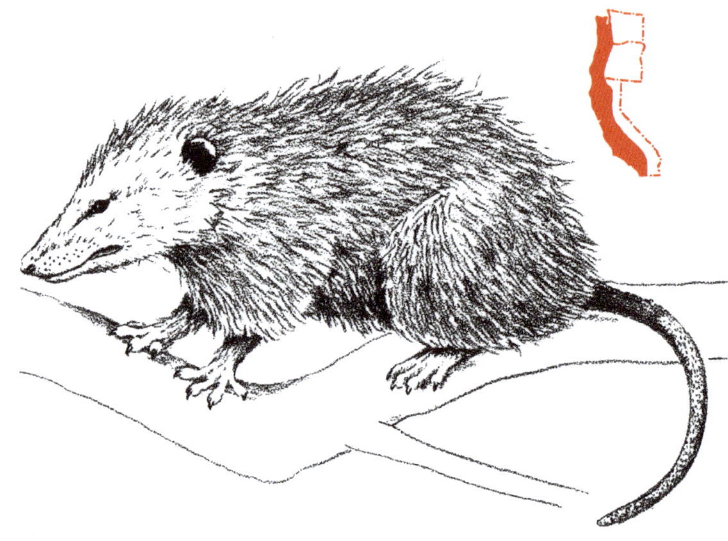

Opossum

Didelphis marsupialis

Only pouched mammal in United States. Scruffy gray body with prehensile tail. Thin, black, hairless ears and part of tail may be missing in north range due to freezing. Opossum feigns death when threatened. Thrives in urban, rural, and woodland areas. Nests in hollow trees, logs, culverts, brush piles, and under houses. Eats fruit, vegetables, nuts, insects, carrion, and eggs. Nocturnal. Solitary. Does not hibernate; stays in den for several weeks in cold weather. Has one to two litters, January to October. One to fourteen embryos crawl out of womb to pouch covering thirteen teats. All could fit in teaspoon. Nurse for two months. Five to seven normally survive to juvenile stage. Lifespan about seven years.

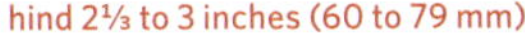

hind 2⅓ to 3 inches (60 to 79 mm)

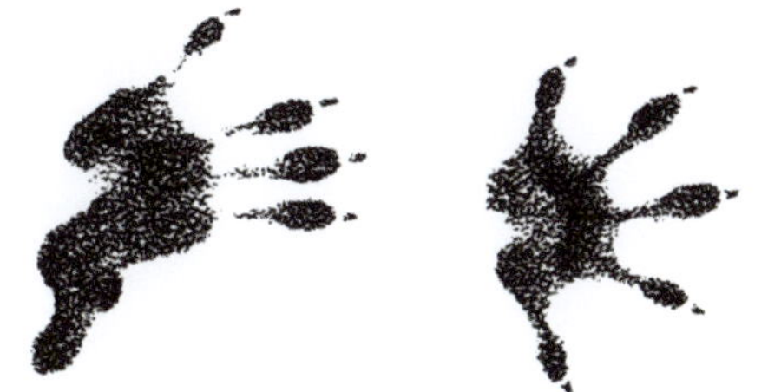

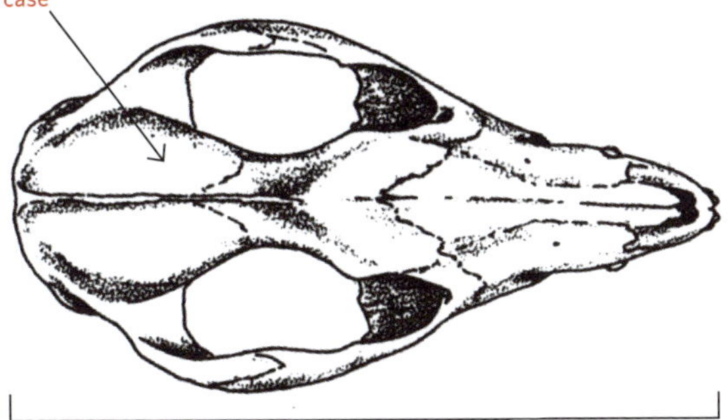

small brain case
4½ to 5 inches (116 to 127 mm)

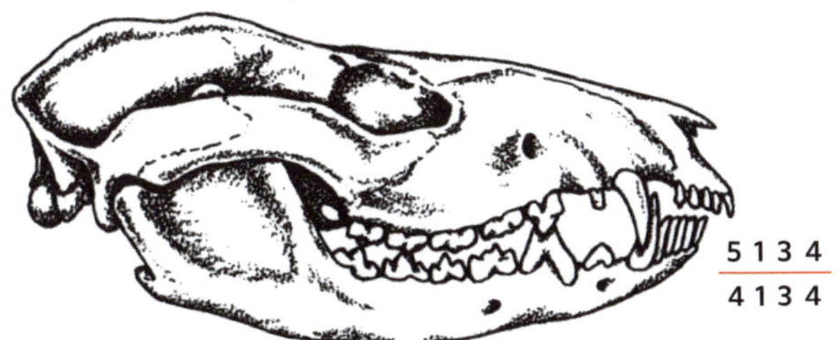

5 1 3 4
4 1 3 4

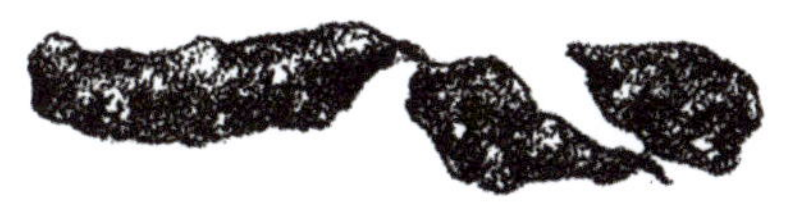

1⅔ (42 mm)

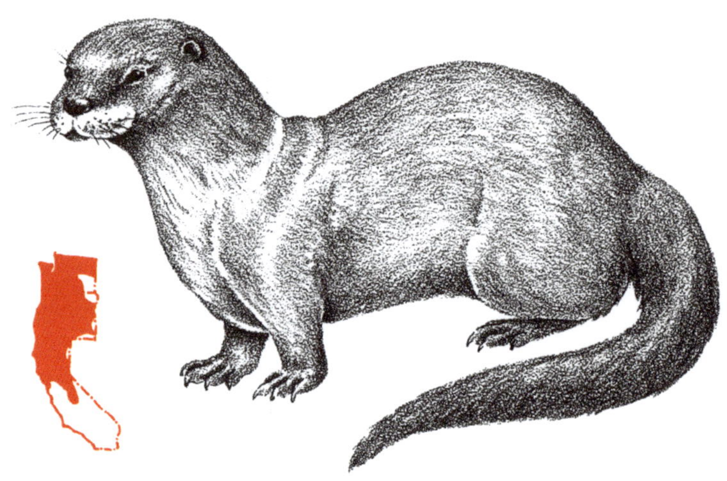

River Otter

Lutra canadensis

Color rich brown above, silvery below. Can stay underwater 2 to 3 minutes because pulse slows and skin flaps close ears and nostrils. Lives in large rivers, streams, sloughs, or along seashore in northern areas. Home range 15 miles (24 km). Travels at night. Enlarges and uses burrows of other animals, such as beaver, muskrat, for dens along water's edge. Makes riverbank slides 12 inches (30 cm) wide. Snow slides show track marks. When on land, spends most of its time frolicking, chasing its tail, or playing tag. Rolls in tules or grass to dry off and mark territory with musk, droppings. Females breed again just after giving birth, in March to May, to one to four blind, helpless, furry pups that must be taught to swim. Eats fish.

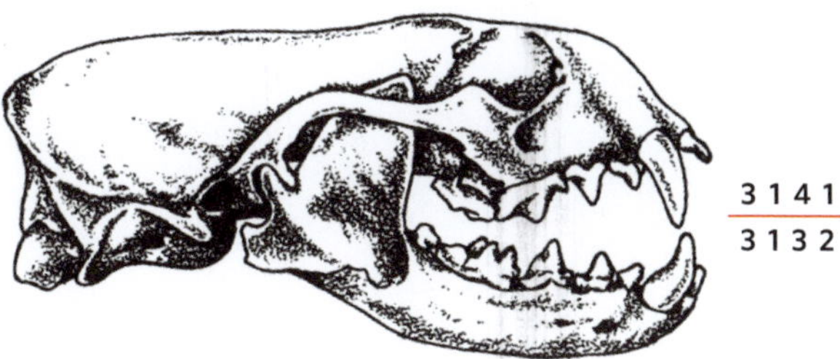

4 to 4½ inches (104 to 113 mm)

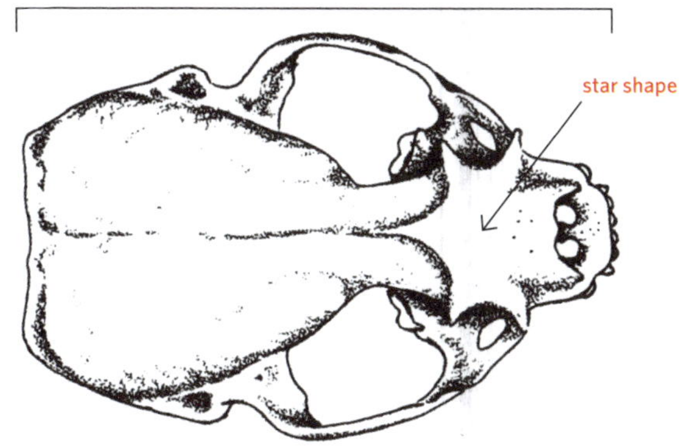

hind 4 to 5¾ inches
(100 to 146 mm)

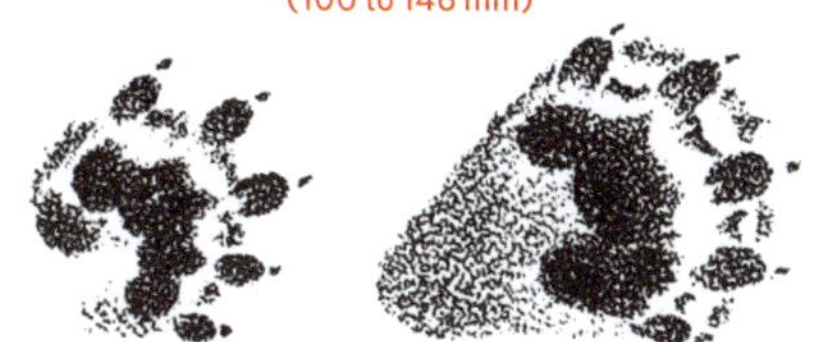

Droppings: 2⅓ to 2½ inches
(60 to 65 mm), rarely found

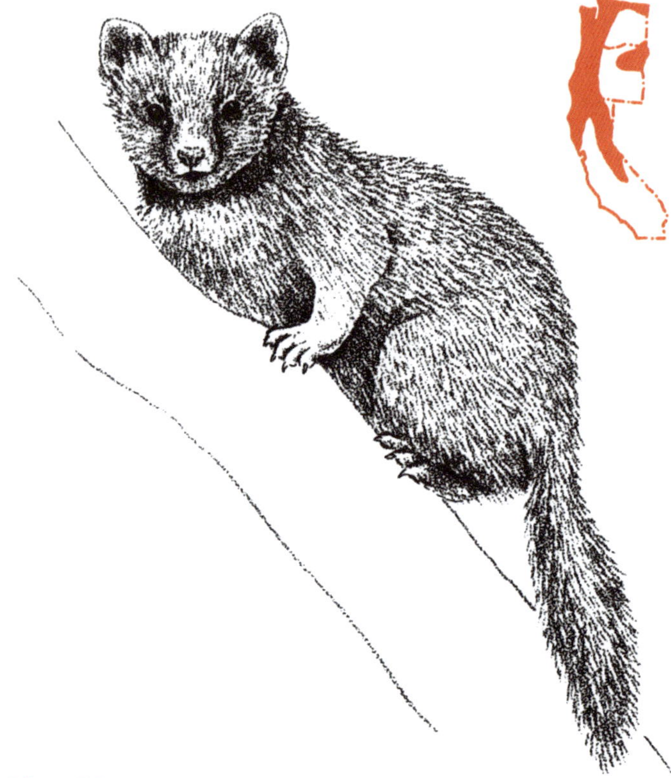

Pine Marten
Martes americana

Brown body with orange-yellow throat and chest. Lives in deep coniferous forests, large high-mountain rock piles. Dens in hollow trees. Active night, early morning, late afternoon, and cloudy days. Spends lots of time in trees. May range 15 miles (24 km) in search of food. Eats tree squirrels, mice, voles, pikas, hares, some berries. Buries surplus meat. Breeds in midsummer; fertilization of eggs by sperm is delayed, embryos form in early winter. Young born in April, one to five in litter. May breed after first year. Males use scent glands under belly skin to mark territory, attract mate. Martens are curious. One male was livetrapped seventy-seven times in a row. Enemies include fisher, large owls.

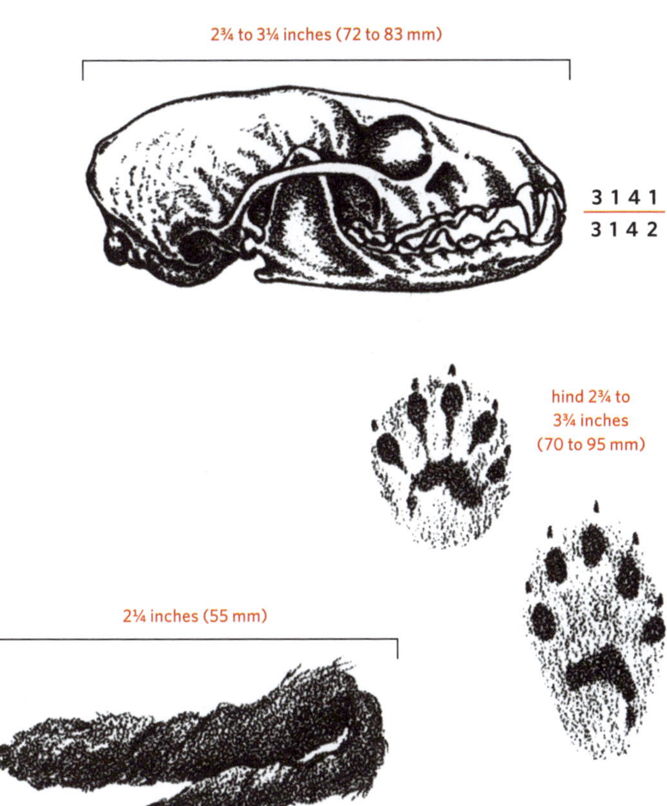

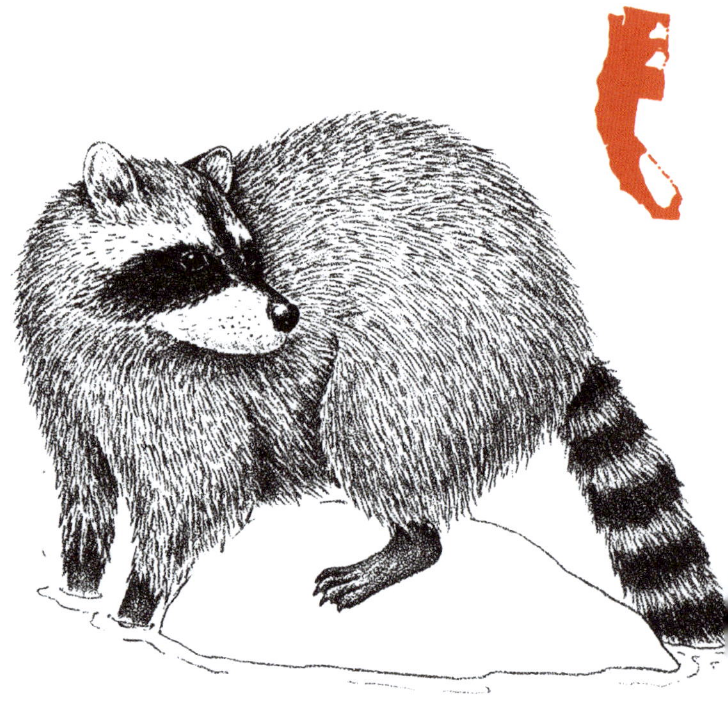

Raccoon
Procyon lotor

Color salt-and-pepper. Playful, curious, good swimmer. Feeds mostly along streams, lakes, ponds, but will wander from water. Dens in hollow trees, logs, rock crevices, or ground burrows. In cold weather, may sleep for several days; does not hibernate. Chiefly nocturnal, but occasionally about in day. Especially active in autumn. Solitary except when breeding, caring for young. Diet varied; fruits, nuts, grains, insects, frogs, fish, crayfish, birds' eggs. Washing food enhances sense of touch in toes, helps raccoon discern nonedible matter. Leaves droppings at base of den tree, on large branches, rocks, logs, across streams. Mates in February or March. Two to seven young born in April or May. In fall, young raccoons may disperse up to 160 miles (260 km), but mostly less than 30 miles (48 km). Chief enemies: dogs, hunters, autos. Raccoon can defend self well against a single dog.

4 to 5⅓ inches (99 to 135 mm)

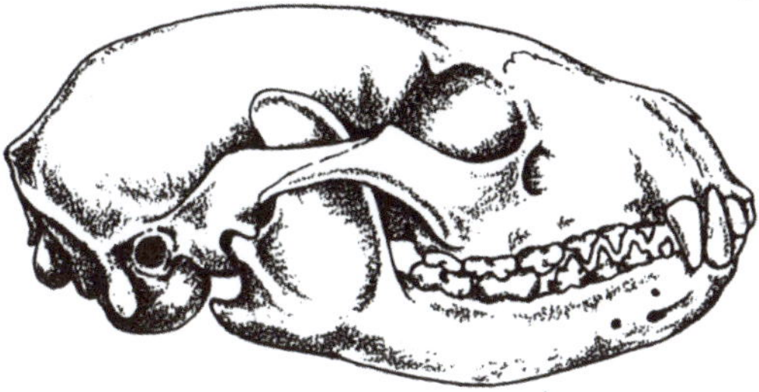

3 1 4 2
3 1 4 2

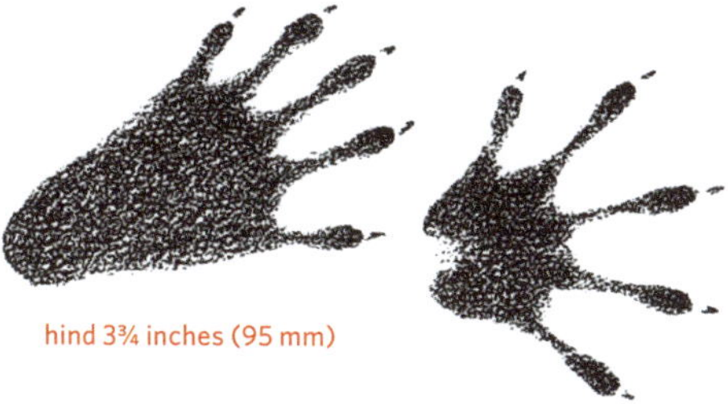

hind 3¾ inches (95 mm)

1¼ to 2 inches (3 to 5 cm)

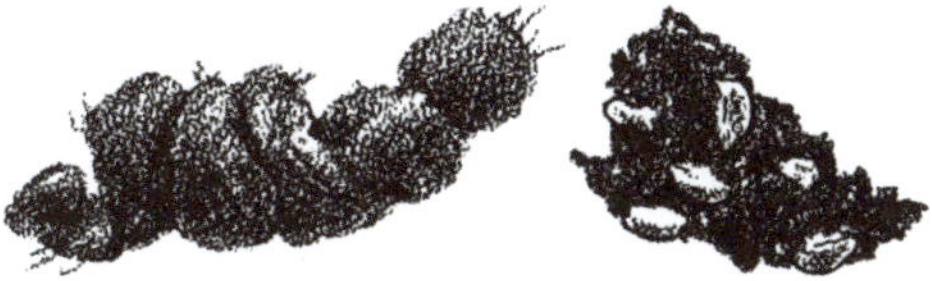

Black Bear

Ursus americanus

Body black or cinnamon. Has keen sense of smell, climbs trees easily. Can run 30 mph (48 kph) in short bursts. Can range 15 miles (23 km). Dens under downed trees, hollow logs or trees, other shelter. Solitary, except when breeding and in garbage dumps. Mainly vegetarian, but also eats fish, small mammals, eggs, carrion, honeycomb, bees, and garbage. Does not hibernate. In fall, bears add thick layer of fat to sustain them during winter sleep; bears without enough fat are active during winter. Mates June to July every other year. Two to three cubs born in winter den. Lifespan thirty years. "Bear trees" have tooth marks as high as bear stands, claw marks above, to mark territory. Bears are dangerous when surprised, hungry, feeding, injured, or with cubs. Use of claws, pancreas as mythical aphrodisiacs is causing decline of bears outside parks.

12 inches (30 cm) (from one specimen)

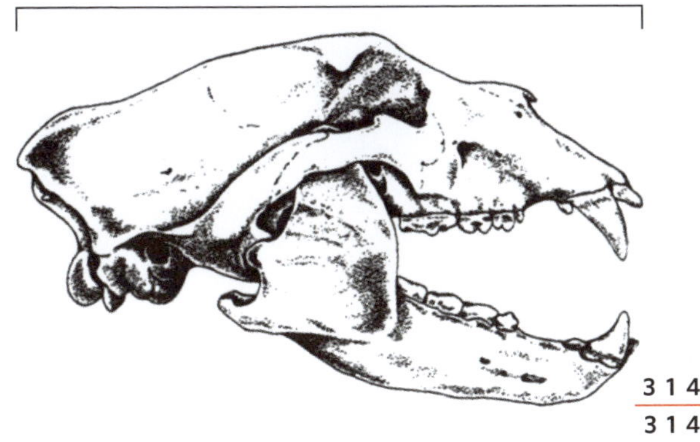

$$\frac{3\ 1\ 4\ 2}{3\ 1\ 4\ 3}$$

3¼ to 4⅓ inches (80 to 110 mm)

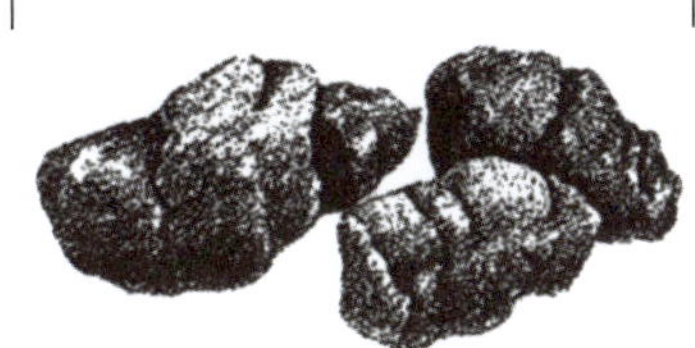

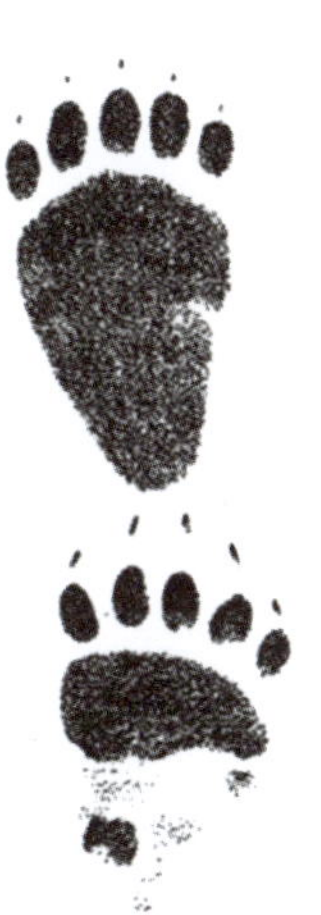

hind 9 inches
(228 mm)

Gray Fox

Urocyon cinereoargenteus

Salt-and-pepper gray with rusty neck, legs, feet. Less vocal than other foxes. Can run, in short bursts, 28 mph (45 kph). Only fox that climbs trees to escape or to hunt. Mainly in woodland, chaparral. Dens in hollow trees, logs, under rock ledges, or in culverts; may have several escape dens nearby. Den area often marked by accumulation of droppings, bones. Gray fox is nocturnal but often seen in day. Eats small rodents, insects, birds, eggs, fruit, acorns; in some areas, diet is largely cottontail rabbits, ground squirrels, and berries. Two to seven pups are born March to April, dark brown, eyes closed. Hunt on their own at four months. Enemies: domestic dogs, bobcats, lions, and people. Poison bait intended for coyotes kills many gray foxes.

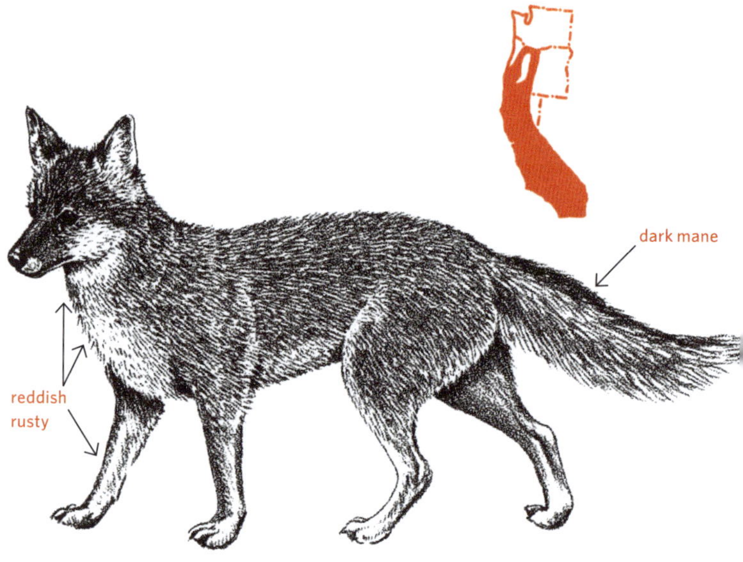

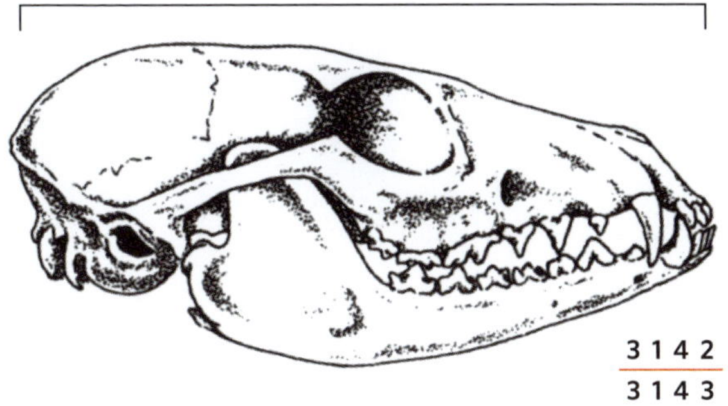

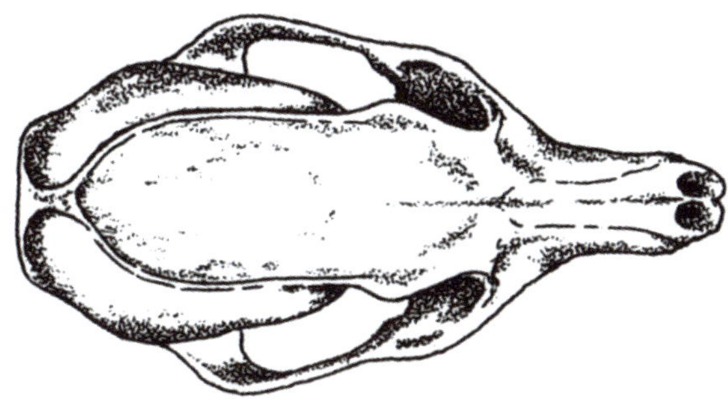

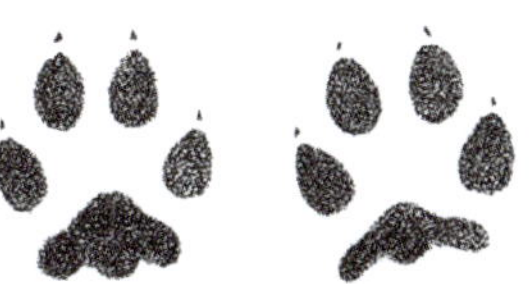

Droppings: 2 inches (5 cm), ⅓ inch (1 cm) diameter, smaller than coyotes, almost always black with stringy ends, berry seeds, fur, etc.

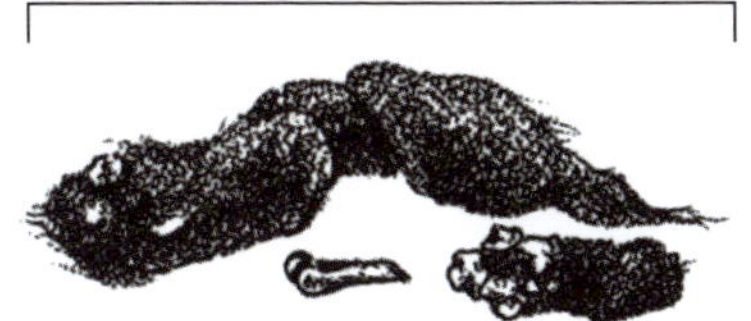

Identifying Select Mammals of the Rocky Mountains

Key to nests, burrows,* dens, mounds, depressions

If you've found a depressed or matted-down area that is less than 1½ feet (50 cm) long, in grassland, sagebrush, or next to rocks or bunchgrass, **see RABBITS, HARES, pages 218–223**

If you've found a nest made of a stack of cut grass piled under rock ledges or among rock piles of high mountain areas, **see PIKA, page 218**

If you've found a hole leading into a tunnel, and the opening of the hole is wider than 8 inches (20 cm), flat across the bottom, solitary, often with smaller holes nearby, **see BADGER, page 226**

*A burrow is the front door and hallway; a den is the whole house.

Key to chew, scratch, browse marks, and other signs

If you've found:

- tips of small shrub branches nipped off at an angle under 3 feet (1 meter) high, **see RABBITS, HARES, pages 218–223**

- a dead animal covered by dirt, snow, leaf or branch litter, or hanging from a tree branch, **see LYNX, page 232, MOUNTAIN LION, page 234.** CAUTION: This could also be a grizzly bear kill, and the bear may not be far away. LEAVE THE AREA IMMEDIATELY.

- chocolate brown to charcoal tufts of hair on shrubs, particularly willow, or on ground, near streams, ponds, meadows, **see MOOSE, page 224**

If you've found other signs like:

- partially buried droppings with vertical scratch or claw marks nearby, **see BOBCAT, page 230, LYNX, page 232, MOUNTAIN LION, page 234**

- tightly bound, twisted, cigar-shaped masses of foil, cellophane, or grass in large, doglike droppings, **see COYOTE, page 228**

Pika, Cony, Rock Rabbit, Piping Hare
Ochotona princeps

Round body, short legs and ears, and dense fur conserve body heat. No visible tail. Common on alpine and fir forest talus slides. Variable color blends with rocks. Territorial. Gives sharp, nasal chirps. Makes "hay piles" of drying vegetation (each one may be a bushel or more). Diurnal. Does not hibernate. Active under snow, using stored hay for food. Re-ingests soft droppings for protein, energy, and vitamins. Stores dried marmot droppings for same use. Concentrates urine to conserve water; leaves distinctive white marks on rocks. Young born May to August, naked, blind, three to four per annual litter. Enemies: weasels, martens, coyotes, and hawks.

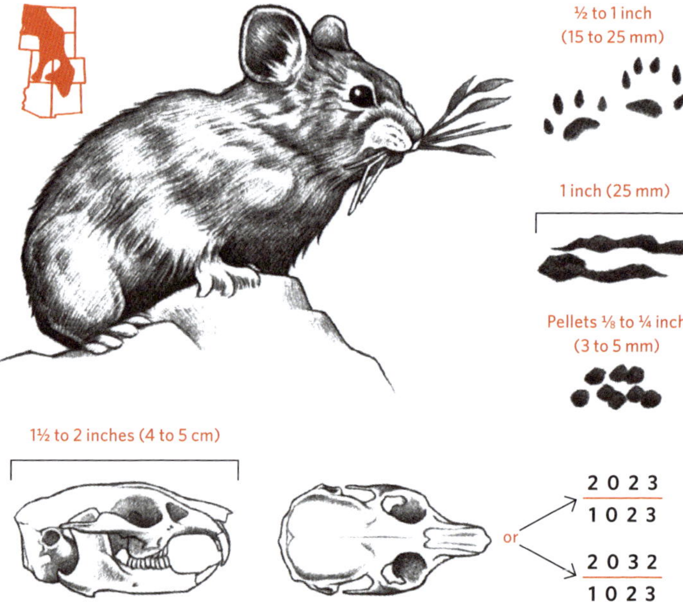

Whitetail Jackrabbit, Whitetail Hare

Lepus townsendii

Gray brown, often turns white in winter, but always has white tail with slight dorsal strip. Ears narrower than those of Blacktail. Common in sagebrush and open areas; does not burrow. Can jump 16 feet (5 meters) on a bound with speeds up to 40 mph (64 kph). Feeds at night on sage, grass, bark of young trees. Young born furred, eyes open, three to six per annual litter. Enemies include: foxes, coyotes, bobcats, lions, owls, hawks, and eagles. Ears of all "Jacks" are long and well supplied with blood vessels to disperse body heat in hot weather.

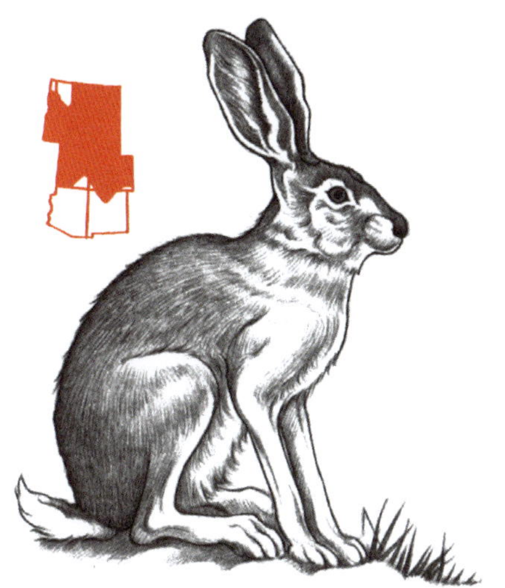

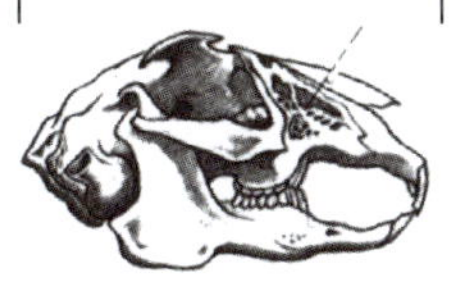

2½ to 3 inches (67 go 75 mm)

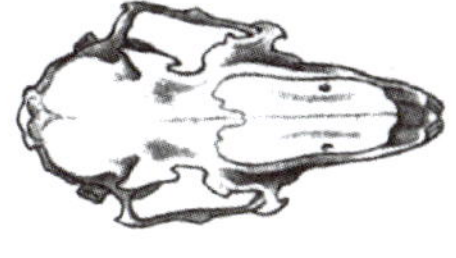

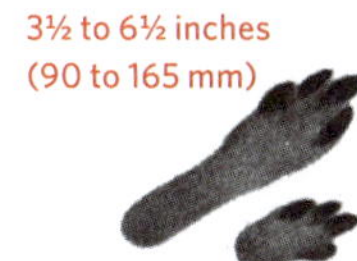

3½ to 6½ inches
(90 to 165 mm)

¾ inch (18 mm)

CAUTION: All hares and rabbits may carry tularemia, or "rabbit disease," which can be transmitted to people. **Avoid handling dead rabbits**, but, if necessary, use rubber gloves.

Blacktail Jackrabbit, Blacktail Hare

Lepus californicus

Gray brown, tawny. Has large, black-tipped ears. Eyeshine is red. Common in sage, cactus, meadow, and open grassland country. Places hind feet ahead of front feet in normal gait. Eats green vegetation, shrubs, and cacti. Most active early evening and morning. Can run 30 to 35 mph (48 to 56 kph). Uses a zigzag running escape pattern. Young are born year-round, furry, with eyes open. Enemies include: coyotes, eagles, hawks, barn owls, and large snakes.

All rabbits and hares form two kinds of droppings: soft, which are re-ingested for vitamin and protein nutrition; and hard, which are not.

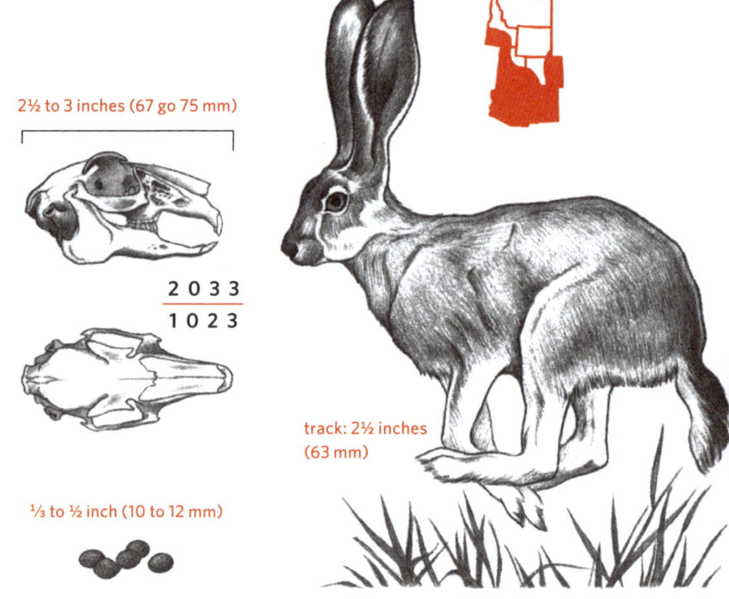

2½ to 3 inches (67 go 75 mm)

2 0 3 3
1 0 2 3

track: 2½ inches (63 mm)

⅓ to ½ inch (10 to 12 mm)

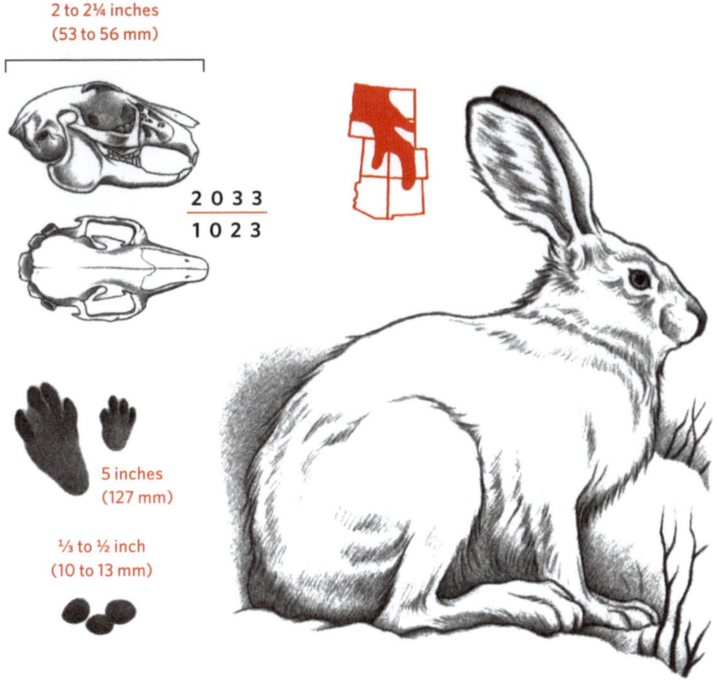

Snowshoe Hare

Lepus americanus

Changes from dark brown in summer to white in winter. Only tips of hairs turn white. Eyeshine is orange. In swamps, forests, and mountain thickets. Home range is about 10 acres (4 hectares). Nocturnal. Eats succulent vegetation in summer; twigs, bark, buds, and sometimes frozen carcasses in winter. Does not build nest. Young born April to August, two to three litters per year, two to four per litter. Populations fluctuate dramatically, with peaks every eleven years. Lifespan in wild is about three years.

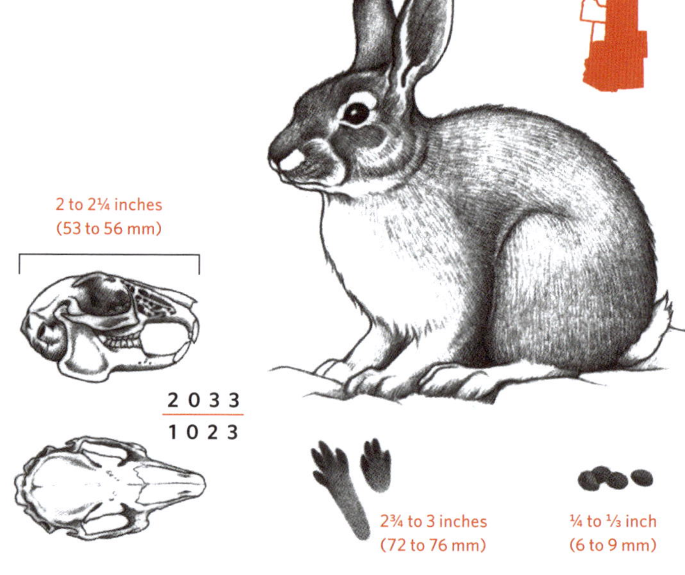

Audubon Cottontail, Desert Cottontail
Sylvilagus audubonii

Long ears that are sparsely furred inside distinguish this rabbit from Mountain Cottontail. In open plains, foothills, and low valleys; in grass, sage, pinyon-juniper, and deserts. Home range is 9 acres (4 hectares) for females, up to 15 acres (6 hectares) for males. Active in late afternoon, night, and early morning. Stays close to thickets. Eats green vegetation and a variety of fruit, tree bark rarely. Young born naked and blind, in fur nests, throughout the year. May live two years in wild. Vulnerable to marauding domestic dogs. Note: Eastern Cottontail, *S. floridanus*, is found in central and southern Arizona, New Mexico, eastern Colorado, southeastern Wyoming.

Mountain Cottontail

Sylvilagus nuttallii

Small, white-tailed. Lives in brushy-rock areas of juniper woodland, pinyon-juniper, and sagebrush desert. Long hairs in uniformly colored ears distinguish it from Audubon Cottontail. Nocturnal, but also active in morning. Eats sagebrush, grass, and tree shoots. Young born naked, blind, in fur-lined nest in early summer, one to eight per litter. Rabbits in some areas produce several litters per year. Enemies same as other rabbits.

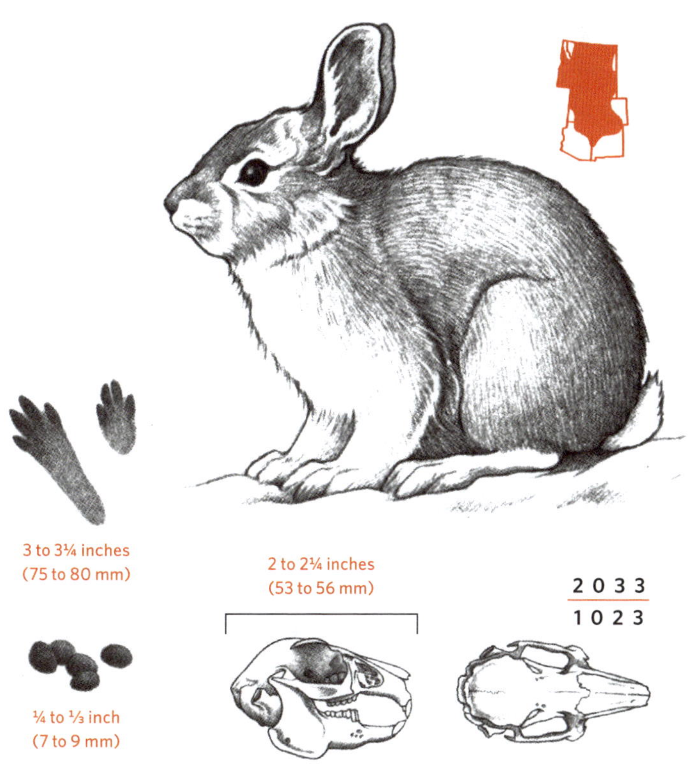

3 to 3¼ inches
(75 to 80 mm)

2 to 2¼ inches
(53 to 56 mm)

2 0 3 3
1 0 2 3

¼ to ⅓ inch
(7 to 9 mm)

Shiras Moose

Alces alces shirasi

Dark brown with gray legs, overhanging snout. Palmlike antlers spread to 81 inches (206 cm). Largest "deer" in North America, with males to 1,400 pounds (635 kg) Found in or near water, in spruce, aspen forests with lakes, swamps, in willow thickets, sometimes in sage flats. Most active at night but seen anytime. Solitary, except for cow-calf pairs, but may herd in winter. Antlers, shed in December to February, are quickly gnawed for calcium by rodents. Moose may submerge almost completely or roll in mud to avoid mosquitos, black flies. Can run 35 mph (56 kph). Eats mostly aquatic vegetation during

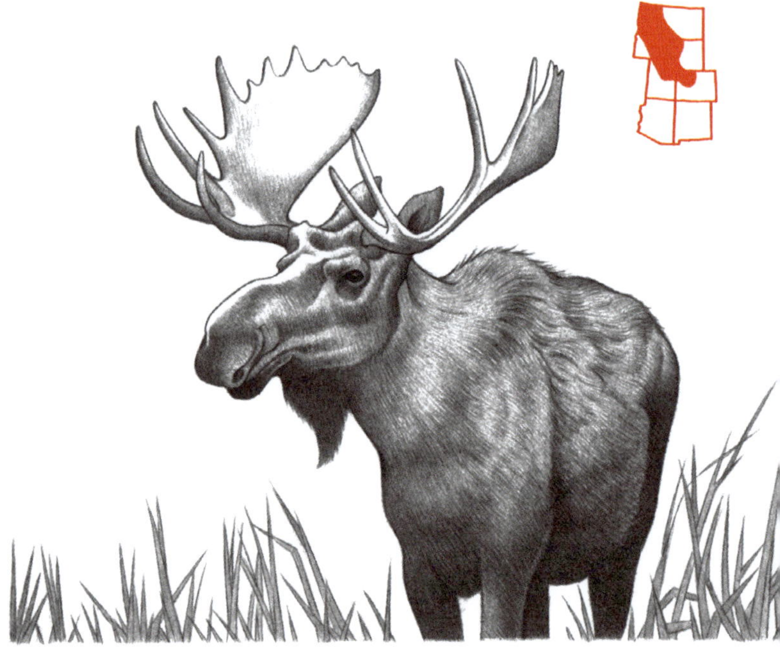

summer. Browses on twigs, buds, bark, especially willow in winter to spring. Known to walk on front knees to reach low-growing plants on sage flats. Breeds September to October. Bulls usually avoid battle with each other but occasionally lock antlers and starve to death. Usually one, but sometimes two reddish-brown calves born May to June. CAUTION: While normally shy and retiring, moose are unpredictable and dangerous. Cows fiercely protect calves and rutting bulls will charge. Enemies: grizzly bears, wolves. Lifespan: to twenty years.

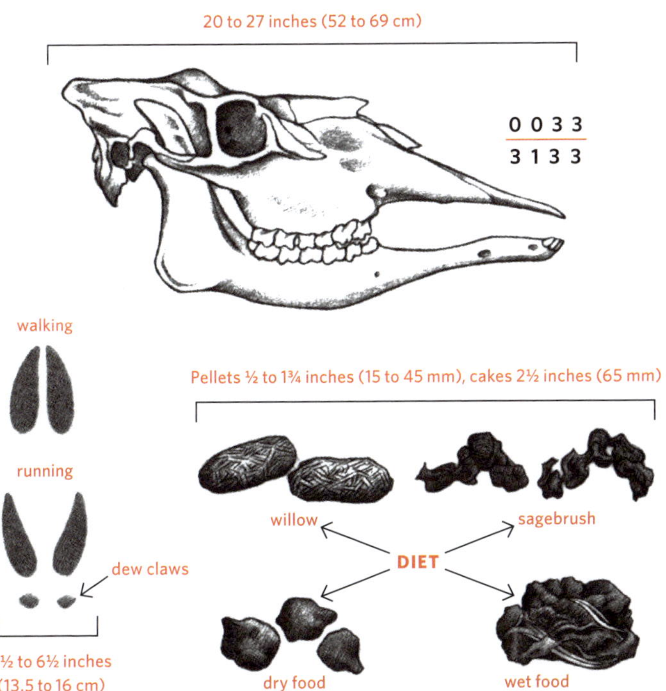

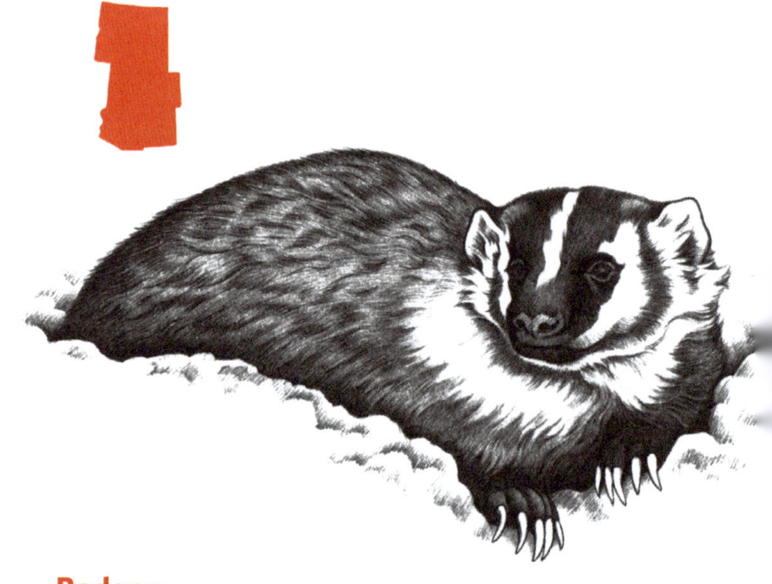

Badger

Taxidea taxus

Coat yellow-grizzled gray. Badger has poor eyesight, strong sense of smell and hearing. Loose skin allows twisting and turning in tight spaces to catch food, defend self. In arid grassland, plains, and deserts. Badger digs to catch food, escape, rest, den, and to bury droppings and extra food. Its burrow entrance is 8 to 12 inches (20 to 30 cm) wide, elliptical, with flat bottom. It eats rabbits, gophers, squirrels, mice, rattlesnakes, and yellowjackets. Coyotes may follow badgers to steal the prey they flush from burrows. Badger is solitary, active day or night. May sleep for weeks during cold weather. Changes dens almost daily in summer. One to five cubs born blind, furred, February to May. Enemies are bears and mountain lions, who do not find badgers easy prey because their squat form, sharp teeth, strong neck muscles are major defensive advantages. Many badgers are killed by cars.

4¼ to 5¼ inches (11 to 13 cm)

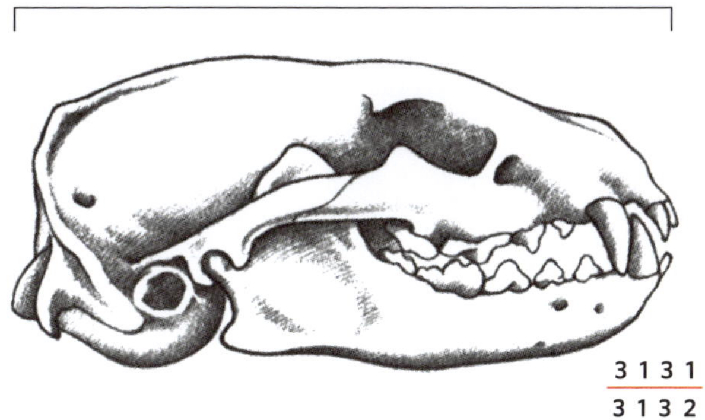

1⅓ to 2 inches
(34 to 49 mm)

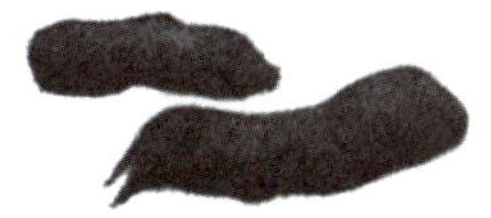

long claws on front foot

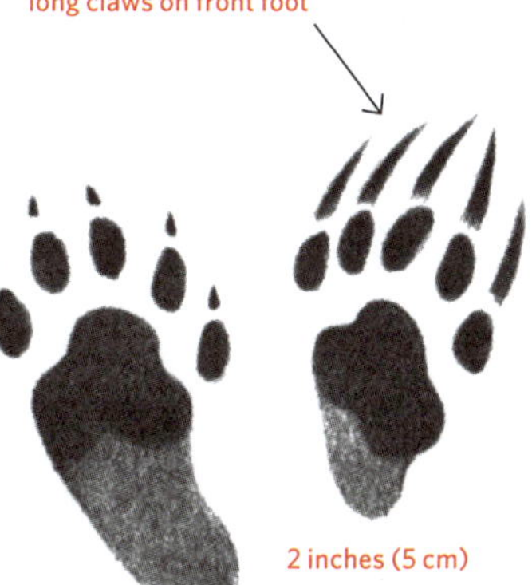

2 inches (5 cm)

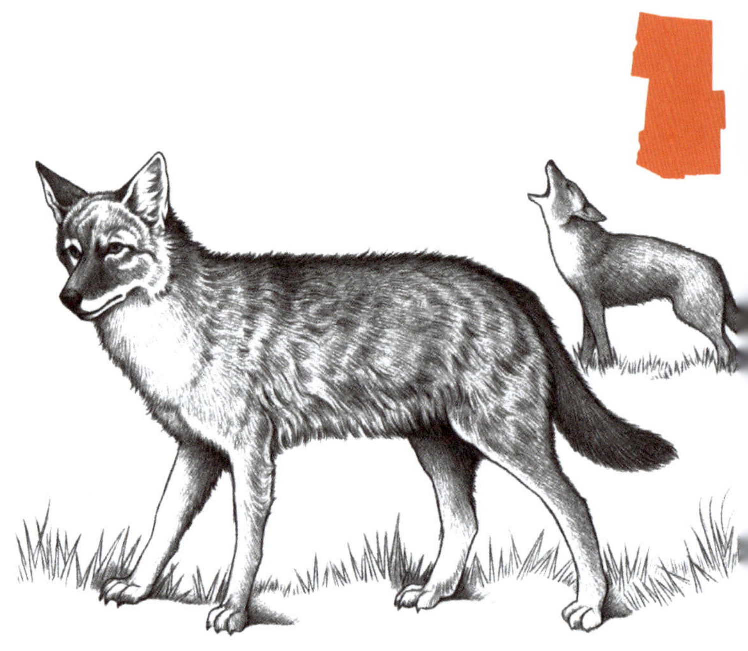

Coyote
Canis latrans

Color and size variable. Mountain coyotes are larger, have longer fur than desert coyotes. Coyote is vocal at night with a series of yaps, a long howl, then short yaps. Holds tail between legs when running. Can reach 40 mph (64 kph). Leaves doglike tracks. Population, range increasing despite hunting, poisoning campaigns. Widespread. Dens along riverbanks, well-drained sides of canyons, gulches. May enlarge badger or squirrel burrows. Chiefly nocturnal but active any time. Often hunts in pairs. Omnivorous, but mostly eats small rodents,

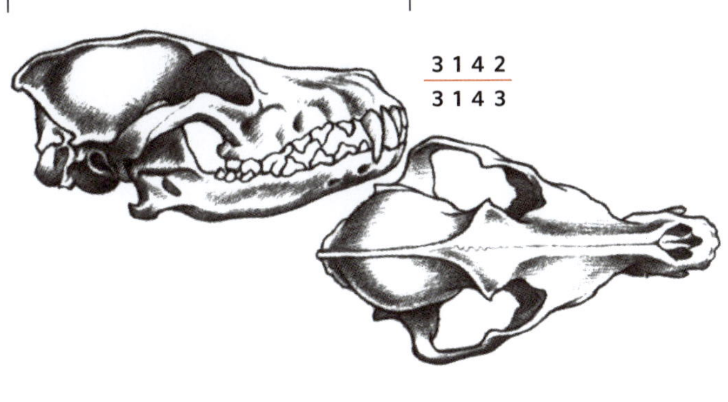

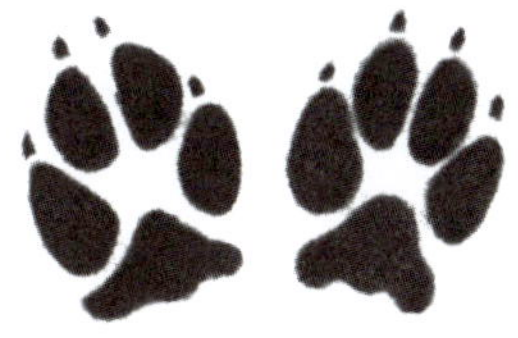

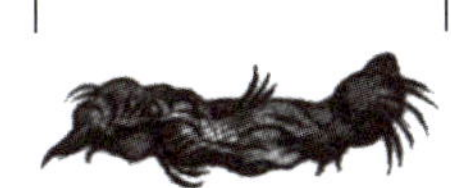

rabbits, squirrels. Droppings gray, with some seeds, but mostly fur, bones, insect parts, reptile skin, feathers; occasionally solid foil, plastic, or grass, which help remove tapeworms. Mates January to February. Six to seven pups born April to May, raised by both parents. Livestock losses blamed on coyotes often the work of dogs. Coyotes kill many grass-eating rodents, earning protection from some ranchers.

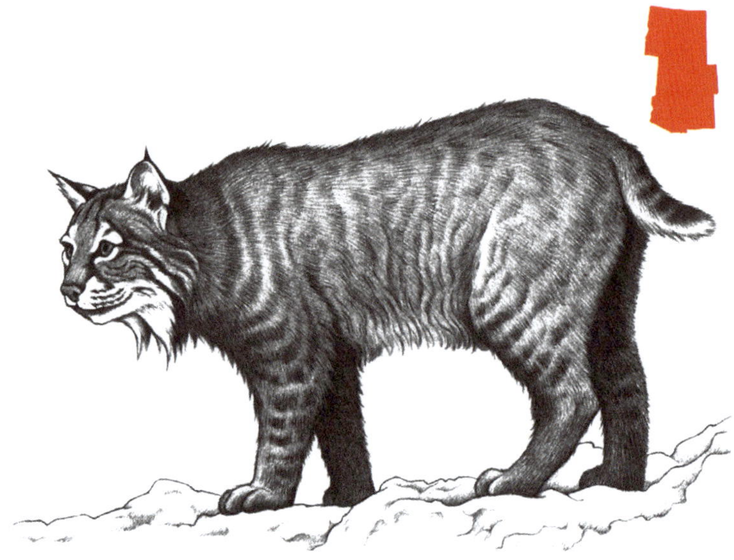

Bobcat

Felis rufus

Gray brown to reddish. Ear tufts used like antennae to aid hearing. Good climber. Gets name from "bobbed" tail. In almost every habitat, life zone. Home range: up to 75 square miles (195 square km). Mostly nocturnal; also seen in daytime. Solitary. Often rests on branches, atop large rocks to watch for passing prey. Eats rabbits, mice, squirrels, and weak deer. Caches large kills. Droppings are like dog's or coyote's but often partially buried, with scratch marks on ground. One to seven (average two to three) kittens born in April to May in den of dry leaves or fallen tree. Southern animals may have two litters per year. Lifespan twenty-five years. Marks territory by urinating on rocks or tree trunks, making scent posts. Uses tree trunk as scratching post. Often killed by poison bait intended for coyotes.

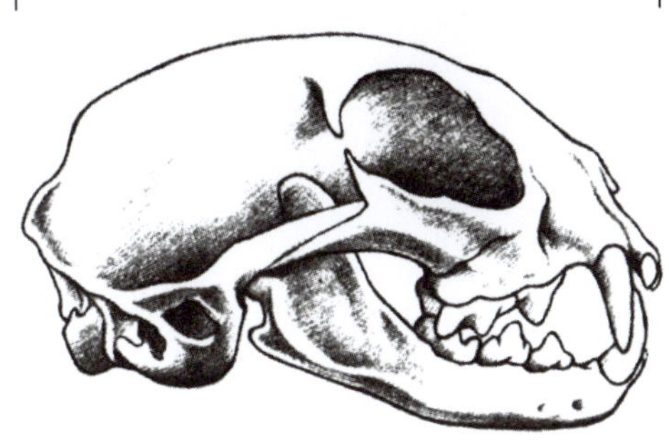

4½ to 5½ inches (11.5 to 13.5 cm)

3 1 2 1
3 1 2 1

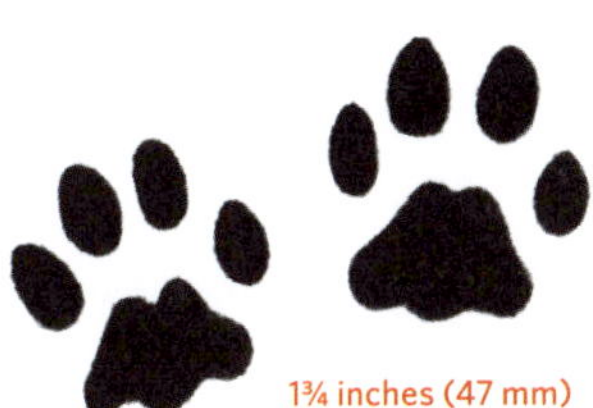

1¾ inches (47 mm)

2 inches (5 cm)

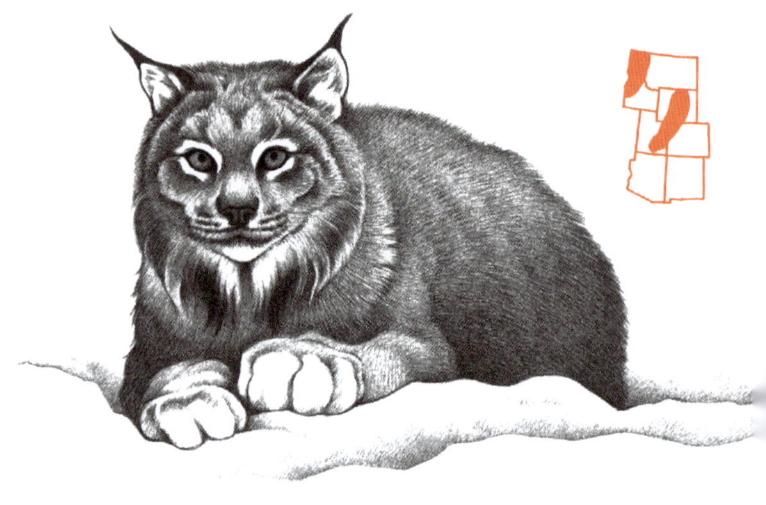

Lynx, Canada Lynx

Felis lynx

Light gray with yellowish wash over back. Underparts gray, buff-tawny with indistinct black spots. Summer fur shorter, reddish. Ear tufts enhance hearing. Large, thickly furred feet allow silent stalking and good footing in soft snow. Primarily nocturnal. Solitary. Found in northern forests, swamps, remote areas. Creates scent posts by urinating on trees, stumps. Dens in hollow logs, beneath roots, other sheltered places. Home range: to 90 square miles (145 square km). Populations peak every nine to ten years coinciding with prey cycles. Lynx frequently rests in trees, waiting to pounce on passing prey. Eats snowshoe hares primarily; also takes rodents, birds, occasionally carrion. Partially buries large prey with snow, litter. Mates in late winter. Two kittens (usually) born May to July. Enemies: wolves, mountain lions, and people. Lifespan fifteen to eighteen years (in captivity).

4½ to 5½ inches (11.5 to 13.5 cm)

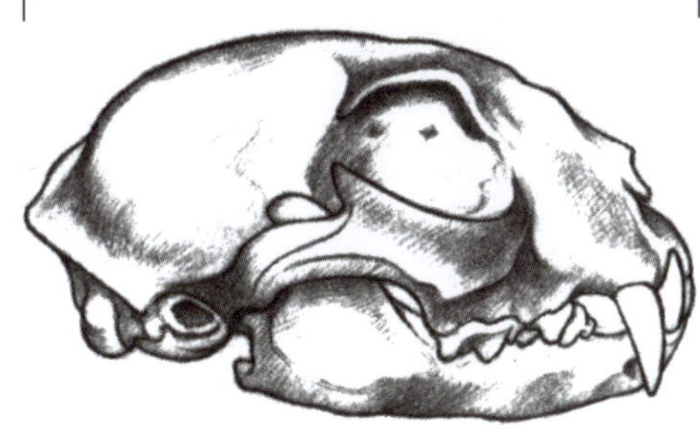

$$\frac{3\ 1\ 2\ 1}{3\ 1\ 2\ 1}$$

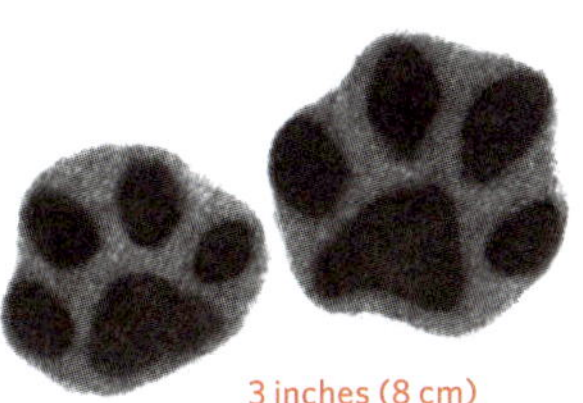

3 inches (8 cm)

2¾ inches (7 cm)

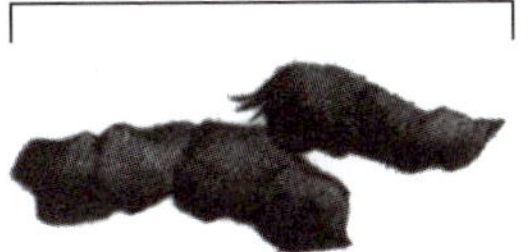

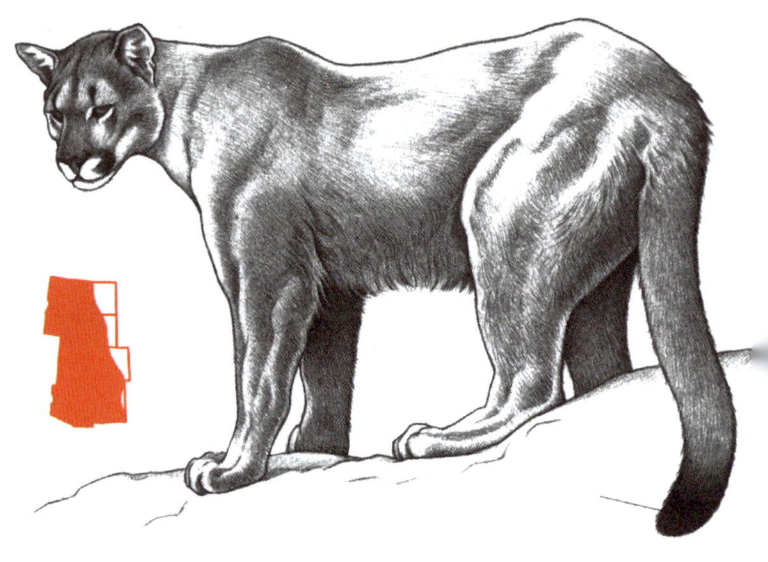

Mountain Lion, Cougar, Puma
Felis concolor

Yellowish, grayish, tawny. Habitat generally wilderness but may hunt in rural areas. Male may travel 25 miles (40 km) in one night. Strongly territorial. Mostly nocturnal. Rarely seen. Has voice like tomcat's, greatly magnified. Uses tree trunks as scratching posts. Solitary. Eats large mammals; one deer per week forms half of diet. Has more success catching old, weak, less alert deer, thus keeps herd healthy. Also eats coyotes, porcupines, beavers, rabbits, marmots, raccoons, birds,

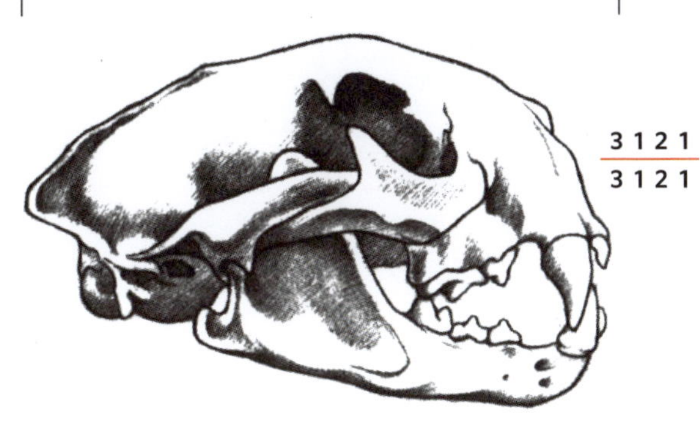

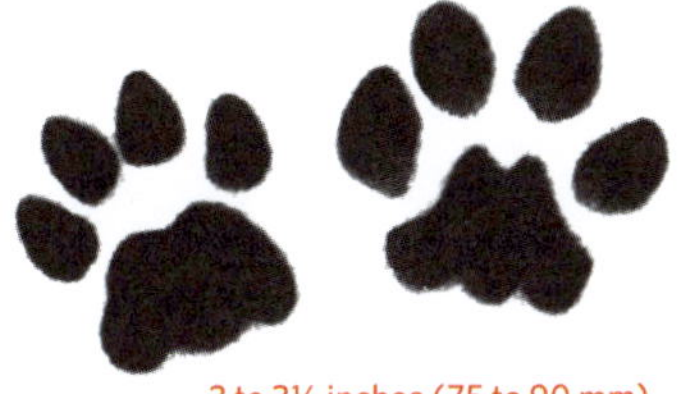

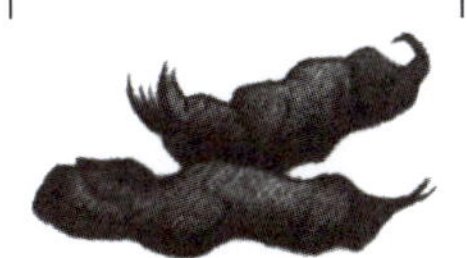

Drawn from pellet droppings in desert. All cat droppings are partially buried.

and sometimes livestock. Covers remains of large kill with branches, leaves. Often partially buries droppings. Adults breed every two or three years. One to six furry, spotted kittens born midsummer, raised by female for one to two years. Enemies: people. Lions are important predators that should be protected from indiscriminate hunting.

Identfying Certain Mammal Tracks in the Eastern United States and Canada

Definitions

PRINT: Impression made by one foot

TRAIL: Series of prints

TRAIL WIDTH: Distance from the outside edge on the left to the outside edge on the right, perpendicular to the direction of travel

TOE-TO-TOE: Measurement of forward movement in zigzag and paired

GROUP LENGTH: Distance from an animal's first to last footfall in grouped patterns

DIRECT REGISTER: Hind foot falls directly into front print

OFF (OR INDIRECT) REGISTER: Hind foot falls partially onto or near front print

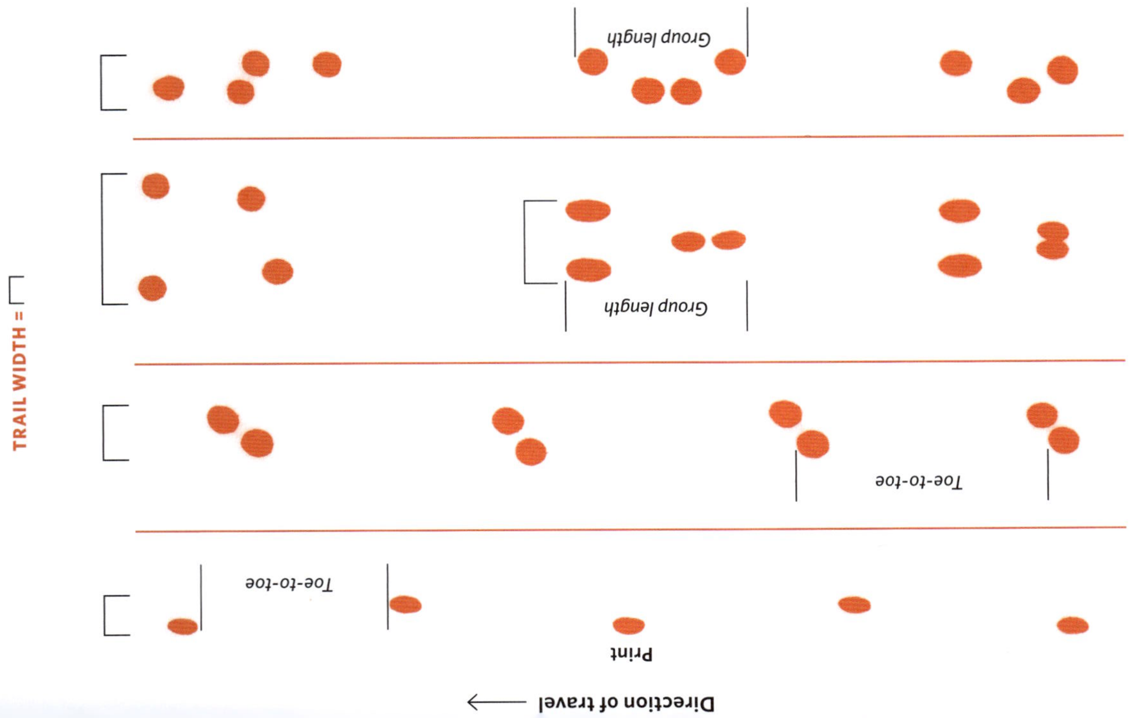

Direction of travel ⟶

TRAIL WIDTH =

Print

Toe-to-toe

Toe-to-toe

Group length

Group length

Pattern Key

A. Zigzag

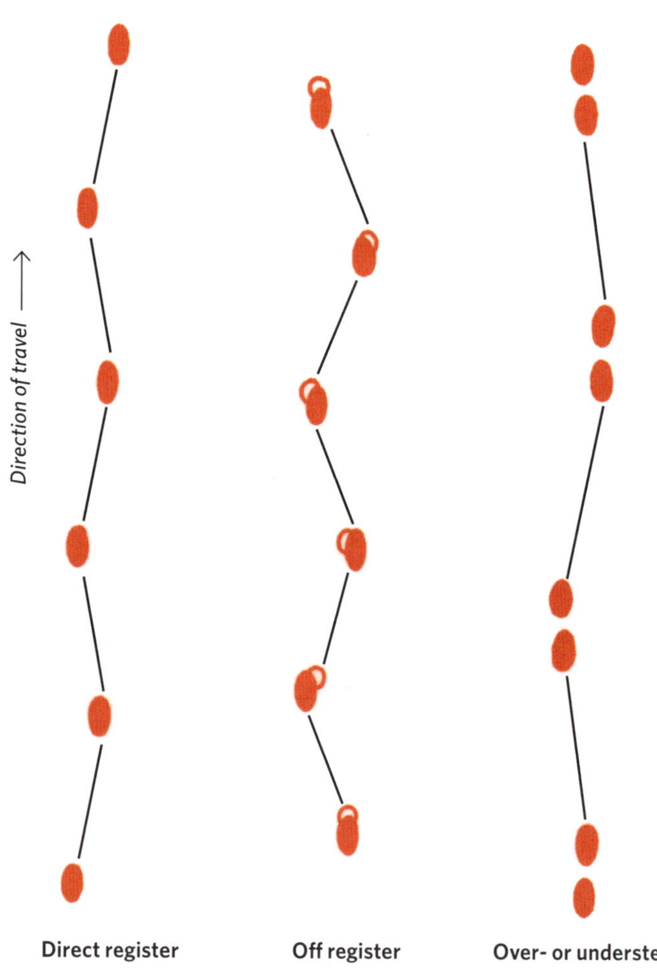

B. Paired Prints

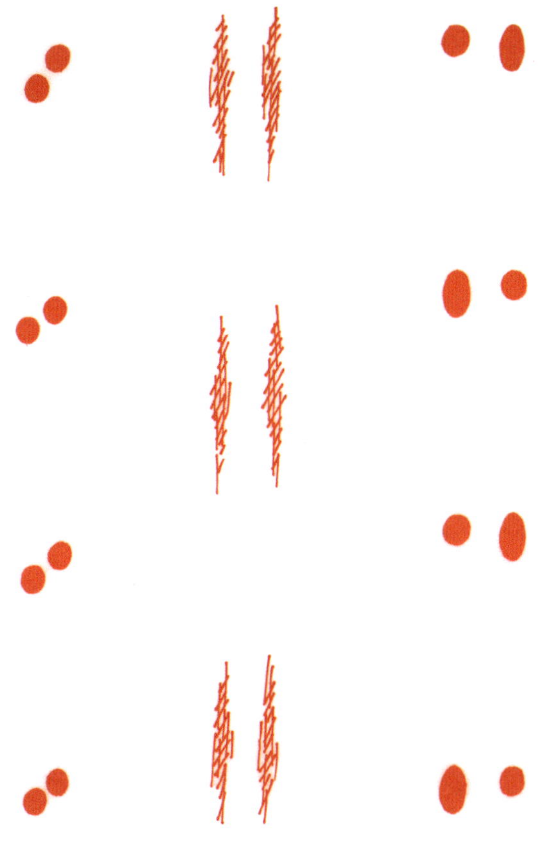

Usually diagonal

Diamond shaped in snow

One large with one small, go to raccoon, page 250

C. Group of Three or Four

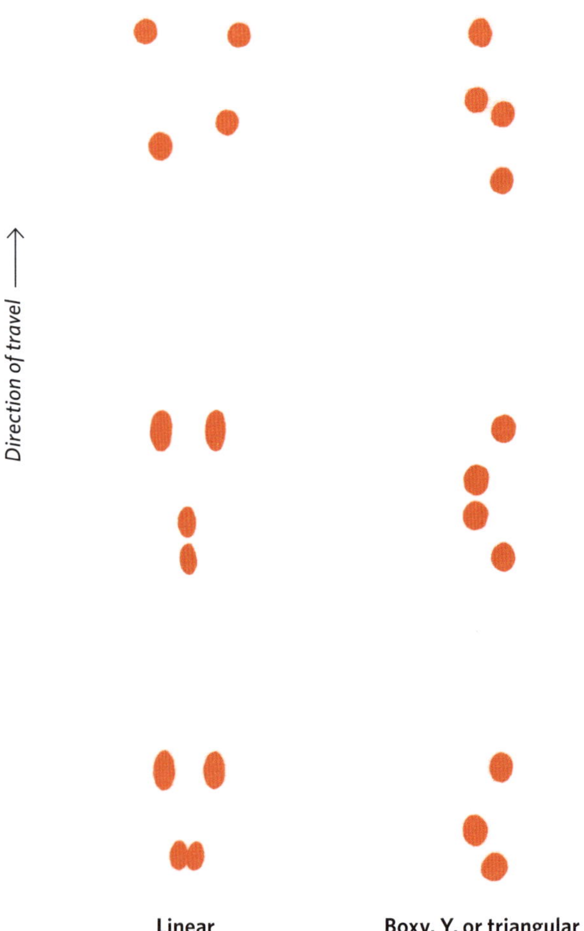

Print Shape Guide

Typical rodent foot plan

Pads alone (left), with foot outline (right)

Includes **mice**, **voles**, **woodrats**, **chipmunks**, **squirrels**, and **woodchucks**

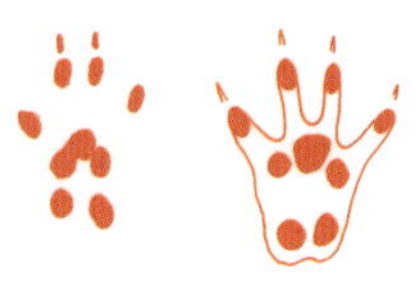

FRONT:
Four front toes (fifth doesn't show) with pads
Three palm pads forming triangle
Two separate heel pads

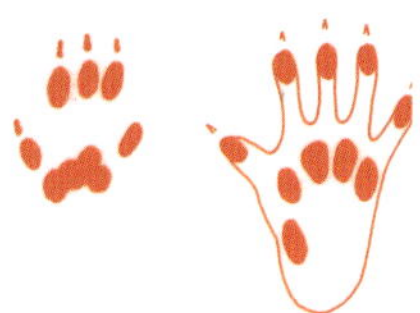

HIND:
Five toes with pads
Four fused palm pads for arc
Two heel pads; pads and heel may not show

A **pad** is a rounded projection from the surface of the sole; the shape and placement of pads can suggest a group or family.

Rodent variations

Prints not to scale

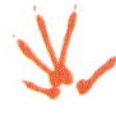

JUMPING MICE, some **RATS:** Long, thin toe prints connect to palm

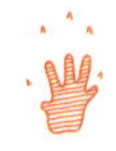

POCKET GOPHER: Front print has five toes with very long nails

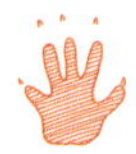

MUSKRAT: Long, wide toes connect to palm; nails extend far beyond toes

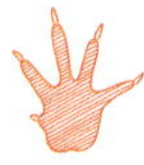

NUTRIA: Front print has five toes; inside toe much smaller; hind print partially webbed

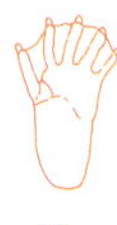
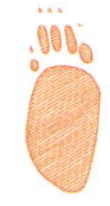

BEAVER, page 244: Print large, V-shaped with long toes; webbing and/or heel may or may not show

PORCUPINE, page 245: Pads fused into one large, textured surface

Other prints

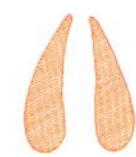

DEER, MOOSE, pages 253–255: Two large toes

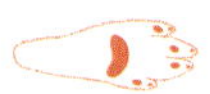

RABBITS AND HARES: Narrow, furry feet with four toes on each foot

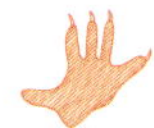

OPOSSUM: Opposable thumb

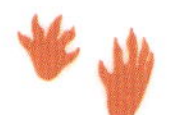

RACCOON, page 250: Five long toes; toe prints connect to wide and/or curved central pad

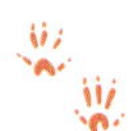

SHREWS: Similar to mouse but five toes on each foot

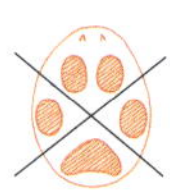

CANINES, pages 248–249: Oval prints, four toes, nails may show "X" between toe pads and central pad

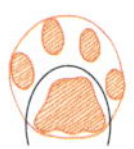

FELINES: Roundish prints, four toes, nails unlikely to show; "C" between toe pads and central pad

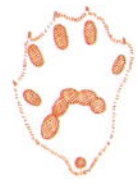

WEASEL, MINK, MARTEN, FISHER, OTTER:
Five toes in an arc (inner, smallest, may not
show); central pad also an arc; furry feet may
obscure print details; print roundish

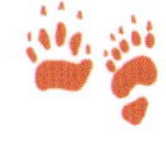

SKUNK: Nails long, sturdy, show in front and
sometimes in hind (print much smaller than
badger)

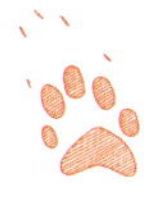

BADGER: Nails long, sturdy, extend far beyond
front pads

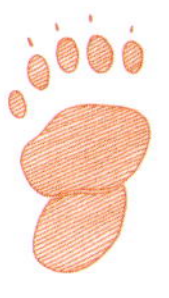

BEAR, page 251: Print large; five toes in a line
(fifth may not show) above wide central pad

Beaver

Castor canadensis

Tail drag usually obscures prints (*below*)

Habitat: Lakes, streams

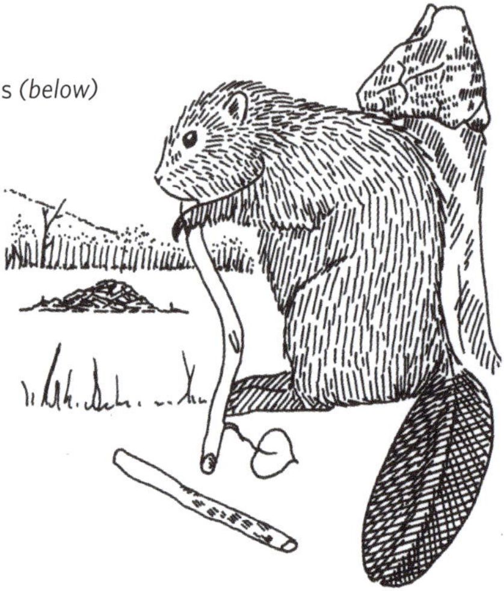

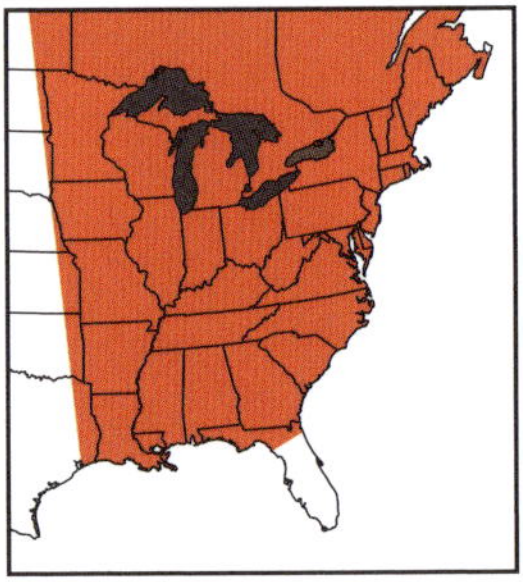

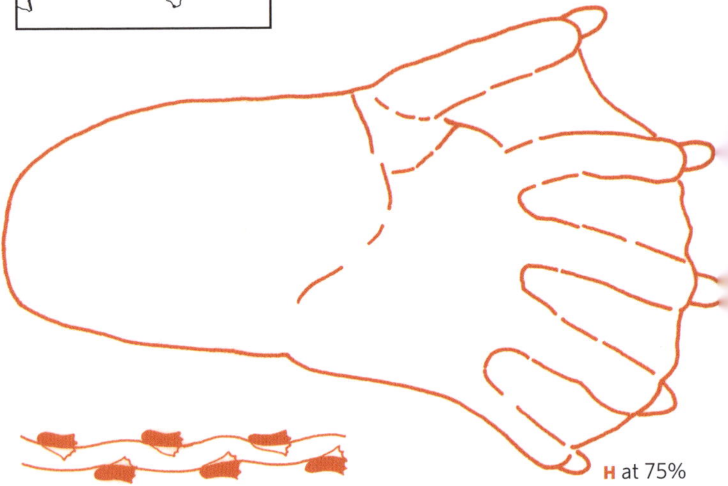

H at 75%

Porcupine

Erethizon dorsatum

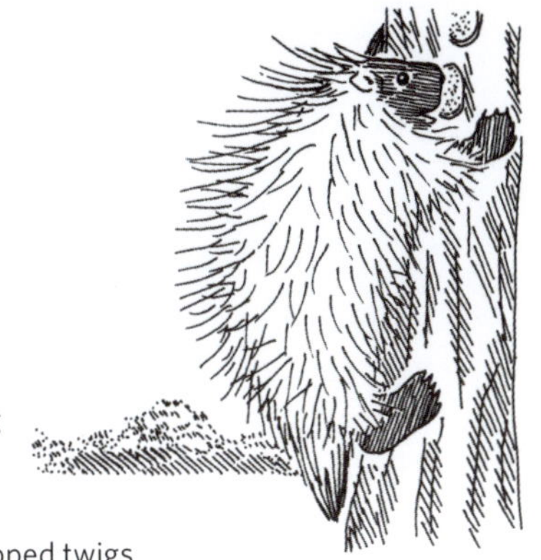

Habitat: Woods

- Toes line up across top of pad
- Nails may show
- Foot pads textured (right) like rubbery dots on slippers; may help feet grip for tree climbing
- In fluffy snow, tail may leave whisk broom sweeps
- *Signs:* Bark gnawed off tree; nipped twigs

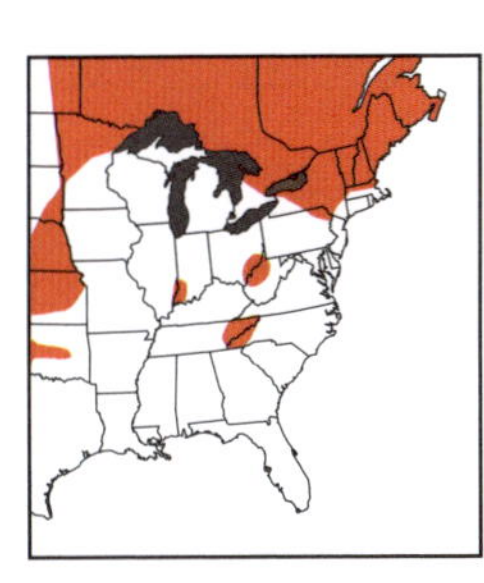

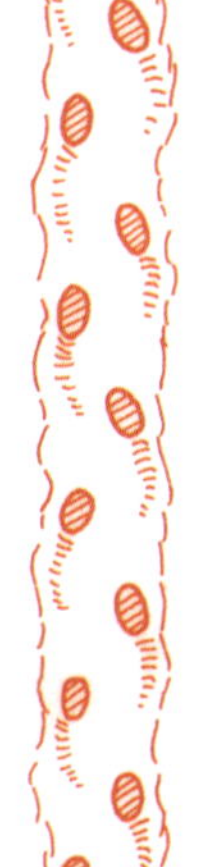

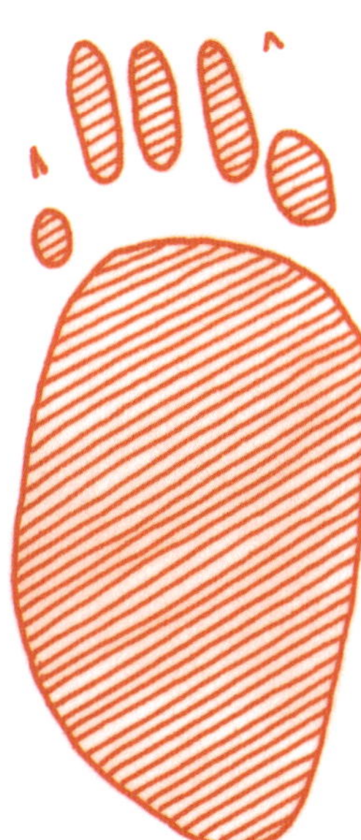

In deep snow, body makes trough; feet leave drag marks and toed-in prints

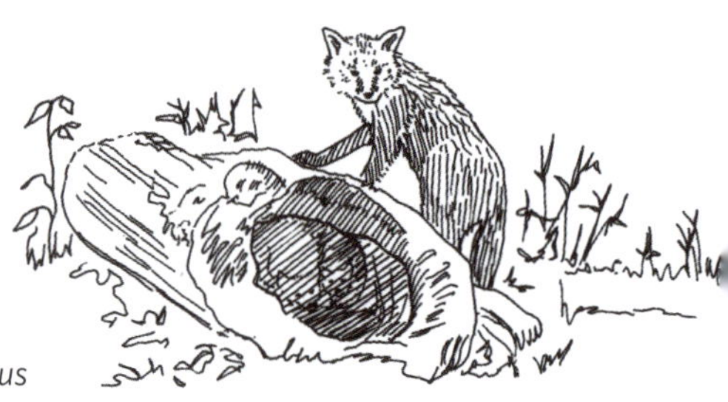

Gray Fox
Urocyon cinereoargenteus

Habitat: Southern pine or deciduous forests, with openings such as farmland or field

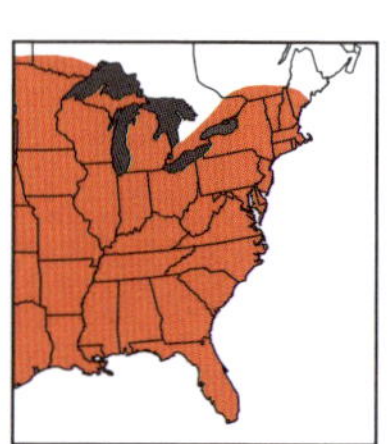

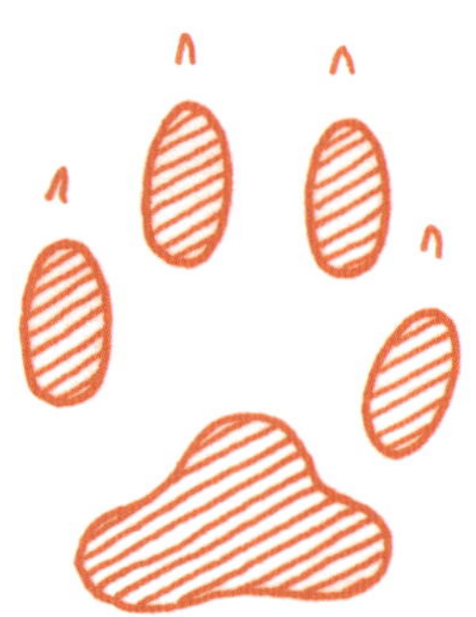

Gray foxes versus red foxes:

	GRAY FOX	RED FOX
Nails	Semiretractable; sometimes show in prints	Semiretractable; sometimes show in prints
Climbs tees?	Yes, including vertical trees	No
Front foot	Not heavily furred	Heavily furred
Heel bar?	No; sometimes base of heel; pad can appear straight	Front has straight/ slightly curved transverse bar
Direct register?	Yes	Yes

Red Fox

Vulpes vulpes

Habitat: Various, especially edge between woods and fields/shrubs

If the hind foot steps exactly in the print of the front foot (direct register), how can the front foot's transverse bar still show? The fox weights toes more than heel; if the heel does not drop, it does not overprint.

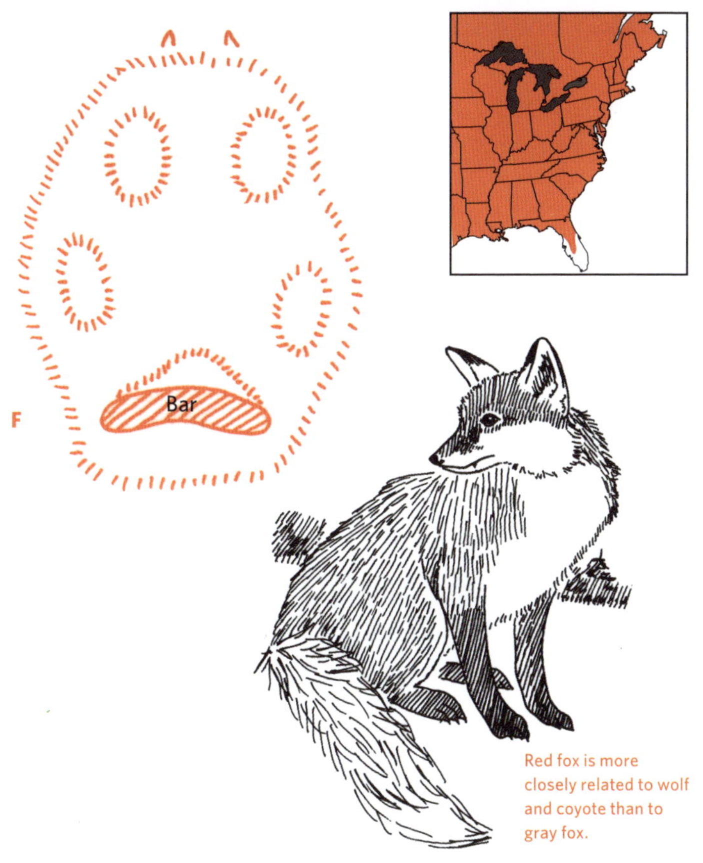

Red fox is more closely related to wolf and coyote than to gray fox.

Coyote
Canis latrans

Habitat: Species highly adaptable, found in variety of natural communities, from rural to urban areas

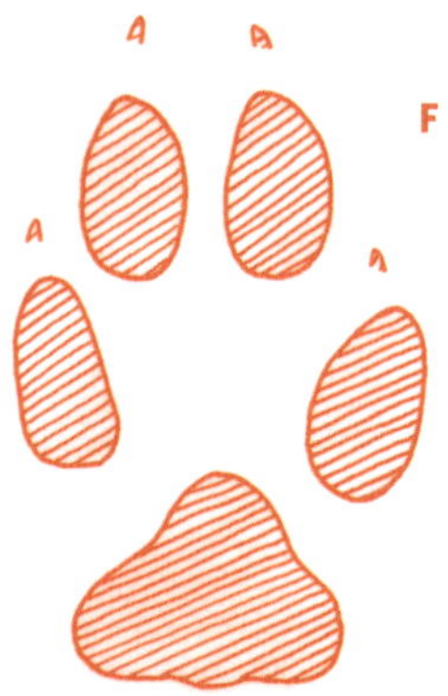

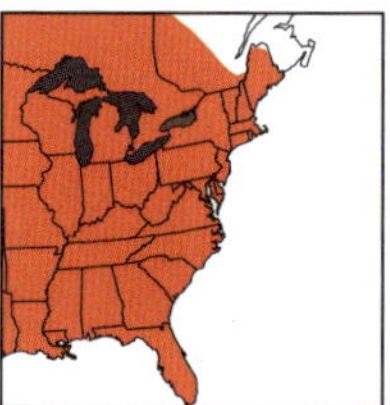

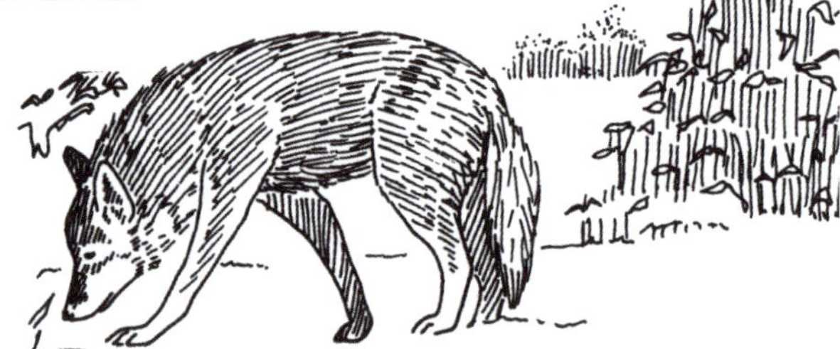

Wild canids versus dogs:

Wild canid: fox, coyote, wolf
- Behavior oriented to hunting
- Direct register when trotting and walking
- Travels at a steady trot; slows to a walk to inspect and speeds to a lope or gallop to cover ground quickly

Dog
- Does not hunt for its food
- Often switches direction and gait; toes often splay
- Walking, trotting trails have off-register prints

Gray Wolf

Canis lupus

Habitat: Wilderness forest

U.S. Fish and Wildlife Service splits Eastern Wolf, *Canis lycaon* (eastern Canada), and Red Wolf, *C. rufus* (being re-established in North Carolina), from Gray Wolf. Other groups classify them all as *Canis lupus*.

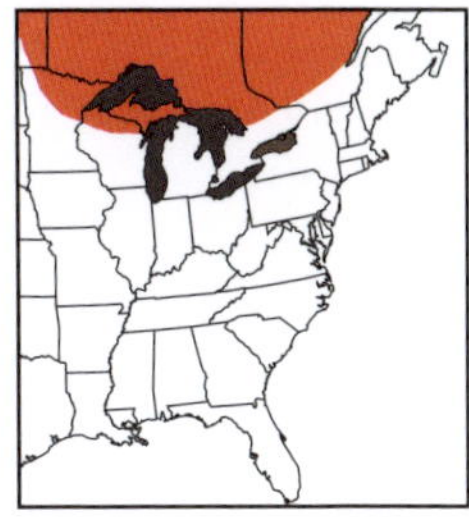

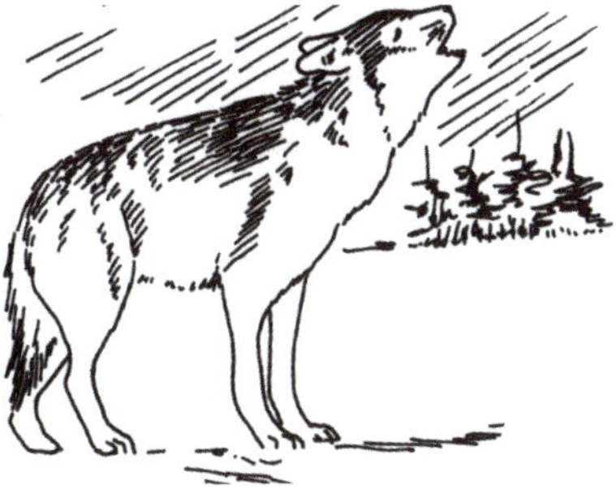

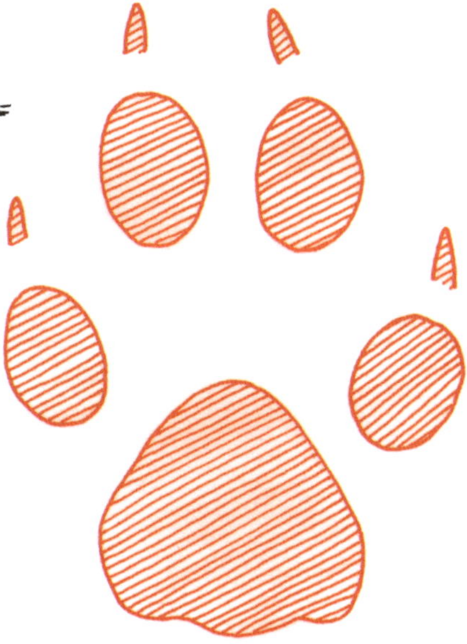

F at 75%

Raccoon
Procyon lotor

Range: Throughout

Habitat: Rural, suburban, and urban, usually near lakes or streams

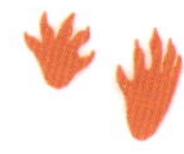

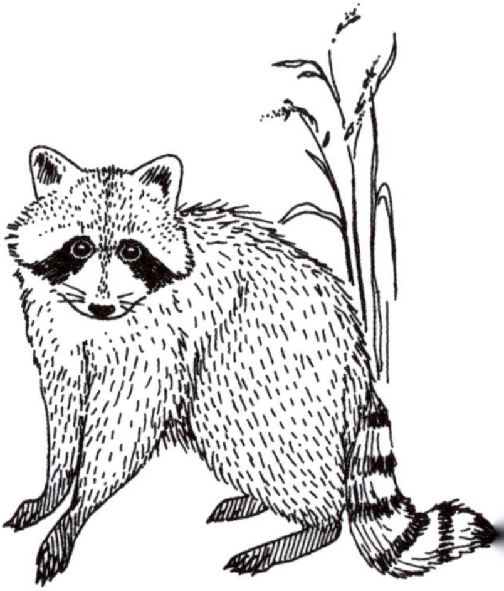

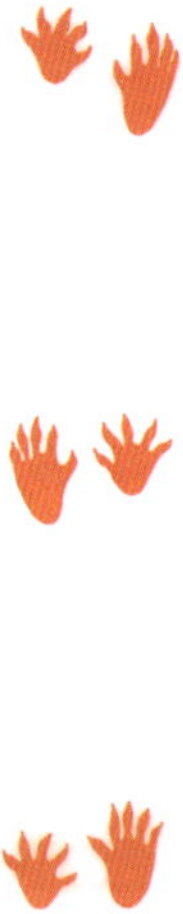

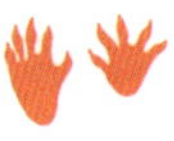

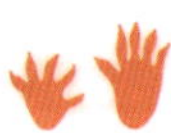

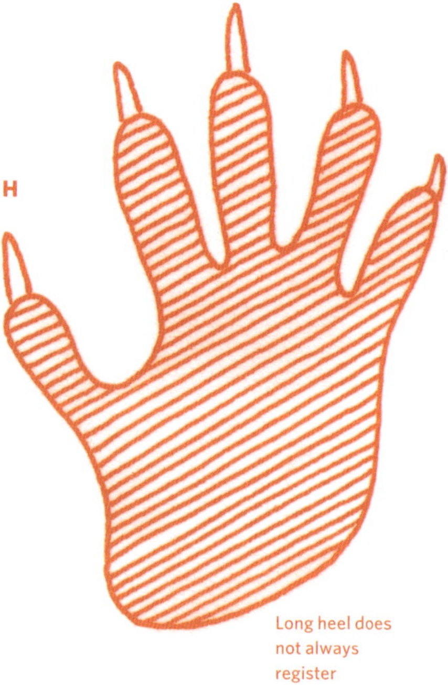

The raccoon's uncommon walking pattern

The animal swings forward both legs on one side, then both legs on the other side, producing pairs of front/hind prints (*above, right*). In deep snow, raccoons make a zigzag pattern.

250

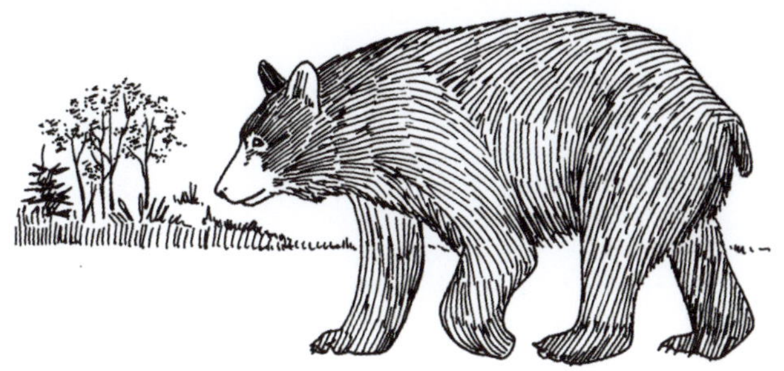

Black Bear

Ursus americanus

- Walks most of the time, either direct register *(lower right)* or overstepping *(upper right)*; uses direct register in snow
- Able tree climber
- Toe pads form a line across top; smallest pad, on inside, may not show
- Nails sometimes show
- Often, hind heel pad does not register

Habitat: Coniferous and deciduous forests

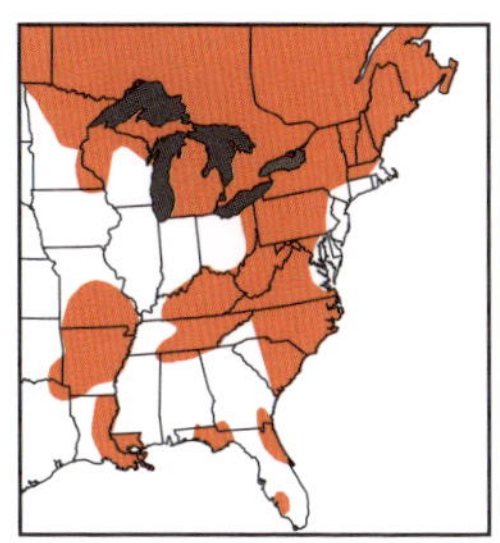

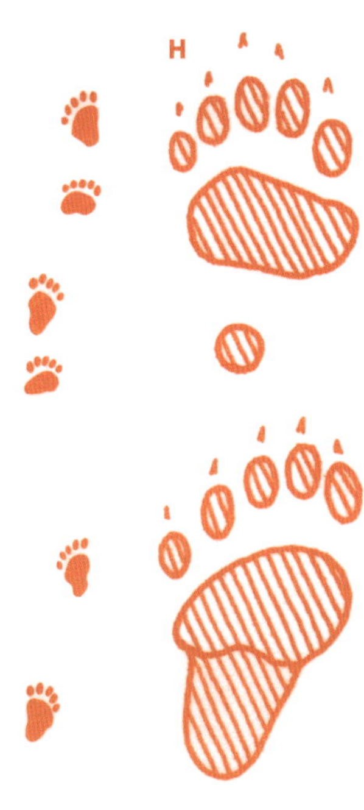

Black Bear

(Description on p. 251)

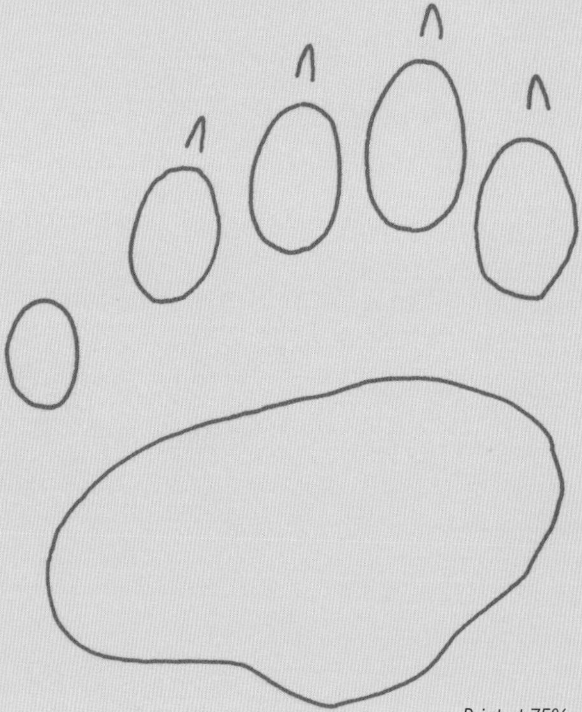

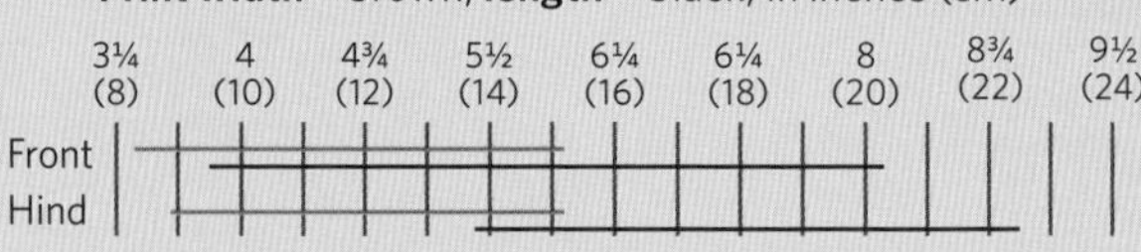

Print at 75%, without h

Trail width: 8 to 14 inches (20 to 36 cm)

Print width = brown, **length** = black, in inches (cm)

	3¼ (8)	4 (10)	4¾ (12)	5½ (14)	6¼ (16)	6¼ (18)	8 (20)	8¾ (22)	9½ (24)
Front									
Hind									

White-tailed Deer

Odocoileus virginianus

Range: Throughout

Habitat: Forests, fields, brushy areas

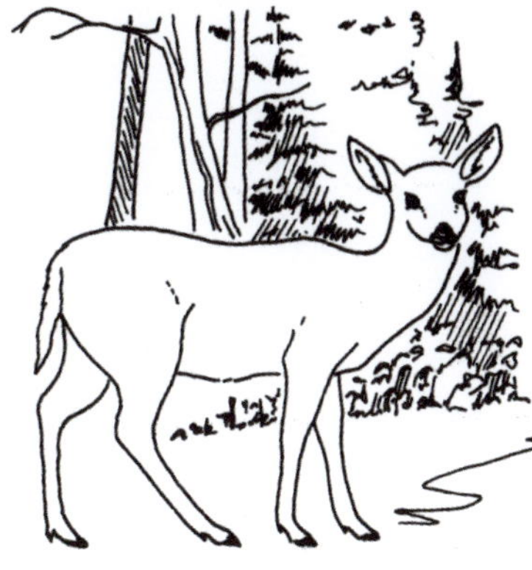

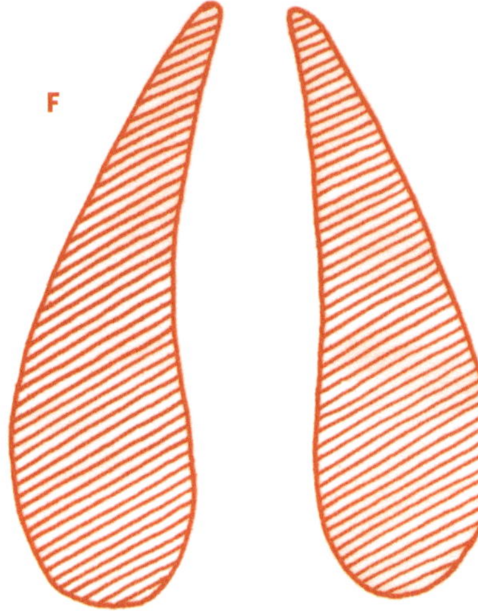

Deer print more tapered at tip than moose print

Deer and moose: Variations in prints

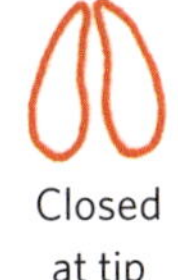

Closed at tip

Splayed

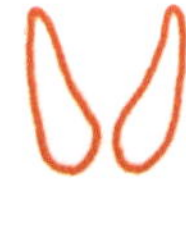

Splayed with dew claws (vestigial toes)

Tips only

Moose

Alces alces

Habitat: Hardwood and coniferous forests in winter, wetlands in summer

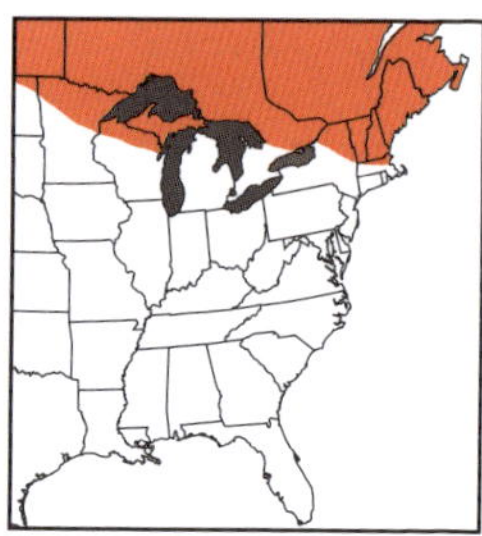

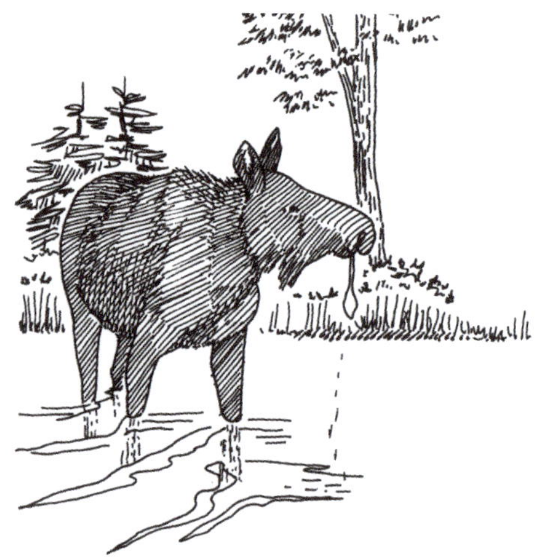

Deer and moose: Variations in trails

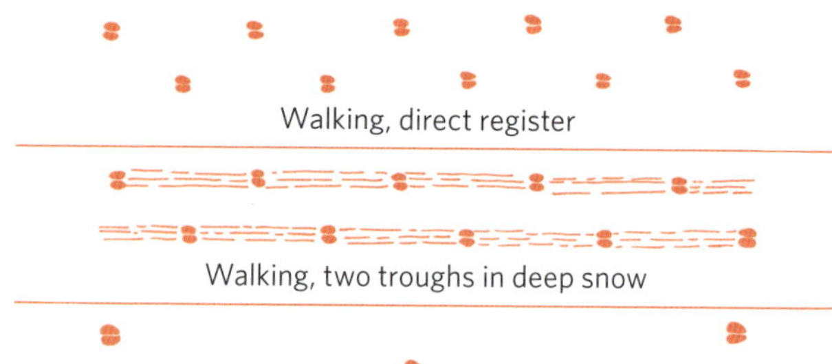

Walking, direct register

Walking, two troughs in deep snow

Trotting, with narrower trail width

Moose

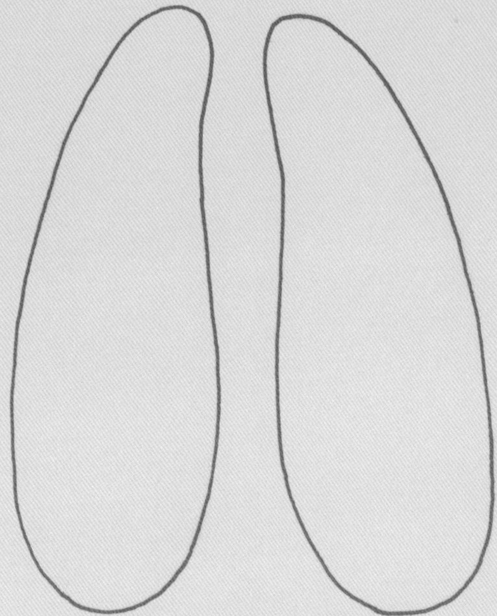

Reduced 10%

Trail width: 8½ to 20 inches (22 to 51 cm)

Print width = brown, **length** = black, in inches (cm)

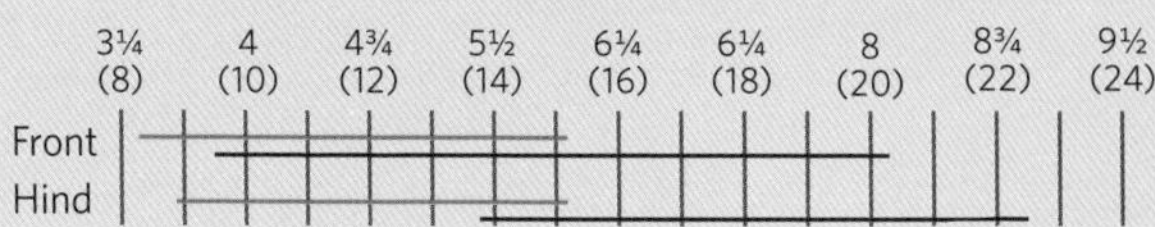

Birds

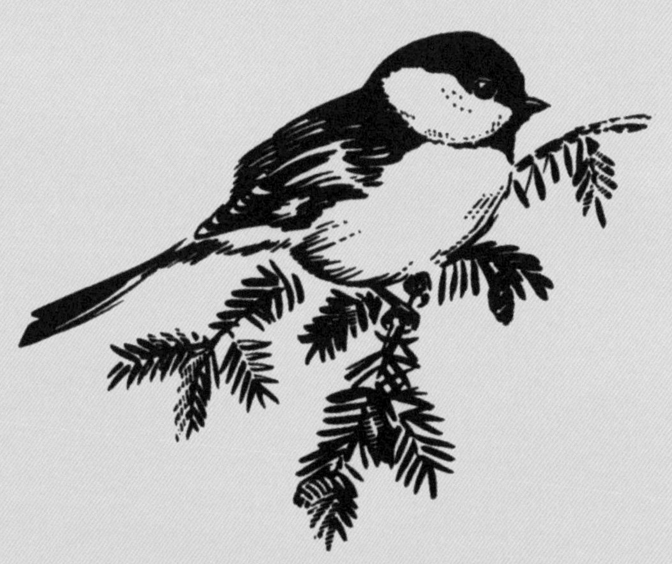

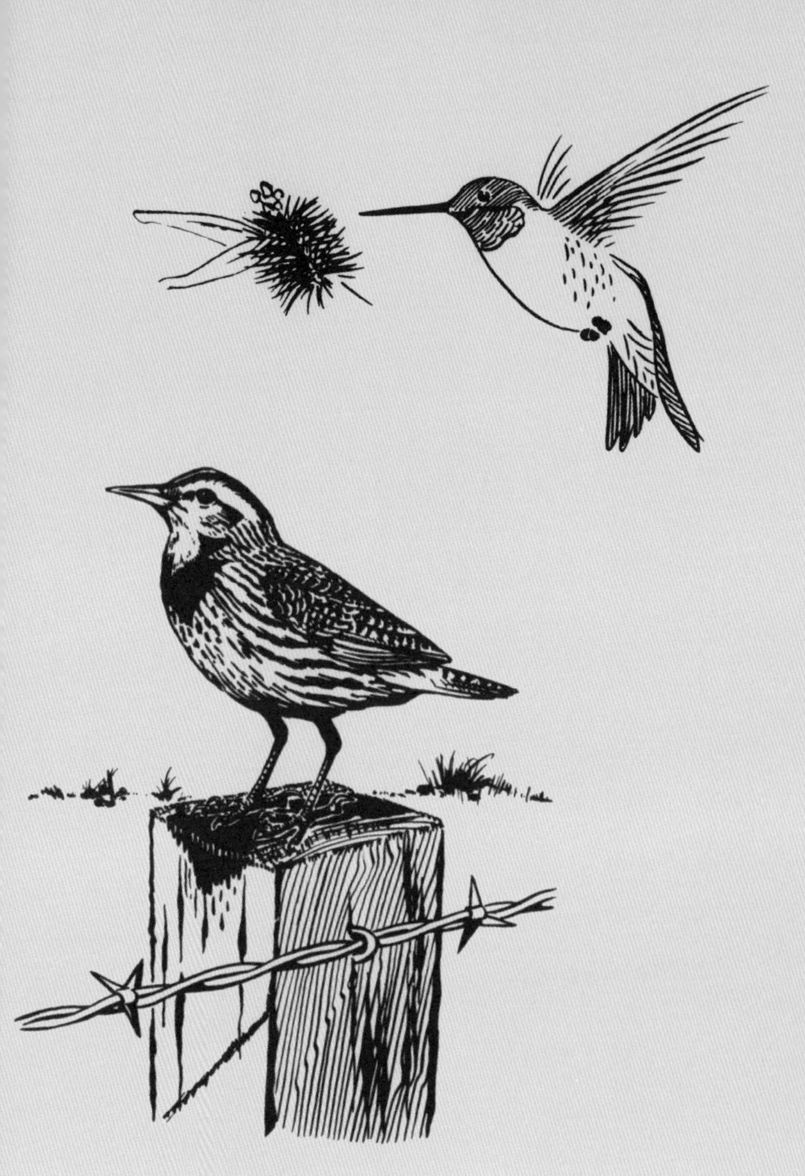

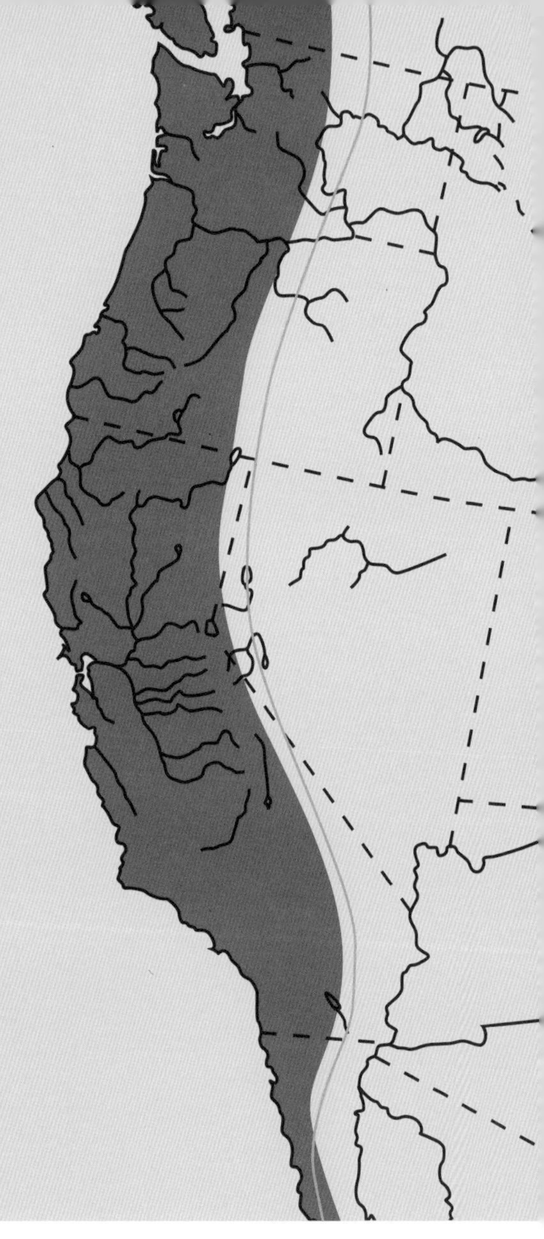

Identifying Select Birds of the Pacific Coast

Habitats

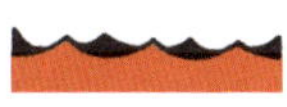

OCEAN
shores of the Pacific (beaches and bluffs)

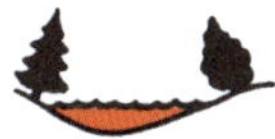

LAKES AND STREAMS
open freshwater (deep or shallow)

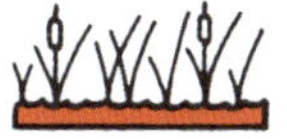

MARSHES
shallow freshwater with tules, cat-tails, etc. (including rice fields)

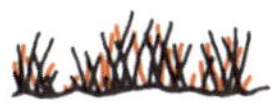

PRAIRIES
grasslands, fields of grain, other crops

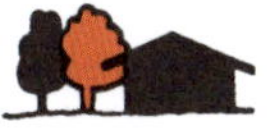

FARMS, PARKS, AND CITIES
areas most affected by civilization

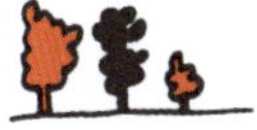

WOODS
open woods, mostly oak or pine

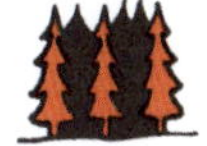

FORESTS
dense, evergreen, conifer forests

Belted Kingfisher

Shaggy, big-headed, blue and white. Gives a long, chattering call in flight. Selects a streamside perch with a good view and waits for fish to dive on. May swim briefly to catch prey. Beats fish on branch to kill it, then swallows it headfirst so fins won't get stuck.

Usually alone except in breeding season. Defends narrow territory along stream. Lays about seven eggs in streamside burrow dug with feet and beak. Eggs are white. Hidden in burrow, they need no camouflage.

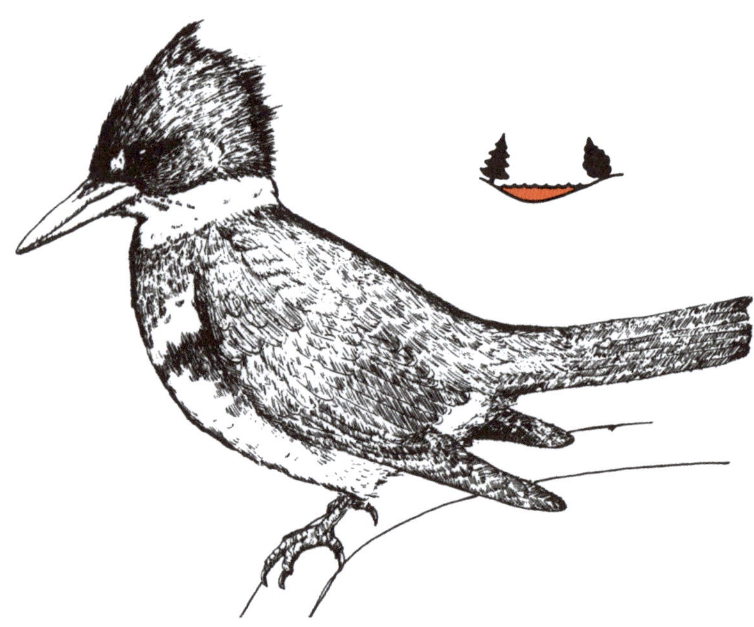

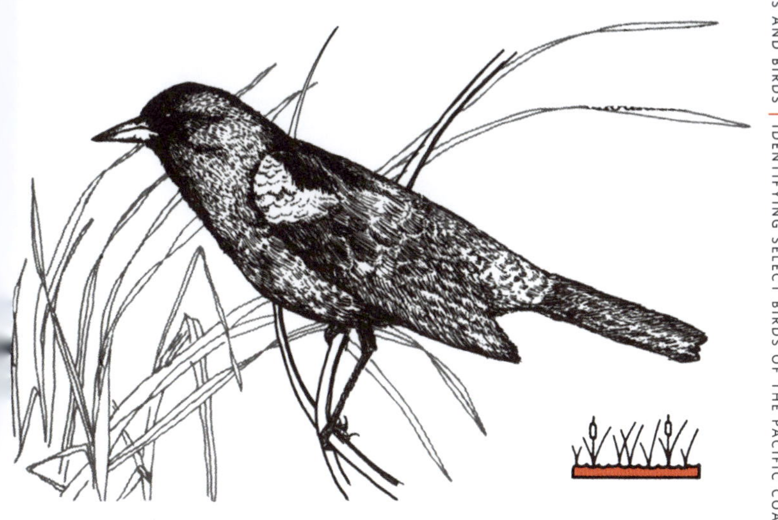

Red-winged Blackbird

Black with bright-red shoulder patch on male. Found in marshes throughout North America. Eats insects. Winters in warmer regions in separate all-male and all-female colonies. Males reach spring breeding grounds first and establish territories from which they drive out other males. Females arrive and select mate with territory—one to three females to each male. Each female builds a nest made of tules and standing tules or cattails. Four eggs per nest.

Red shoulder patch is essential for warding off competing males. Males with artificially darkened shoulders can't gain territory or mate.

Wood Duck

Male has striking multicolored head, red eyes, flowing nape feathers. More common in eastern United States. Unlike most other ducks, it nests in tree cavities (or artificial nest boxes) but has, nevertheless, some waterfowl habits that seem more appropriate to ground nesting. Adults fly into nest headfirst, without perching. Fluffy young ones are precocial. They climb from the nest before learning to fly and tumble softly to the ground. (Female is reported to carry young to ground on back.)

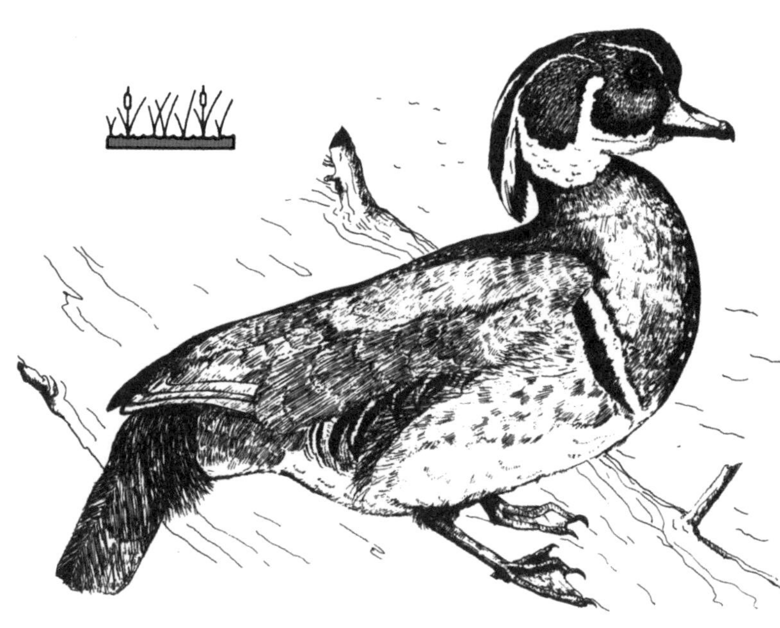

Mallard

Males have green head, white neck ring, and cinnamon chest. Female is brown-streaked. Both sexes have purple-iridescent wing patch bordered by white. The best-known duck. Has a resounding, classical "quack." Many barnyard ducks are domesticated versions of Mallard. Can take off almost vertically from water.

Lays ten to fifteen eggs in down-lined nest on ground, usually well concealed by vegetation.

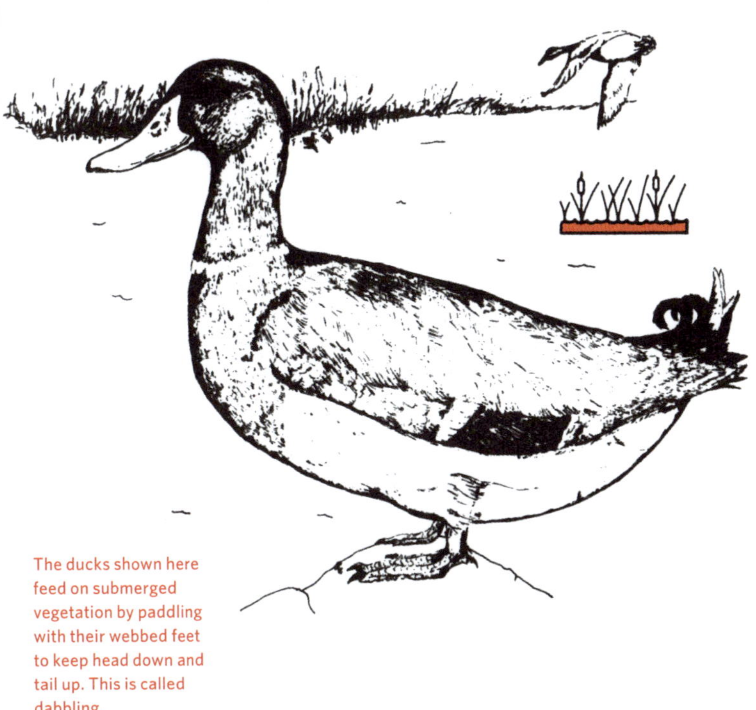

The ducks shown here feed on submerged vegetation by paddling with their webbed feet to keep head down and tail up. This is called dabbling.

Pintail

Male has white chest,
strip of white running up
neck, and long pointed tail.
Female is drab, mottled brown.
Lays seven to twelve eggs in
simple grass nest on ground.
Winter flocks of thousands
gather in wildlife refuges of
California Central Valley. Prized
by hunters who call it "sprig." Whistles.

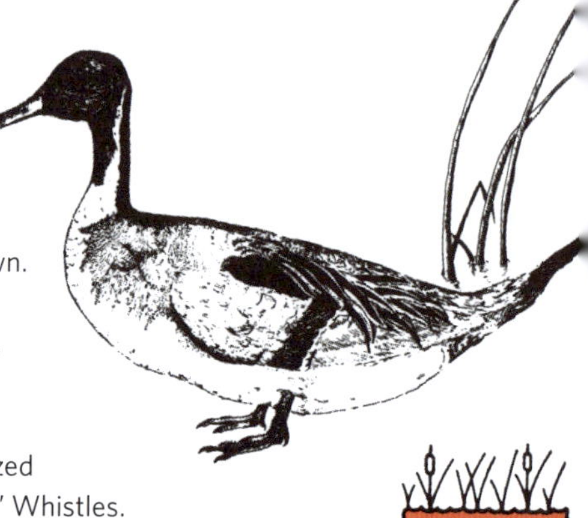

Loggerhead Shrike

Gray, white, and black—like Mockingbird, but stockier with shorter tail.
Wingbeat very rapid. Shrikes and their relatives are all in the large group
of small, brightly colored, perching songbirds. But shrike habits are hawk-
like. They catch sitting birds, small mammals, large insects in hooked
beak. Their songbird feet are too weak to serve as talons, so they impale
prey on twigs, rushes, barbed wire, earning the name "butcher-bird."

Shrikes often sit in open on fences, wires, tops of small trees, especially
near farms. Lay four to seven eggs in bulky nest of twigs placed in dense
shrubbery.

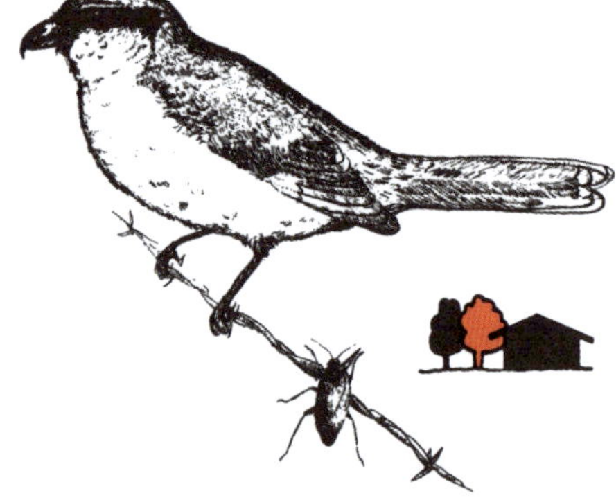

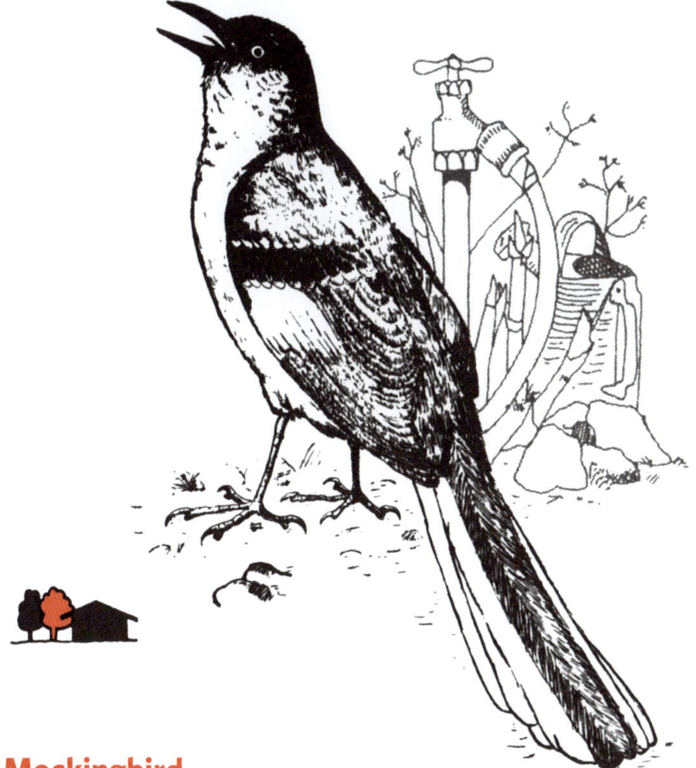

Mockingbird

Gray, robin-sized, with white wing patches. Sings at night. Mimics and "mocks" other birds with a wide repertoire. Lays four pale blue eggs with brown splotches in bulky, twiggy nest in dense shrubbery. Eats berries, seed, insects. Rare north of California.

Many birds sing to clear their territories of competing members of their own species. Mockingbird imitations may be accurate enough to fool and drive away other species as well. Experiments show imitations are learned, not instinctive. Flashing displays of white wing patches may also drive off intruders.

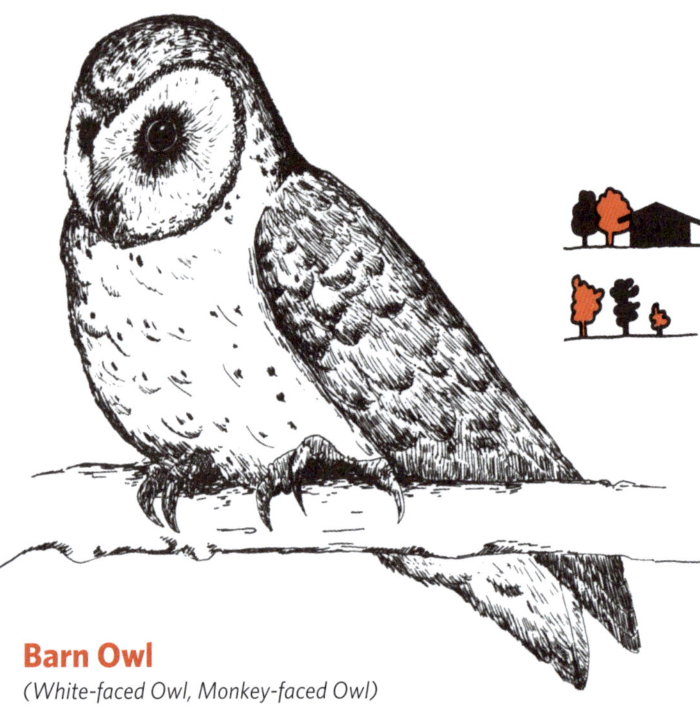

Barn Owl

(White-faced Owl, Monkey-faced Owl)

The most commonly seen owl. Light brown and white with heart-shaped face. Like distantly related hawk, preys mostly on rodents. Searches at night over prairies and farms using night vision that far exceeds that of humans. Feather-covered, asymmetrical ears, which look human under feathers, can sense direction and pinpoint source of sound. Head can turn almost full circle. Fringed wing feathers produce silent flight. Unlike most birds, which hatch young in even-aged groups, owl nests contain young of varying ages. Young may leave nest early and require care until they can fly. Young found on ground have seldom been abandoned and are extremely hard to raise in captivity. Look under roosting trees for compacted, regurgitated "owl pellets," which contain complete skeletons and fur of one or more small animals—a record of owl diet.

Acorn Woodpecker

Back of head is red. Displays white wing patches and rump in flight. Lives in large groups that raise young cooperatively and share "granaries"—old trees, poles, buildings they riddle with holes to hold acorns and nuts tightly, safe from squirrels, crows, etc. They transfer drying, shrinking nuts to tighter holes and eventually crack shells to extract nutmeat and insects feeding on it. Abundant in California oak woods and many urban areas.

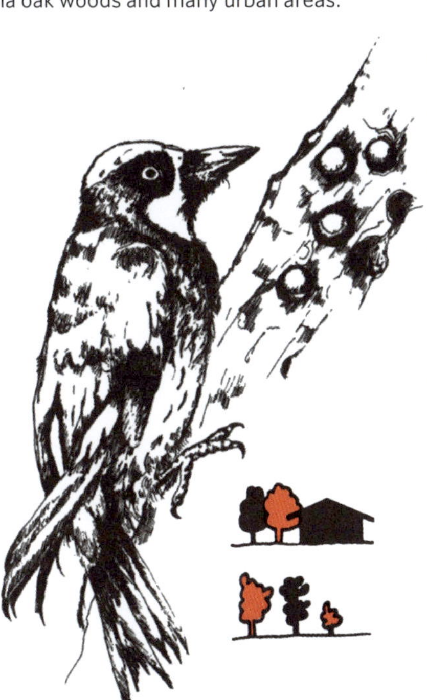

Woodpeckers

Woodpeckers clutch tree bark with extra backward-pointing toe, lean on still tail to peck, and explore crevices with stiff, extensible tongue for insects and their eggs. Special jaw suspension absorbs shock of pecking. They chisel nesting cavities in trees, crack nuts, and proclaim territorial rights by pounding resonant branches, trunks, even tin roofs. Call is raspy. Flight is undulating.

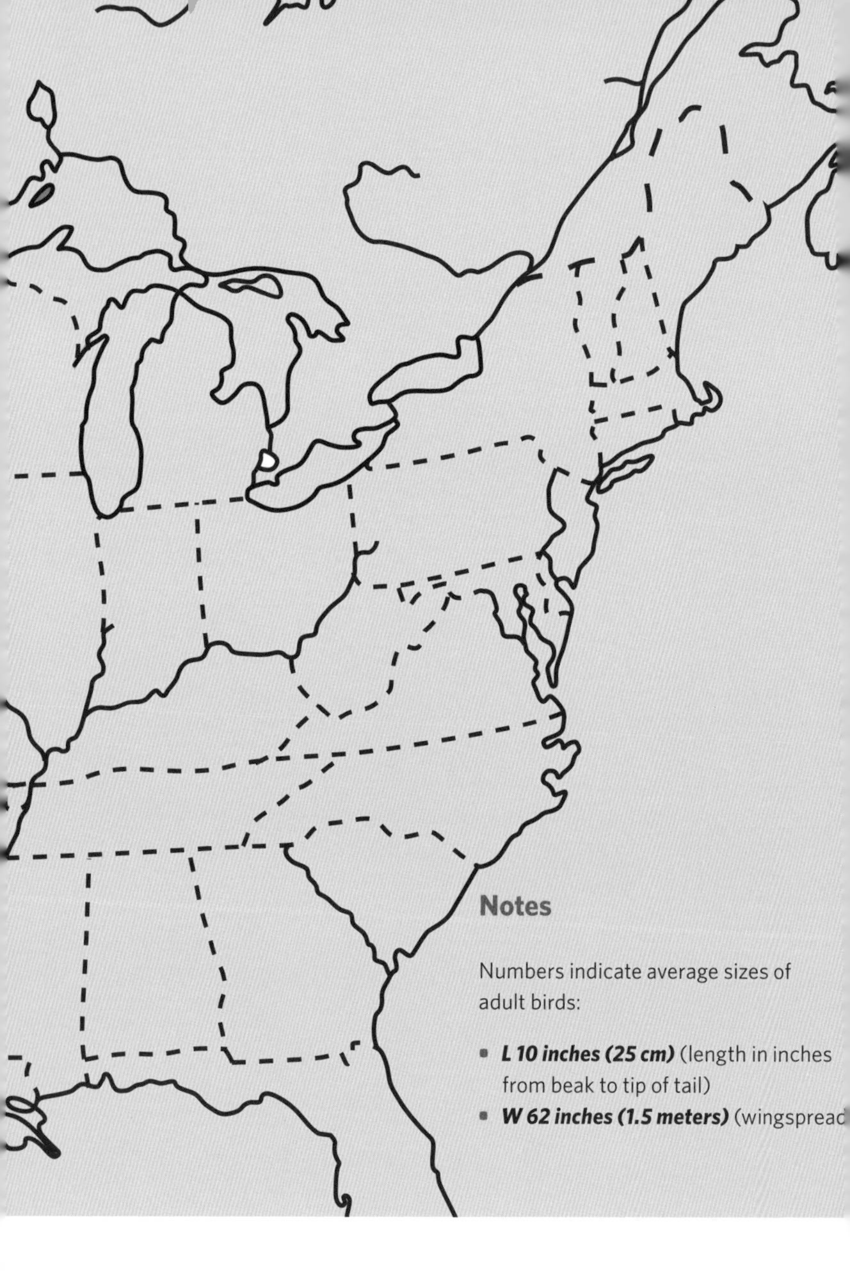

Notes

Numbers indicate average sizes of
adult birds:

L 10 inches (25 cm) (length in inches
from beak to tip of tail)
W 62 inches (1.5 meters) (wingspread

Identifying Select Birds of Eastern North America

Habitats

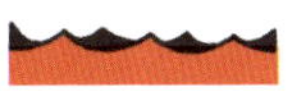

OCEAN
shores of the Atlantic (beaches and bluffs)

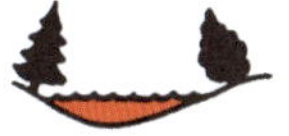

LAKES AND STREAMS
open freshwater (deep or shallow)

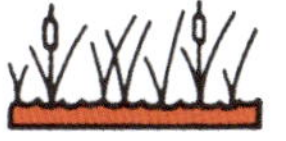

MARSHES
shallow freshwater with tules, cattails, etc.
(including rice fields)

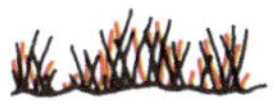

PRAIRIES
grasslands, fields of grain, other crops

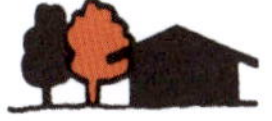

FARMS, PARKS, AND CITIES
areas most affected by civilization

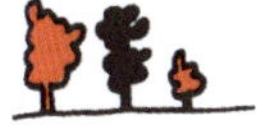

WOODS
deciduous woods and forests

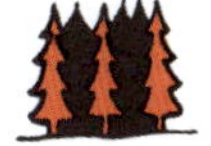

FORESTS
evergreen forests

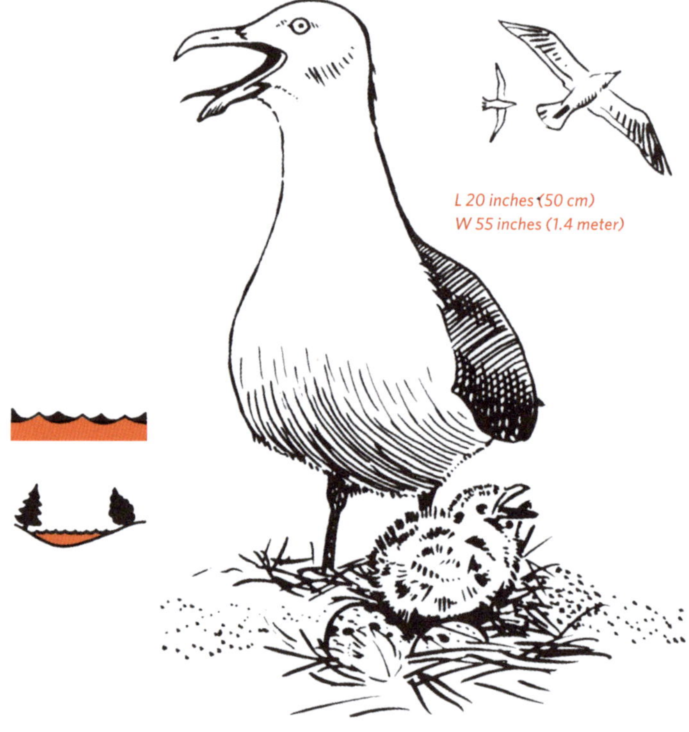

Herring Gull

White body and head, gray wings with black tips, pink legs, and red spot on bill.

Two other common gulls are Ring-billed (similar, but yellow legs and black ring around bill), and Bonaparte's (smallest of three, bright orange legs and feet, black head).

Gulls eat almost any food—fish, crustaceans, insects—living or dead. Often at garbage dumps, beg for scraps in parks, follow ships. Make simple nests of vegetation on ground or cliffs, in colonies. Both parents incubate the two to four spotted eggs.

Killdeer

Has dark brown cap; wide brown face band; white underparts except for two dark breast bands; brown back; pointed wings; rusty tail and rump. Call is high, shrill "kill-deer, kill-deer." A common freshwater shore bird. Also found in almost all open habitats. Finds invertebrates to eat along shore and on ground.

Nest on ground is often only a depression in gravel. Arranges four eggs with small ends inward to prevent rolling. May replace missing egg with rock to keep others in place.

Young are speckled like eggs for camouflage. On approach of intruder they sit still and inconspicuous while parents scream and run with drooping wing to feign injury and divert intruder's attention from nest.

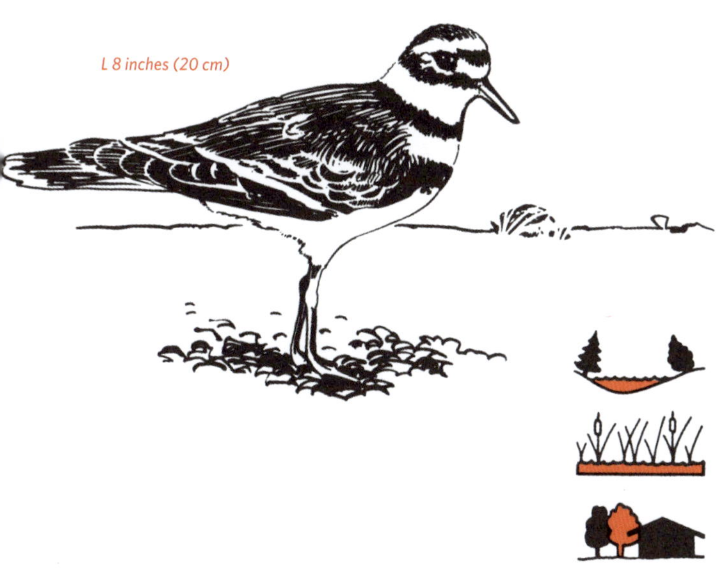

L 8 inches (20 cm)

Marsh Hawk

Males are gray, females tan. Both display distinct white rump in typical low flight over marshes, meadows, and grasslands seeking small rodents. Lay four to six eggs in nest of grass and reeds built on twig and tule platform in marsh or meadow. Female incubates eggs for a month while male brings her food, often passing it to her in flight.

Adapted to hunting over tall and dense marsh vegetation. Other hawks searching over dry land vegetation can profit by soaring high to view a wide area from many angles, but Marsh Hawk's effective field of vision is narrowed to a single, vertical channel by tules, high grass, cattails, etc. Since prey is visible only when directly underneath, nothing would be gained by high soaring. Hovering allows careful examination of ground to detect scurrying of prey.

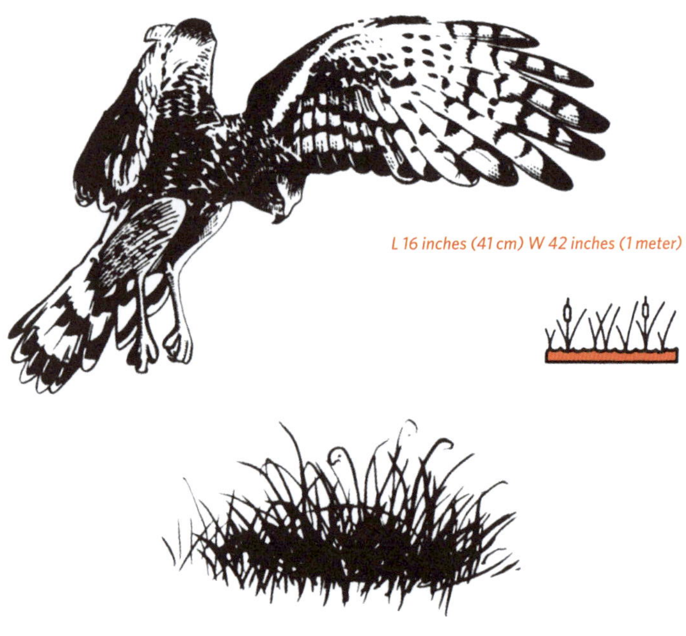

L 16 inches (41 cm) W 42 inches (1 meter)

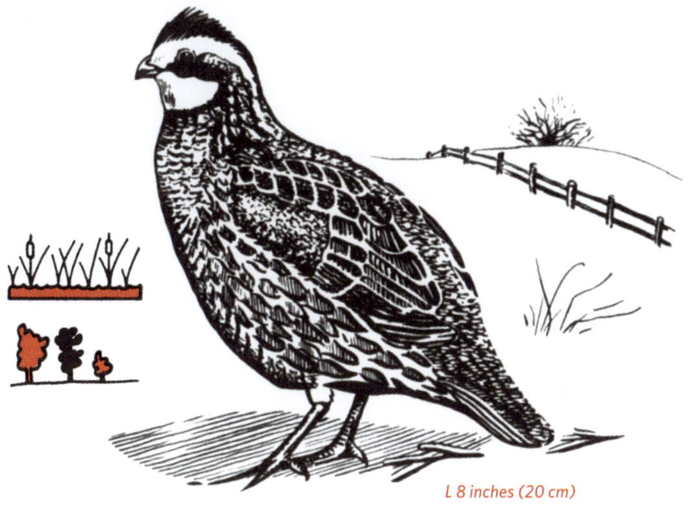

L 8 inches (20 cm)

Bob-white Quail

Male has white stripe over eye, down side of neck, and white throat. Wings and back are brownish, belly barred with dark brown markings. Female is similar but has pale brown face and head. "Bob-white" call is distinctive.

Nests on ground in dense cover. Lays up to twenty eggs. Hatchlings fly in a few days. Quail form winter flocks (coveys) and roost together in circles, facing outward, ready to scatter at first sign of predator.

Eat seeds, grain, fruit, berries, and insects. Live year-round in brush, forest edges of eastern and midwestern United States.

Grain Eaters

Quails, pheasants, turkeys, and chickens are related, and have equipment for digesting hard seeds and grain. Lower neck has expansible crop to store quickly gobbled food for slow grinding in muscular part of stomach (gizzard), which contains swallowed gravel (grit). They run fast, fly laboriously.

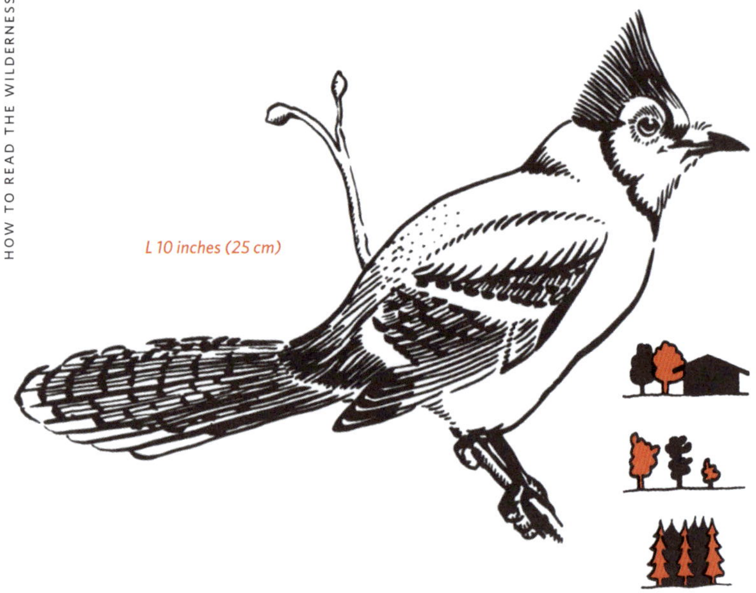

Blue Jay

Has crest, white face with black necklace, blue and white back and tail, and whitish underparts. Raucous call like squeaky pump.

May prey on eggs and young of smaller birds. Otherwise eats berries, insects, seeds, nuts. Common at feeders in winter, where it sweeps smaller seeds aside with bill to find sunflower seeds.

Inhabits deciduous and coniferous forests, wooded suburbs, and parks. Nonmigratory but may move several miles in flocks in search of food or better weather. Some oak trees owe existence to a jay that could not relocate an acorn pushed into the ground for storage.

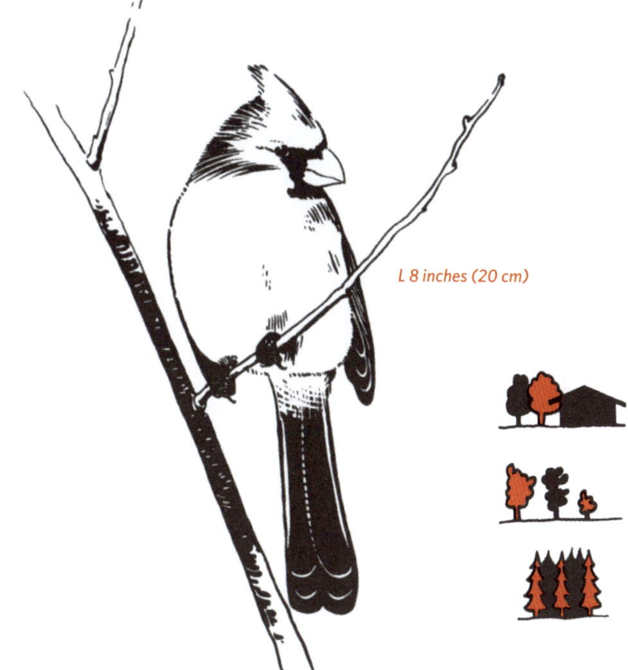

Cardinal

Bright red body and topknot and black face of male Cardinal are well known. Female dull red and greenish. Inhabits woods, city parks, and backyards, where its loud, clear whistle is often heard.

Nonmigratory resident of almost entire eastern half of United States. Range has been expanding northward for the past few decades, although severe winters may temporarily reduce its numbers. Backyard feeders allow Cardinals to survive in areas otherwise devoid of food. But if seed supply stops abruptly, as when human seed provider leaves for weekend, birds dependent on it may die. Fond of large seeds. Powerful beak can inflict painful bite.

State bird of Illinois, Kentucky, Indiana, North Carolina, Ohio, Virginia, and West Virginia.

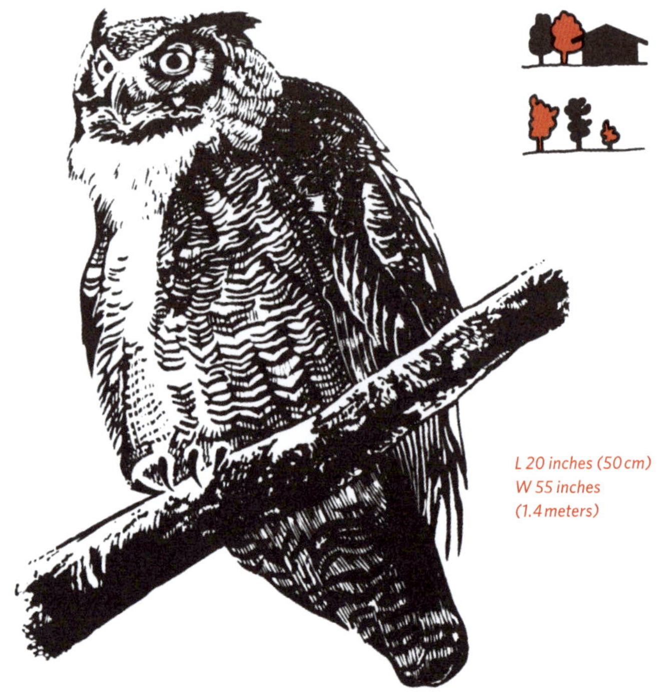

Great Horned Owl

Lives year-round in most of North America. Brownish with white throat, earlike feather tufts on head. Four-foot wingspread. Familiar hooting call.

Preys on gophers, mice, rats, snakes, fish, squirrels, and skunks. May attack sitting ducks, occasionally chickens, but does far more good than harm.

Often mobbed by groups of crows or blackbirds seeking to drive this large predator away. Owl decoys are used to both lure crows and blackbirds into traps, and to scare them away.

Nests in tree cavity as early as February.

Mourning Dove

Gray brown, plump bodied with pointed white-edged tail. Takes off noisily, flies swiftly with rapid wing beats. Has plaintive "mourning" call.

Like others of pigeon family, can drink by sucking, while other birds must tilt heads back to swallow.

Eats seeds and grain. Found throughout United States and southern Canada. Most doves migrate south from their nesting areas in winter.

Two eggs are incubated by female during day and by male at night in simple saucer-shaped nest on or near ground. Naked, helpless hatchlings are fed secretions from the parents' crops called "pigeons' milk." Later eat seeds digested and regurgitated by parents. May have several broods per year; invariably lay two eggs.

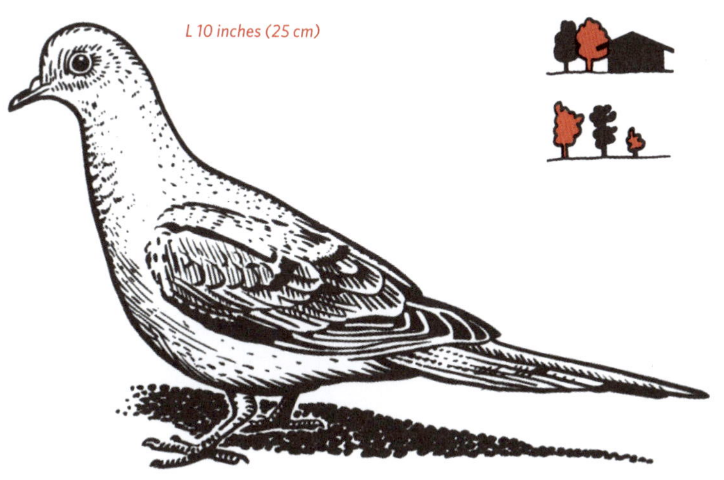

L 10 inches (25 cm)

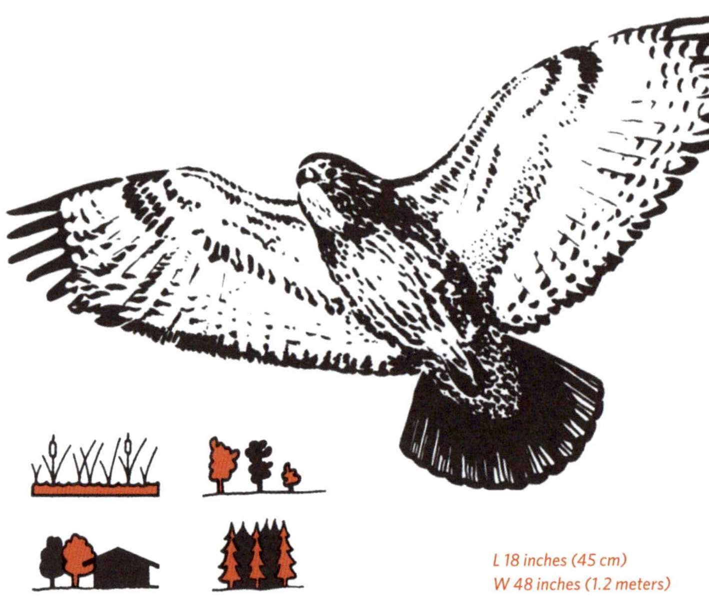

Red-tailed Hawk

The most commonly seen hawk. Only adults have rust-colored tail, but all ages have dark, contrasting band across belly. Body darkness varies. Often seen soaring over meadows and woods when sun makes thermal updrafts, or waiting on posts and poles in cloudy weather. Uses excellent vision to spot rodents to grasp in powerful talons and tear apart with hooked beak. Young may cover captured prey with outspread wings to avoid sharing with nestmates, etc. Build large, open tree nests.

Ecological value in rodent control vastly outweighs rare chicken taken by hawks, but ignorant gunning kills many "chicken hawks" despite federal and state protection.

Warblers

Warblers are small, flighty birds with thin bills. Sing complex, melodious songs. Adult males in breeding season are brightest—later, males resemble drabber females. Eat insects, larvae, and eggs. Build small cup-shaped nests of grass, bark strips, leaves, and moss, in trees or on ground.

Warblers winter in South and Central America. Many breed in northern United States and southern Canada. Forty species east of Mississippi. Common warblers and their obvious colors include: Black and White (black, white stripes); Yellow (yellow undersides with red streaks); Yellow-rumped (yellow rump, flanks, and cap); Chestnut-sided (yellow cap, white underparts, rusty flanks); Common Yellow–throat (yellow underparts, black face mask); American Redstart (black with pink patches on tail, wings, shoulders).

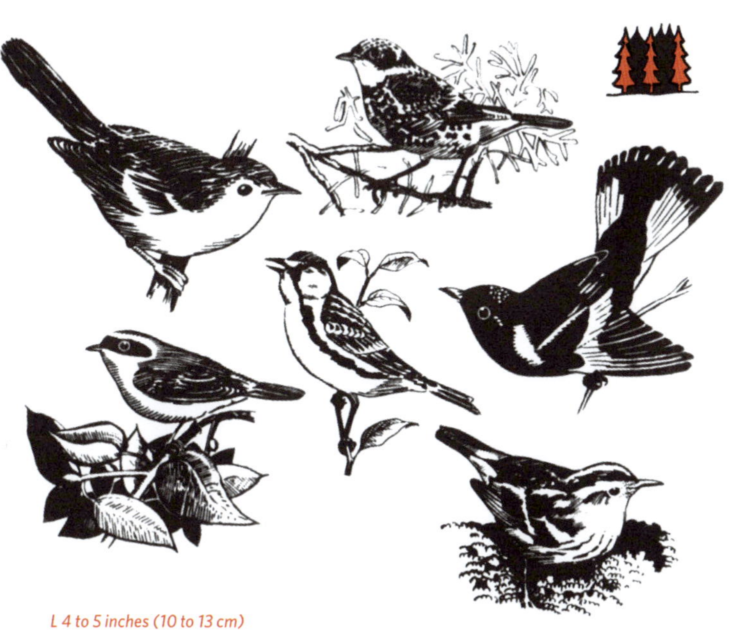

L 4 to 5 inches (10 to 13 cm)

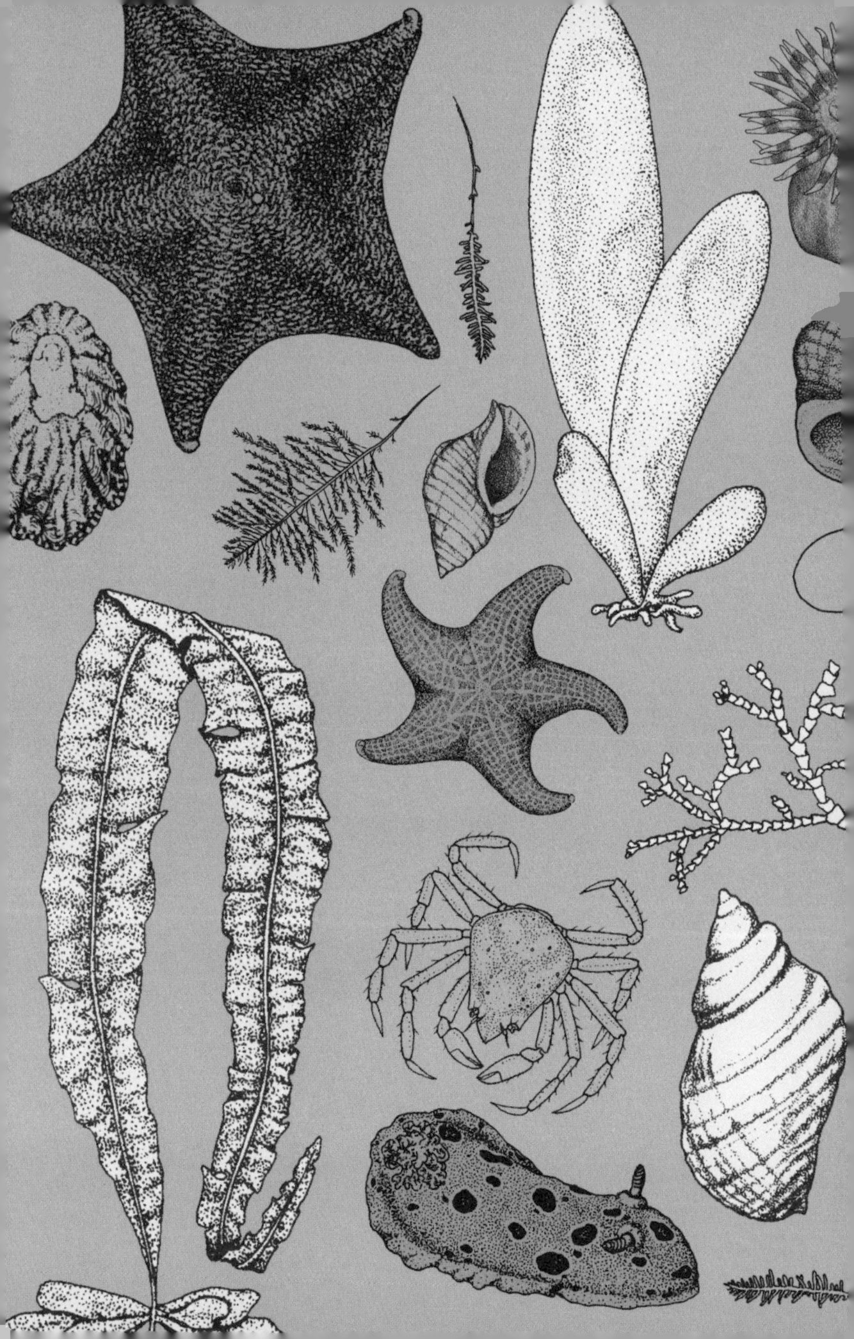

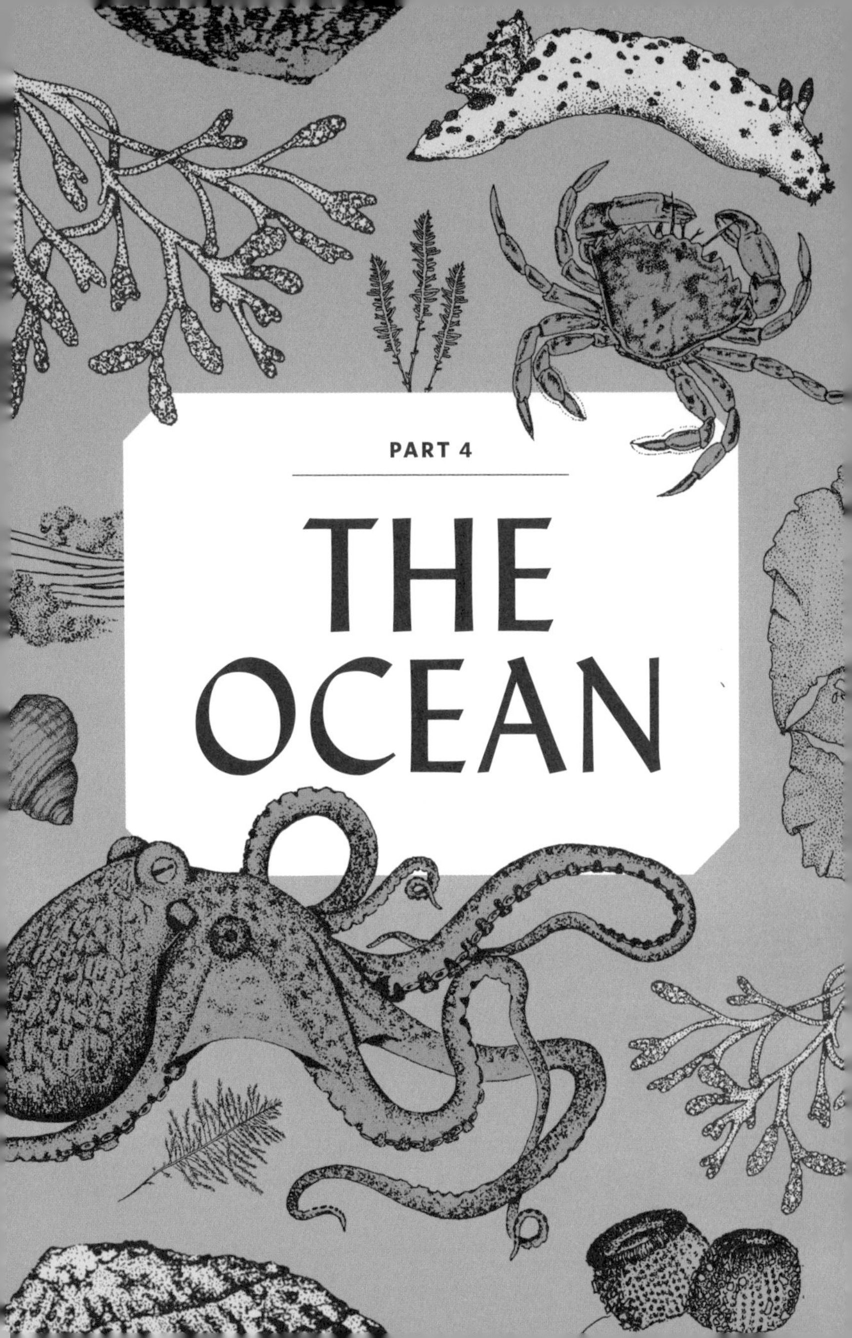
PART 4
THE
OCEAN

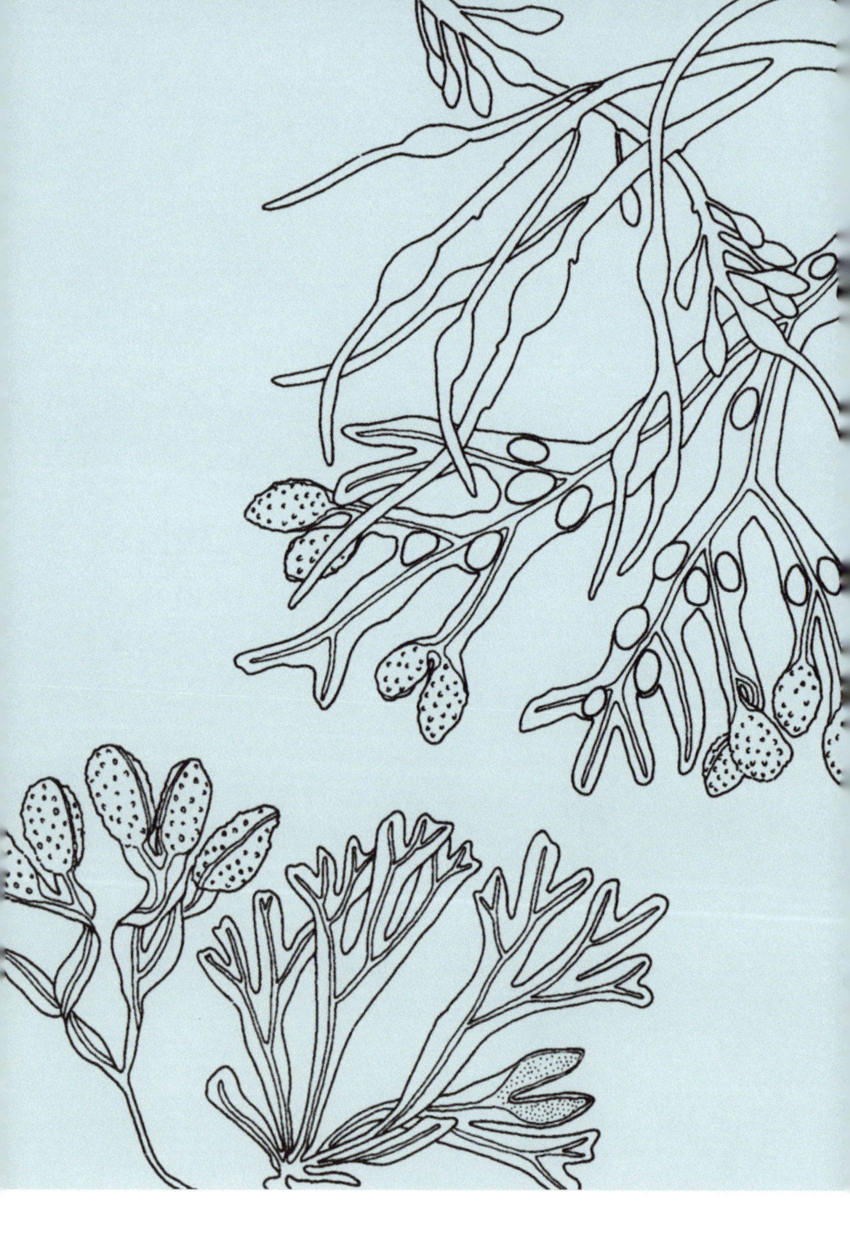

Intertidal Life on the Pacific Coast

CAUTION: Watch your footing! Accidents are common on slippery or unstable rocks. Don't let the rising tide leave you stranded. And watch out for waves in exposed areas.

Intertidal Ecology

Rocky reefs along the Pacific coast have some of the world's richest intertidal life. The secret behind their abundance lies with winds and currents that drive surface water away from shore, causing colder water to rise from below. The upwelling water is rich in the nutrients plants need. So plant life—from microscopic diatoms suspended in the water to giant attached seaweeds—flourishes. There's plenty to eat and an incredible variety of eaters. Animals subsisting on the vegetation support, in turn, a horde of predators, parasites, and scavengers.

Intertidal organisms use each other for much more than food, however. Some attach themselves to, hide in, or ride about upon others. Scale worms may live harmlessly on the undersides of sea stars and urchins, or in the shells of hermit crabs. A dozen or more different species may live on or under an abalone's shell. Some species are experts at mimicking and use particular animals or plants for camouflage. Tidepool sculpins alter their color to match the algae where they rest. Interaction among the species is complex.

The lives of intertidal creatures are further complicated each time an outgoing tide exposes them to a radically different environment: pounding waves, drying wind, heat and ultraviolet radiation from the sun, rain that dilutes salinity, and a whole new set of predators. It's a hard life. The reef dwellers you find are the tough survivors—shaped by ages of evolutionary change that has gradually adapted them to these kinds of adversity.

Although the natural environment may be harsh, the greatest present dangers to intertidal life are man-made. No species has had time to evolve a defense against oil spills or chemical pollution. Nor can intertidal animals resist either the unintended damage done by careless visitors or the deliberate killing done by inconsiderate collectors.

You can help preserve reef life if you look under rocks, then return them to their exact original positions. Make sure that people with you do the same and know why it's important: sponges, hydroids, anemones, and other animals attached to the undersides of rocks will die if you leave them exposed to air, sun, and predators. You can use a jar to observe free-swimming animals, but put them back. Above all, don't try to remove permanently attached animals, and don't take any animals home with you. They will soon die. Besides, it's usually against the law to collect these animals. So don't do it. Be a careful visitor so that others can see the intricate variety of life that exists on rocky reefs.

Intertidal Zonation

The intertidal zone extends from the highest wave-splashed rocks down to levels that are uncovered only by infrequent, extreme low tides. Within this zone are many different kinds of places to live.

That's why you'll find different plants and animals as you move from one level to another. Species with the least tolerance to atmospheric exposure are found at the lowest part of the zone or closest to the sea. More tolerant ones live near the tops of rocks or higher up the reef. Species that can't take heavy wave action occupy protected crevices of the lee sides of rocks.

On flat reefs, these changes in the kinds of plants and animals present are subtle. But where there are large, deep pools, or rock walls and overhangs, the changes are more abrupt and dramatic. You can often find sponges and tunicates at the lowest levels of a rock wall that has mussels and goose barnacles at the top. The following chart shows three tide zones used as a guide to where you're most likely to find the species described in this chapter.

About Tides

The highest and lowest tides, called spring tides, occur every two weeks near the times of either full moon or new moon (dark of the moon). Between periods of spring tides there are less extreme tides called neap tides.

The low spring tides are usually the best times for exploring, especially in the lower part of the intertidal zone. To determine the tides for a given date, consult the daily tide information in your newspaper or online. Yearly tide tables are available where fishing supplies are sold.

CAUTION: When exploring a reef at low tide never let the incoming tide flood your route back to shore.

Intertidal Zones

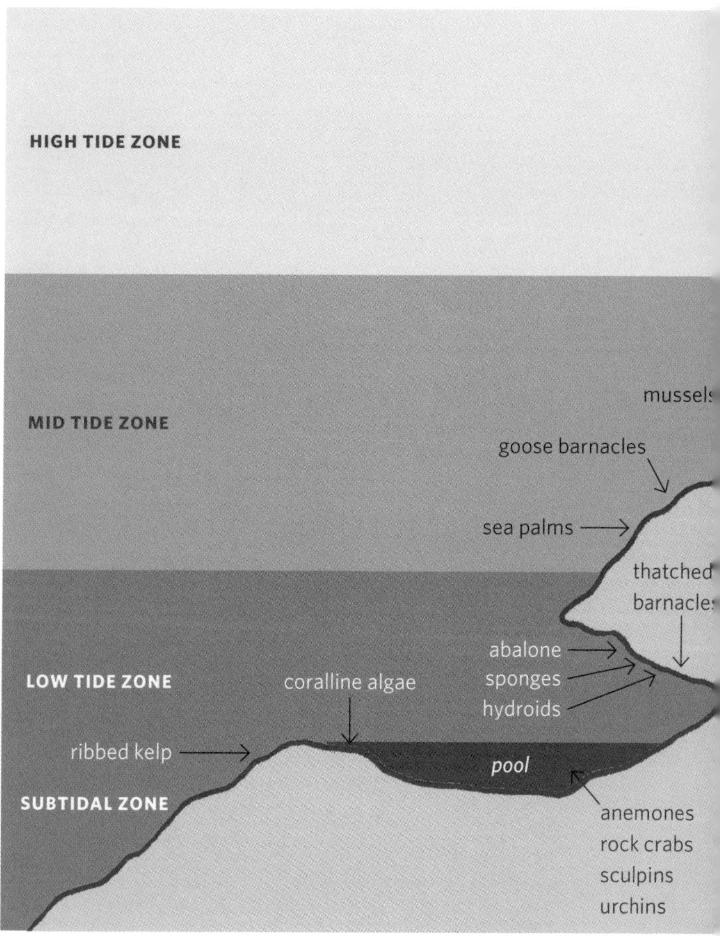

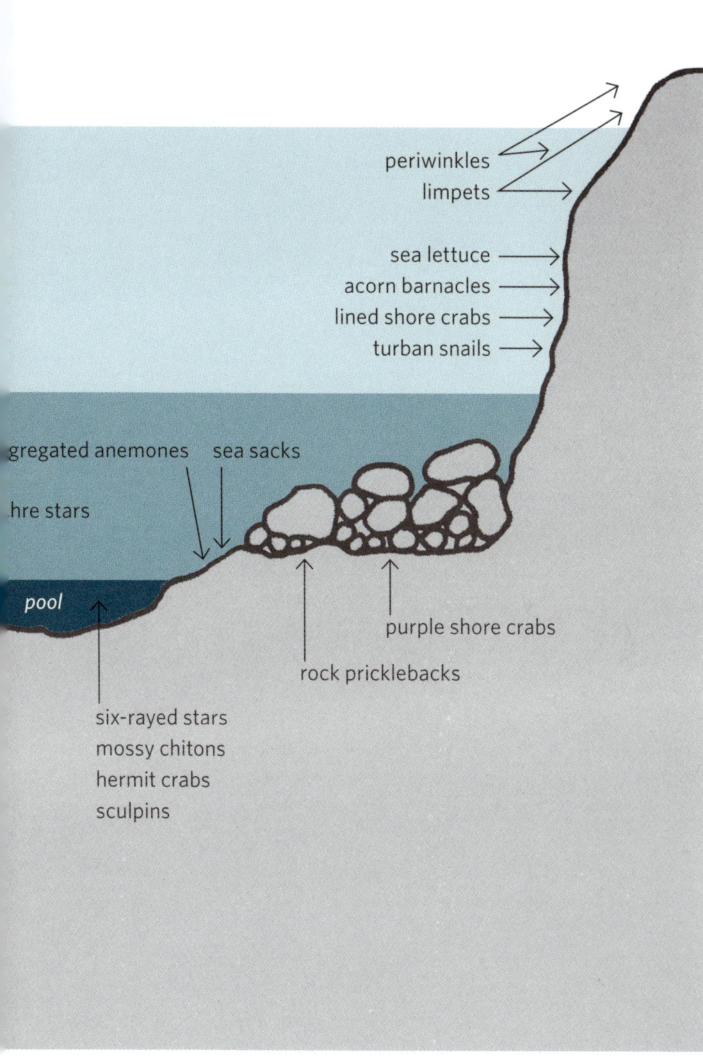

periwinkles
limpets
sea lettuce
acorn barnacles
lined shore crabs
turban snails
gregated anemones
sea sacks
hre stars
pool
purple shore crabs
rock pricklebacks
six-rayed stars
mossy chitons
hermit crabs
sculpins

Identifying Octopi and Sea Stars

If an organism is neither a plant nor a plantlike animal, **go to this symbol** ⟶ 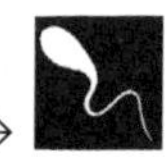

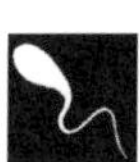 If the animal is not covered with a one- or two-piece shell, nor with several hard plates, nor with long spines; if it has neither jointed legs, nor black rubbery flesh, nor eight white, exposed plates; and if it is not shaped like a sphere, **go to** ⟶ 

 If the animal has rays or arms radiating outward like this , ⟶ **it is a SEA STAR**

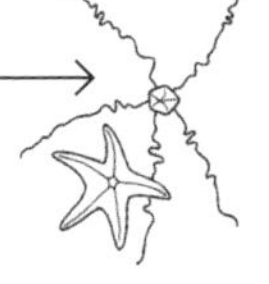

If not, **go to** ⟶ 

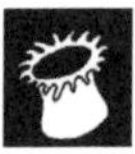 If the animal has a soft, saclike body and eight arms with suckers, **it is an OCTOPUS**

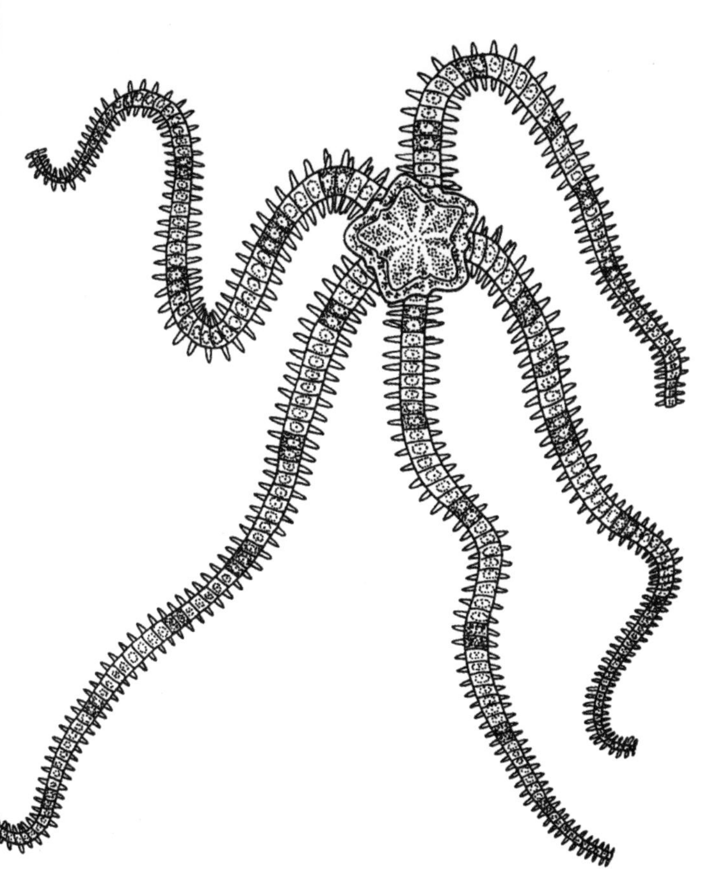

Octopi

Octopi have eight arms that converge into a funnel with a beak at the center. Some species can bite painfully, but most are shy, retiring masters of disguise. Pigment cells in the skin are manipulated by muscles to alter color instantly for camouflage and to startle predators. They escape by squirting out water to propel themselves backward. They also spread inky fluid to blur vision and confuse predators such as Moray Eels. Octopi eat crabs, clams, mussels, scallops, worms, and fish. They're like vertebrates in the complexity of their eyes and brain. An internal remnant of a shell shows the relationship of the octopus to other mollusks. A valve system passes water over gills hanging inside the body. The sexes are separate; fertilization internal. Females lay eggs under rocks and in small caves, then forgo eating during the four-month hatching period while they steadfastly guard the eggs. Octopi taken into laboratories and classrooms have an uncanny ability to crawl out of aquaria and die on the floor. Intertidal octopi are becoming scarcer as thousands are turned into trophies for ignorant collectors and sold as curios.

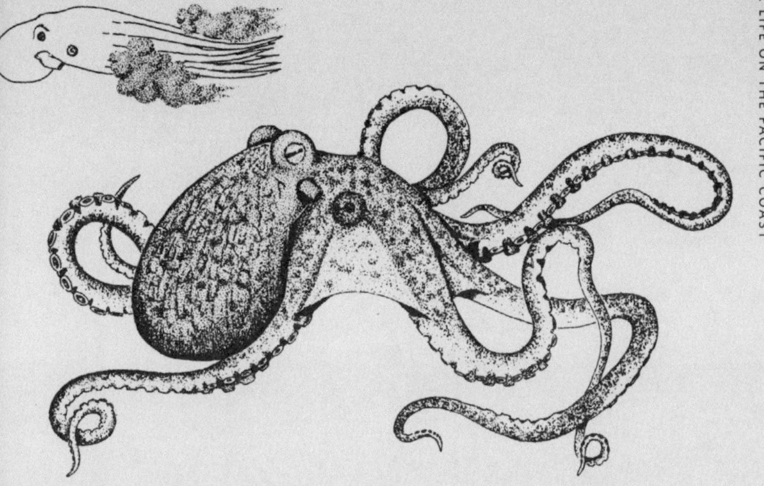

Two-Spot Octopus

Octopus bimaculatus

Body including arms to 24 inches (60 cm) long, but most about
12 inches (30 cm). Gray brown, yellow brown, red brown with two
dark, false eye spots ringed in blue. Under rocks and in pools. Common
in low tide zone, central to southern California. Young Pacific Giant
Octopus, *O. dofleini*, occurs intertidally from Alaska to Oregon with a
subspecies to central California.

Sea Stars

Sea stars are echinoderms with knobby spines on their backs. Their skin also has soft, fuzzy clumps of oxygen-absorbing tissue and tiny pincers that preserve the breathing ability of skin and prevent suffocation by routinely cutting up the larvae of barnacles, sponges, and hydroids seeking attachment there. Hold an Ochre Star gently against your arm. When you remove it, you'll feel pincers tugging at your hair. Just off center on the star's back is a small strainer for taking water into its internal hydraulic system, which operates hundreds of flexible, elastic tubes on its underside, called tube feet. Each foot is tipped with a tiny suction cup, which helps a star to hang tight when a wave hits. Whichever way stars go is "forward." Having neither heads nor tails, they change direction without turning around. Tube feet can pry open clams or mussels just far enough for the star to insert its inside-out stomach into the shell and digest the victim. When finished, a star retracts its stomach and moves on. Some species have free swimming larvae; others brood eggs until they hatch into little stars. Stars can usually regenerate lost arms. In some species severed arm tips can regenerate whole new animals. Sea stars' greatest enemies are human curio collectors.

Bat Star
Patiria miniata

To 4 inches (10 cm) across, webbed between arms, scaly. Color variable: solid or mottled red, orange, brown, yellow, green, or purple. Eats sea stars, tunicates, and algae. Worm lives in arm grooves. Low tide zone, but more often subtidal, Sitka to San Diego. Some to Mexico.

Ochre Star
Pisaster ochraceous

To 12 inches (30 cm) across. Rows of white-tipped spines resemble pentagons near center. Harsh to touch. Purple or orange. Arm tips have light sensors. Slow growing. Needs only a 4/1000-inch (.1-mm) opening to insert stomach into mussel shell. Also eats barnacles, limpets, certain snails. Mid- to low tide zones, Sitka to Point Conception. Subspecies to Mexico.

Knobby Star
Pisaster giganteus

To 16 inches (40 cm) across. Large, blunt, club-shaped spines surrounded at base by blue rings; show no pattern. Blue gray, sometimes with brown tones. Common in low tide zone, but more often subtidal, British Columbia to Mexico.

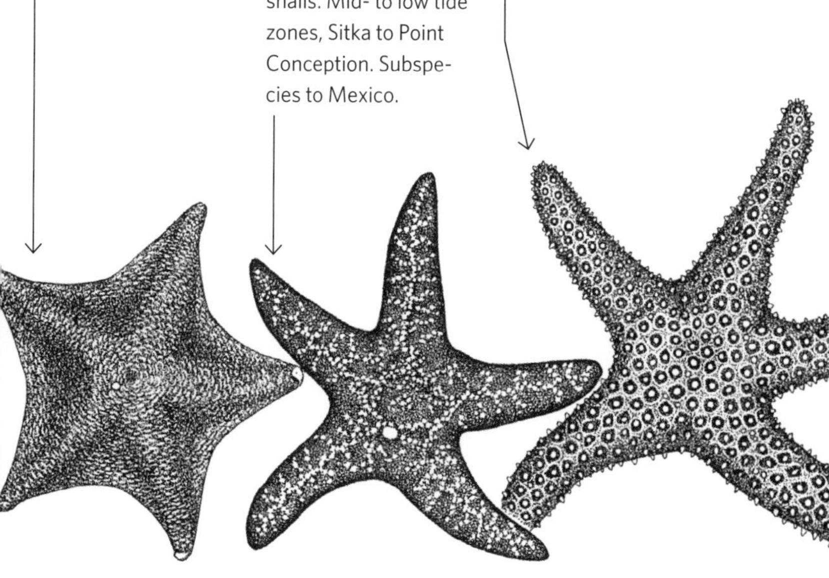

Leather Star

*Dermasterias
imbricata*

To 4¾ inches (12 cm)
across. Texture is soft
and slippery. Lead
gray with patches of
red, brown, or purple.
Often has garlic or sul-
fur odor. Eats urchins,
sponges, anemones,
bryozoans, and sea
stars. In open rocky
pools, more abundant
in sheltered places.
Low tide zone, Sitka to
San Diego.

Sun Star

Solaster stimpsoni

To 10 inches (25 cm)
across. Texture gritty.
Orange or rose with
streaks of gray blue
or purple from center
to arm tips. Arms
taper uniformly. One
of the most beautiful
intertidal stars, low
tide zone, Bering Sea
to Eureka.

Sunflower Star

*Pycnopodia
helianthoides*

To 16 inches (40 cm)
across. Starts life with
six arms, develops
up to twenty-four.
Orange, purple, gray
blue. Soft and limp out
of water, arms break
off easily. Moves rap-
idly. Largest intertidal
star on the Pacific
coast. Eats urchins,
bivalve mollusks, and
dead fish. Low tide
zone, Unalaska to
Monterey, occasion-
ally to San Diego.

Six-Rayed Star

Leptasterias hexactis

Body to 3 inches (8 cm) across, usually smaller. Lead gray, olive green, blackish, or orange with pink tones. Female groups eggs in clusters, broods them for about sixty days, apparently doing without food. Eats barnacles, chitons, snails, small sea cucumbers. Low tide zone, Puget Sound to Channel Islands. Young often confused with *L. pusilla* of central California midtide zones.

Brittle Star

Amphipholis pugetana

Body to ¾ inch (2 cm) across. Black and white, often nearly all white. Does not drop arms easily. Often found in large groups under rocks for protection and to avoid suffocation. Fast moving. Cannot extrude stomach. Feeds on detritus. Smallest brittle star on the Pacific coast. Mid- to high tide zones, Alaska to San Diego, sparsely to Mexico.

Brittle Star

Ophiothrix spiculata

Body to ¾ inch (2 cm) across. Arms have tiny, fuzzy spines. Color variable, orange margins on arm bands. Drops arms easily. Eats detritus and small animals. Usually solitary in algal holdfasts and under rocks. Low tide zone, Moss Beach to Peru.

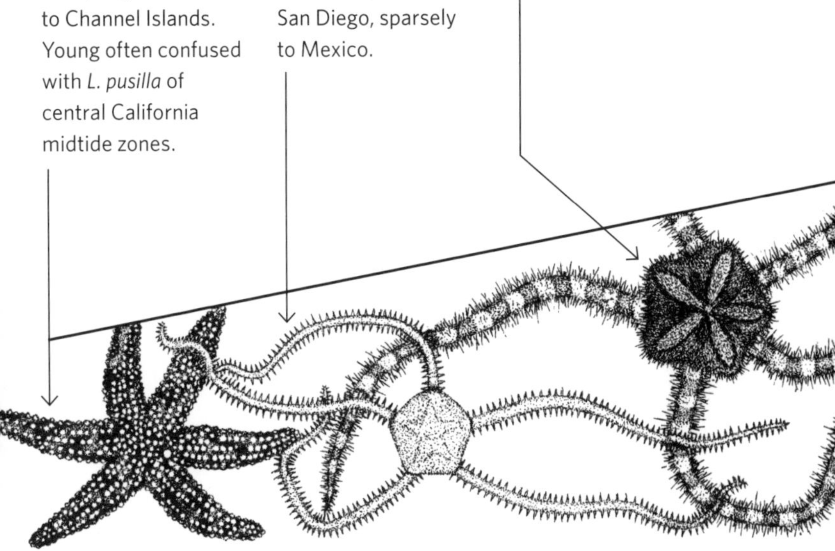

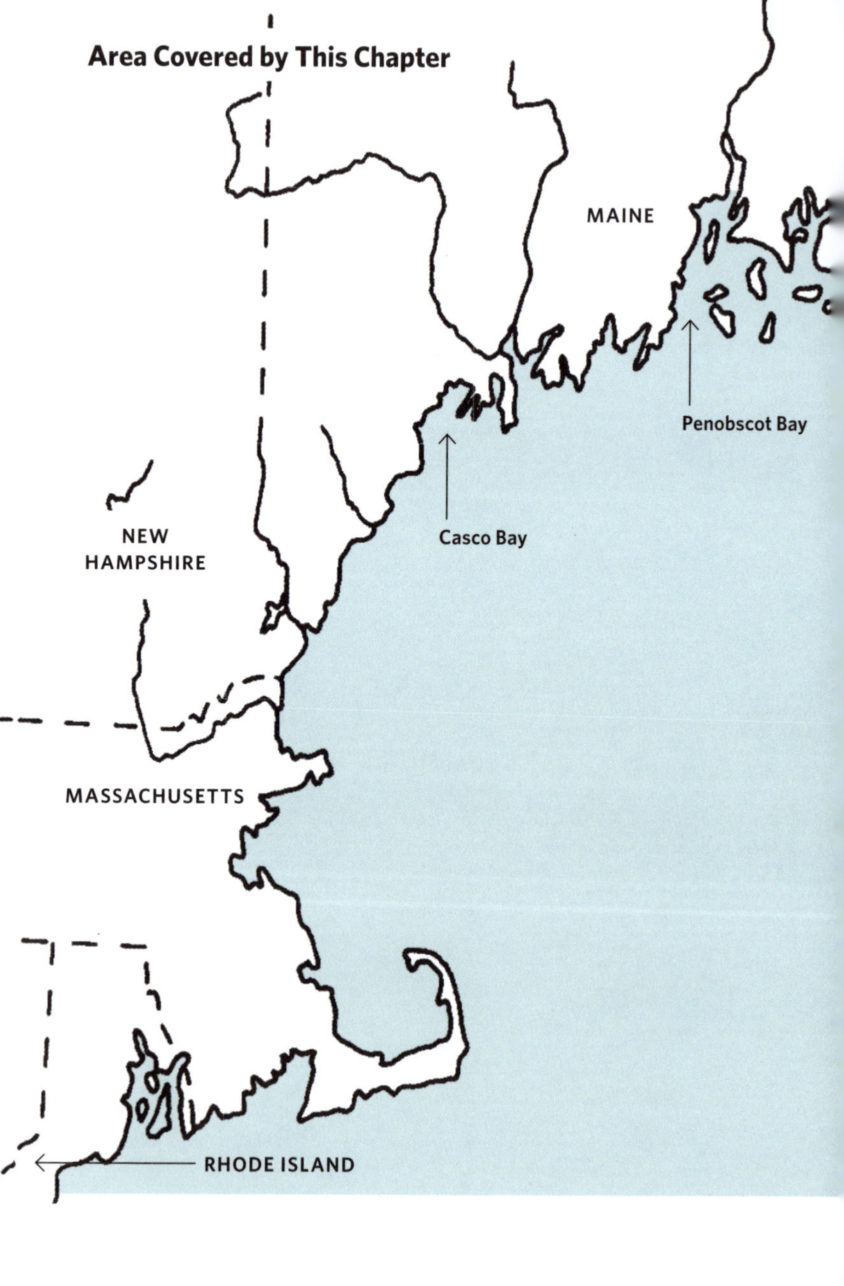

Area Covered by This Chapter
MAINE
Penobscot Bay
Casco Bay
NEW
HAMPSHIRE
MASSACHUSETTS
RHODE ISLAND

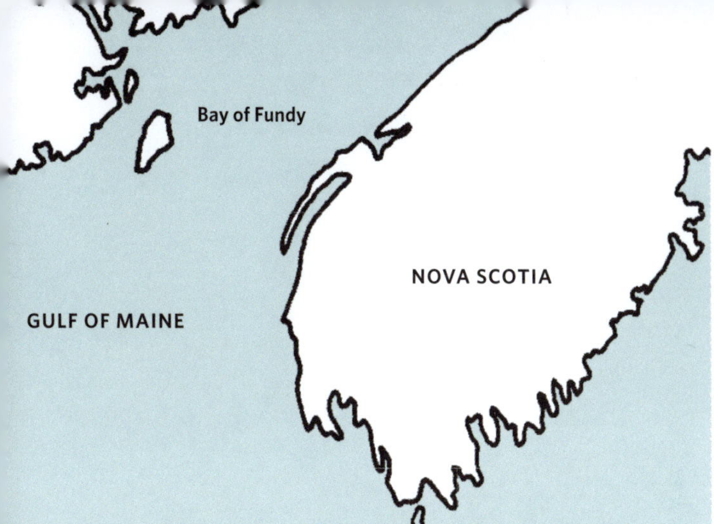

Intertidal Life on the Rocky North Atlantic Coast

CAUTION: Watch your footing! Accidents are common on slippery or unstable rocks. Don't let the rising tide leave you stranded. And watch out for waves in exposed areas.

Formation of the Rocky Coast

The rocky coast of the Gulf of Maine, with wave-beaten headlands; narrow sheltered bays; steep, rocky cliffs; and many islands, was shaped by Ice Age glaciers. Sheets of ice up to a mile thick advanced as far south as Cape Cod, scouring away sediment. As they retreated, the melting glaciers exposed the granite bedrock and left behind boulders and gravel. This made an intertidal habitat with many places for organisms to hide, and a firm substrate for them to grip—a rocky habitat quite different from the shifting beach sediments of the Atlantic coast south of Cape Cod.

Water Temperature

Water temperature determines where a plant or animal lives. Each intertidal organism has a range of temperatures in which it can flourish. Beyond this optimal range, it can survive only by adapting to local conditions. The hook of Cape Cod juts into the Atlantic Ocean and acts as a barrier between cold-temperate waters in the Gulf of Maine and warmer waters to the south. Some species occur only in the Gulf of Maine. Others, more abundant south of Cape Cod, have adapted to the cooler waters by moving into higher, warmer levels of the intertidal zone or by completing their life cycles during the warmer months. Some are confined locally to sheltered bays. Conversely, intertidal species common in the Arctic may be found in the Gulf of Maine at lower intertidal and subtidal levels.

Adaptations

Intertidal plants and animals must contend with a wide variety of environmental stresses. Twice daily they tolerate submersion and exposure to air and sunlight. Storms with crashing waves threaten to

rip organisms off their substrates.
Predators are a constant menace.

The seasons bring dramatic
changes. Ice scrapes creatures off
exposed rocks and freezes shallow
tide pools, trapping the inhabitants.
Some animals can escape winter cold by
moving to deeper water, but attached organ-
isms die back or become dormant, waiting for spring. In summer heat
animals can suffocate since warm water holds less dissolved oxygen.

Salinity (salt concentration) changes as rivers swollen with melted
ice and snow dilute coastal waters in the spring. Shallow tide pools
have the widest range of salinity, as rain dilutes or the sun evaporates
the water. Few intertidal creatures can tolerate the brackish water of
estuaries, where salt water intrudes only at high tide.

Despite all these difficulties, intertidal rocks are crowded with
creatures competing for food and shelter.

Food

Intertidal plants manufacture their own food by photosynthesis. Animals, unable to make their own food, must gather or capture it—and
avoid getting caught themselves.

The greatest source of food for animals is rarely seen: plankton,
the tiny, often microscopic plants and animals that float in the ocean.
Various feeding structures have evolved that grasp, snag, or strain
the plankton from seawater. Filter-feeding mussels strain seawater
through their gills, barnacles scoop up plankton with their feathery
feet, and sea cucumbers lick off food trapped by their bushy tentacles.
But plankton is not the only food. Intertidal herbivores eat plants,
carnivores consume other animals, and omnivores have a mixed diet.

Scavengers clean up whatever dies. Whatever remains is recycled into the seawater as mineral nutrients.

The Greatest Threat

Intertidal life is endangered by humans who pollute the intertidal environment, or who ravage it for food or fun. Intertidal animals almost never survive removal from their habitat. Enjoy them, but do not disturb or destroy intertidal life.

Marine Plants

A vast number of microscopic one-celled plants drift in sunlit water. They are mostly diatoms and dinoflagellates known collectively as phytoplankton. Seasonal fluctuation in salinity, temperature, the upwelling of nutrient minerals, and dissolved gases trigger spring and autumn blooms of these algae. Despite their size, they are the primary food for a rich diversity of animals, from the microscopic zooplankton to many of the intertidal creatures.

The more familiar "seaweeds" grow attached to intertidal rocks and pilings, and form distinctive bands of green, brown, and red algae. The colors are from pigments sensitive to different light intensities and thus mark the gradual transition from the sunny surface to the darker depths. In general, blue-green algae and green algae grow near the upper zone; the larger brown algae drape the middle zone; and the delicate red algae flourish below the canopy of brown algae in the dim light of the lower zone and the subtidal zone. Some species have different colors at different exposures. Irish sea moss, a red alga, grows olive green in sunny locations.

Some algae, such as kelps, have distinctive body regions: a leaflike blade, a stalk or stipe, and a rootlike holdfast. But they lack the complex vascular tissue, woody support, flowers, and seeds of land plants. Their forms can be much simpler—a single cell, a filament, a broad sheet, or an undifferentiated mass of cells—because they depend on the sea to support, water, and nourish them, and to disperse their reproductive spores and gametes (eggs or sperm). Others have air bladders to float their leafy parts up toward sunlight. Their junglelike growth provides food, oxygen, anchorage, and hiding places for other organisms.

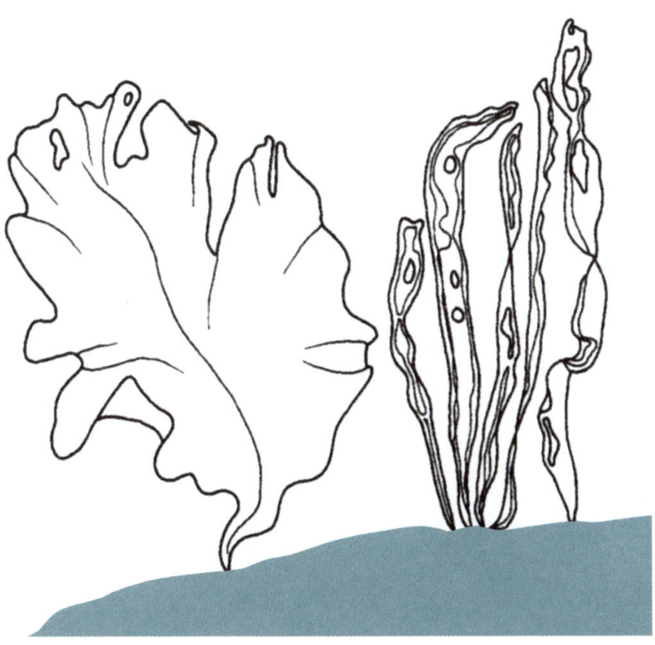

Intertidal Zones

The twice-daily rise and fall of the sea results from the gravitational pull of the moon and, to a lesser extent, the sun. Twice a month, near full and new moons, the spring tides occur. They have the highest high tides and the lowest low tides (best for exploration). Neap tides, when high and low tides differ least, come during the first- and third-quarter moons. A high or low tide will be fifty minutes later the following day.

SPLASH ZONE — wet from wave splash but never submerged

level of spring high tide

UPPER ZONE — submerged only during high tide

level of neap high tide

MIDDLE ZONE — submerged and exposed during every tide ←

level of neap low tide

LOWER ZONE — usually submerged, exposed briefly during most low tides

level of spring low tide

SUBTIDAL ZONE — always submerged

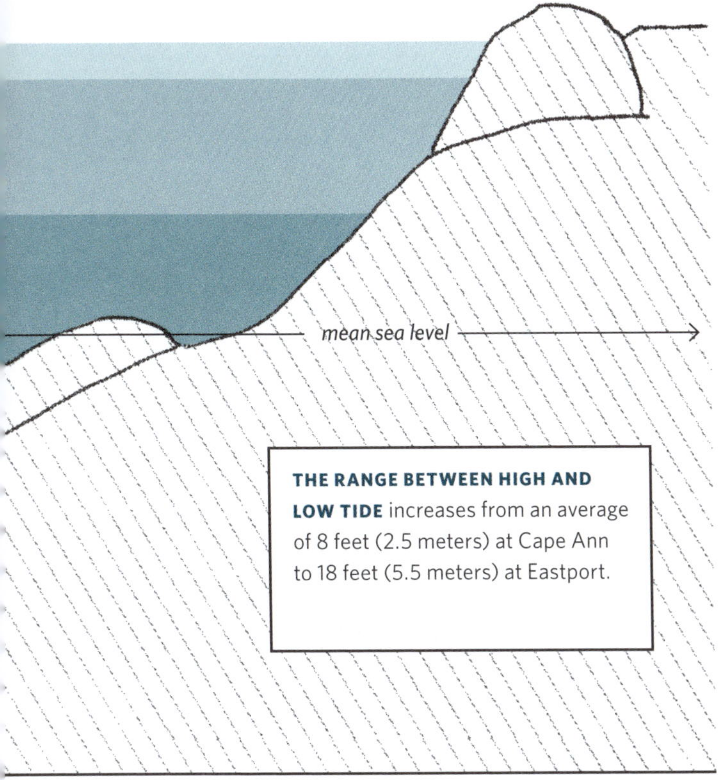

THE RANGE BETWEEN HIGH AND LOW TIDE increases from an average of 8 feet (2.5 meters) at Cape Ann to 18 feet (5.5 meters) at Eastport.

Dominant Intertidal Species

Intertidal organisms occupy the zone with the pattern of exposure and submergence to which they're best adapted. Some attached species tend to dominate a certain zone. By recognizing them, you can tell which zone you're in.

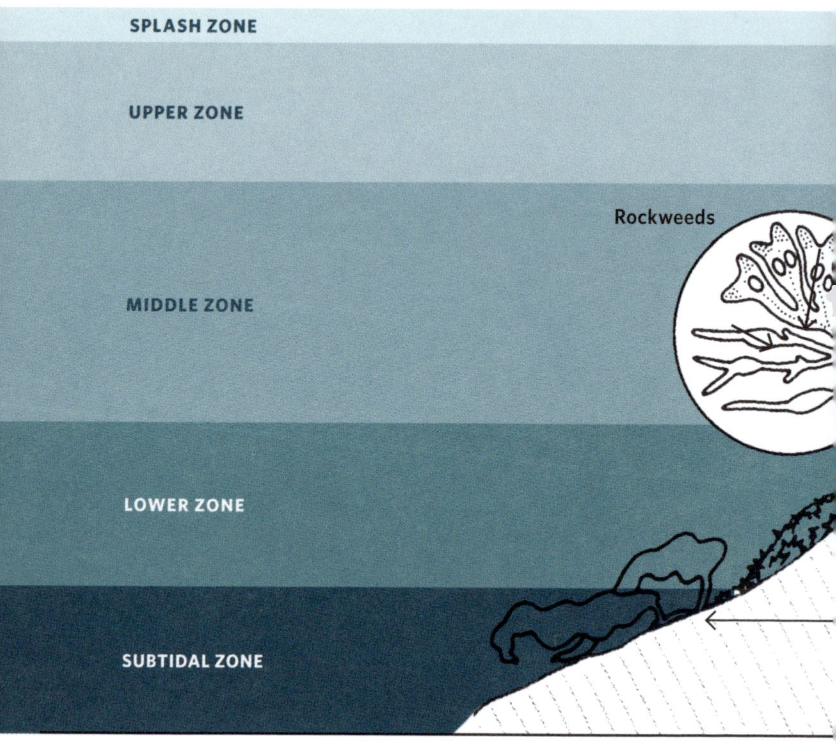

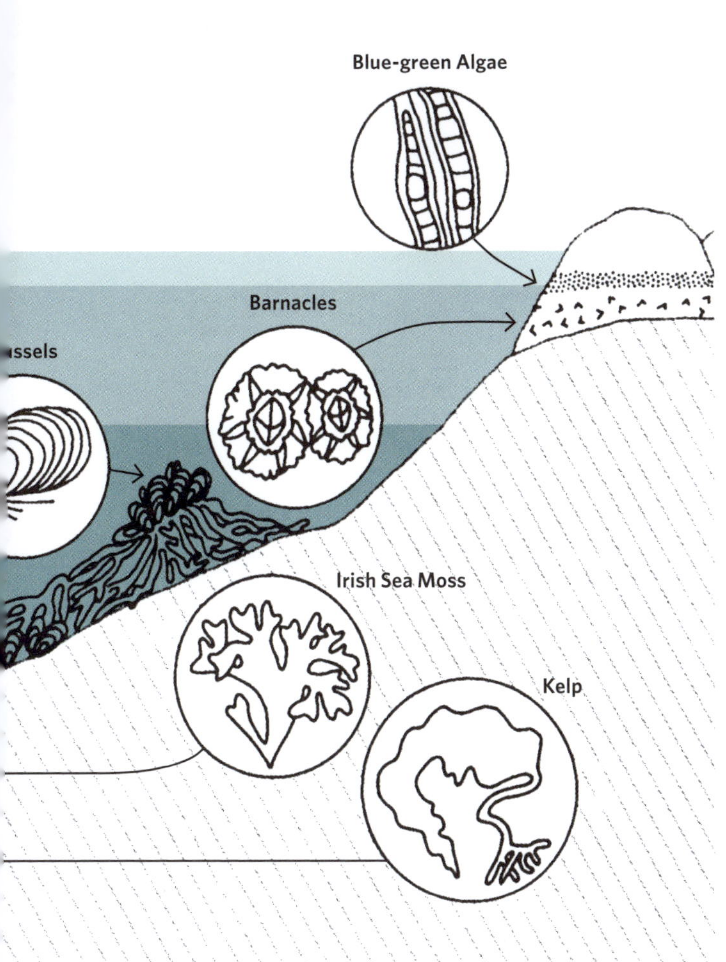
Blue-green Algae
Barnacles
ssels
Irish Sea Moss
Kelp

Associated Species

Once you're familiar with the dominant species, look more closely for the associated animals and plants attached to, feeding on, or lurking among the dominants.

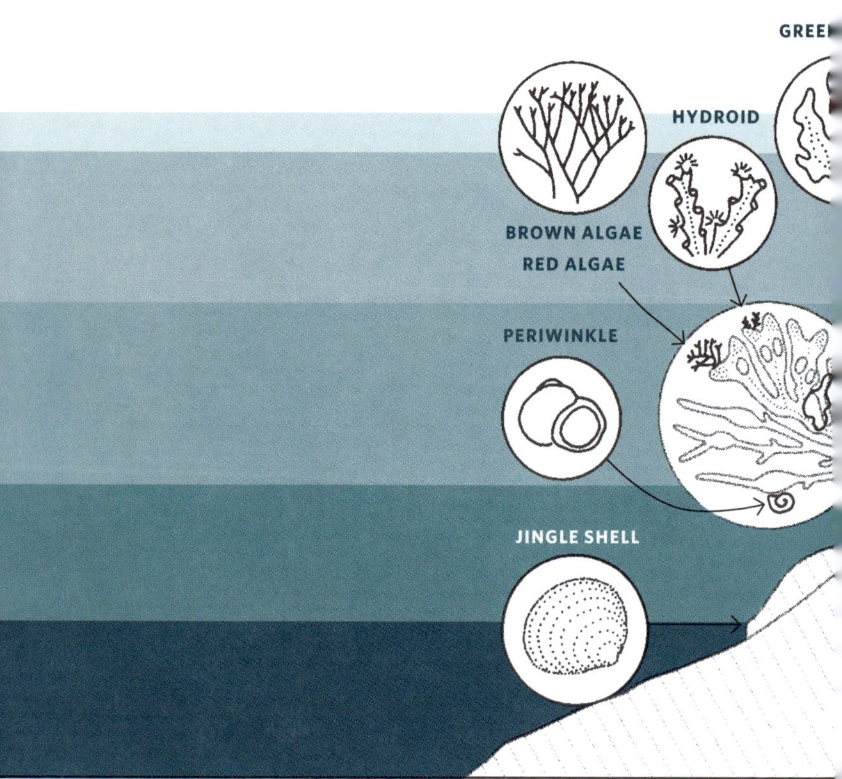

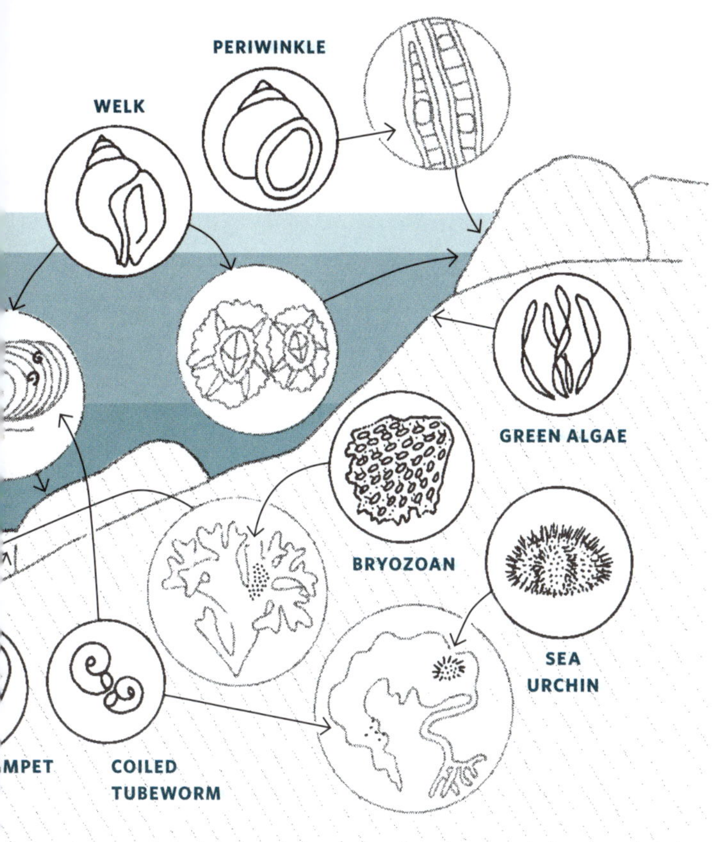
WELK
PERIWINKLE
GREEN ALGAE
BRYOZOAN
SEA
URCHIN
MPET
COILED
TUBEWORM

Tide Pools

As the tide recedes, rock basins become calm pools where you can observe underwater life. Wait quietly for the creatures to adjust to your presence and resume their activity.

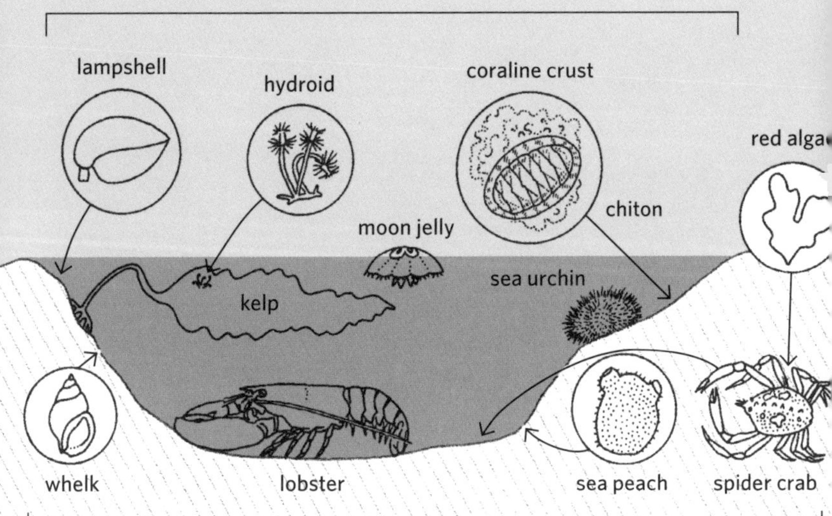

Lower zone pools often serve as nurseries for subtidal animals.

Middle Zone Pools

Middle and especially upper zone pools are cut off longer from the sea. The sun's heat may warm them and deplete their oxygen or even evaporate them. Or rainwater may lower their salinity. This adversity limits their population.

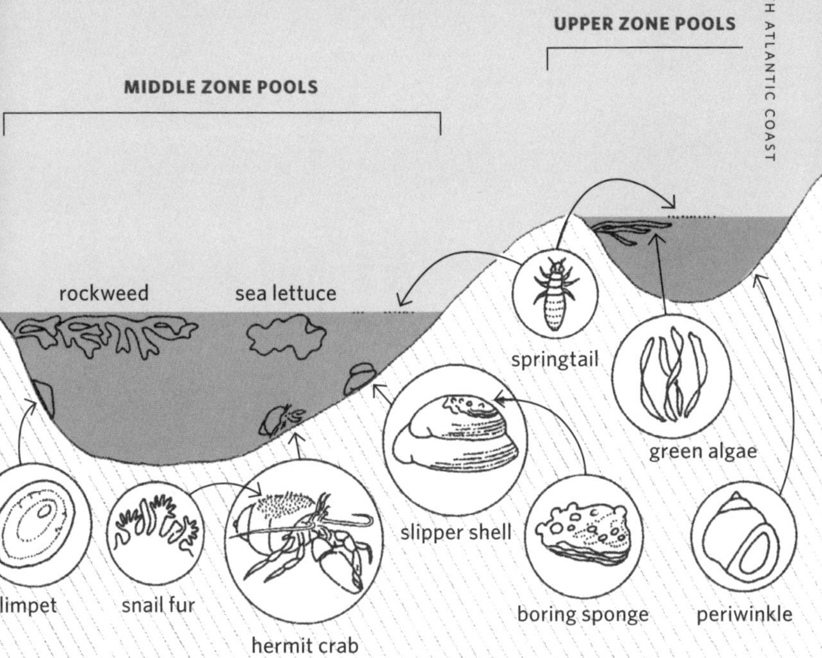

Under Rocks and Between Boulders

During low tide many creatures hide under rocks and between boulders. Rockweed mats add further protection. Sediment and organic debris collect here. After exploring, carefully restore rocks to their original positions.

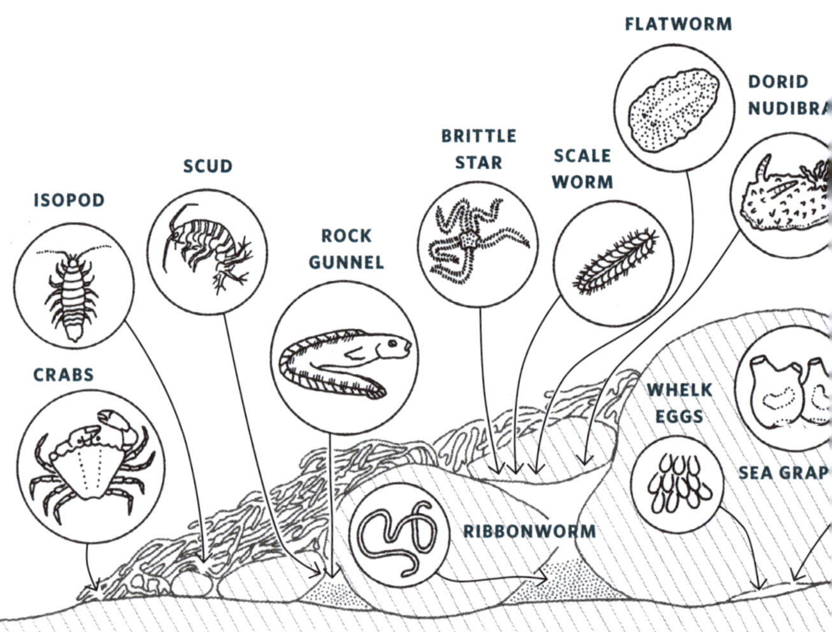

Caves and Crevices

Indentations provide a haven from hot sun and drying wind. Slowly circulating water allows sediment to settle.

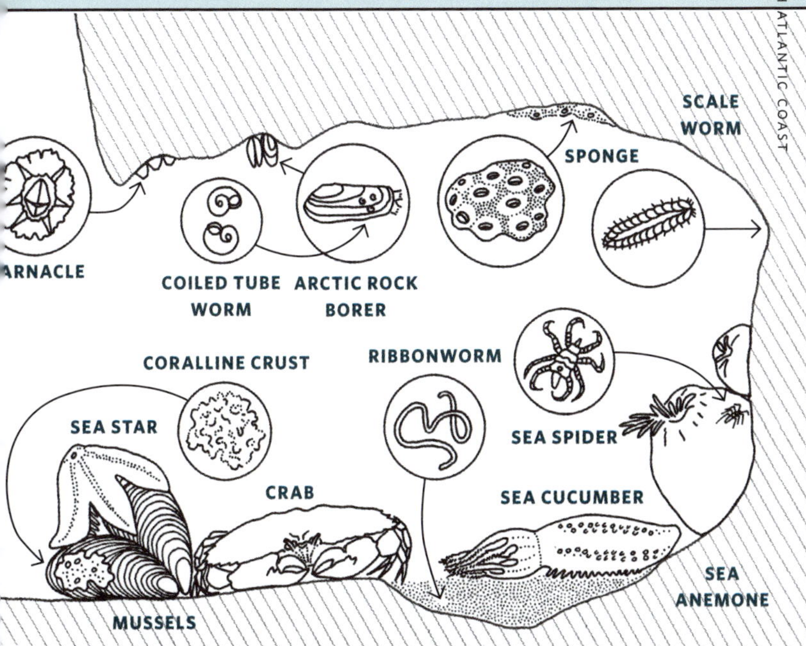

Pilings and Wharves

Away from crashing waves, pilings and wharves provide a hard wooden substrate and shade. Life in harbors, although protected, must contend with increased fresh water and pollution.

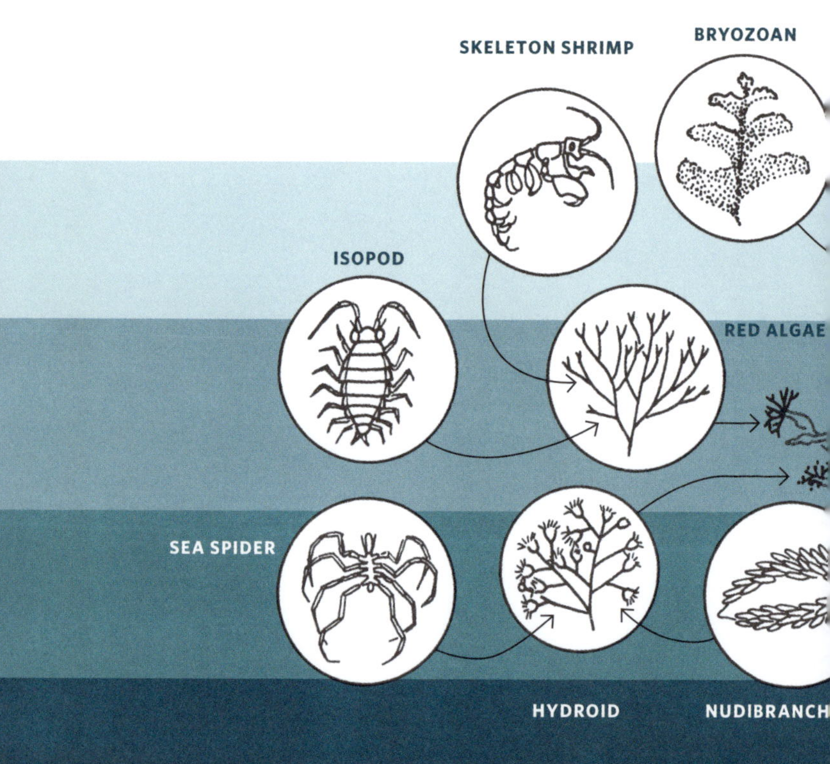

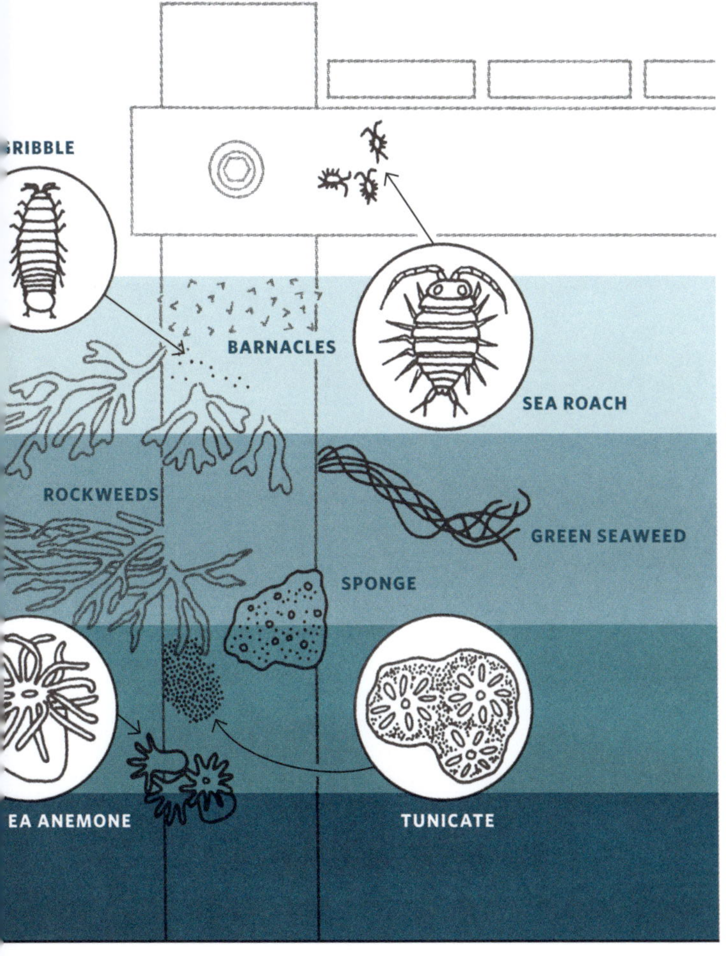
GRIBBLE
BARNACLES
SEA ROACH
ROCKWEEDS
GREEN SEAWEED
SPONGE
SEA ANEMONE
TUNICATE

Identifying Kelps

THESE LARGE BROWN ALGAE of lower and subtidal zones have noticeable body regions: blade (A), stalk or stipe (B), and holdfast (C). Kelps are harvested for alginates to stabilize paints, soaps, cosmetics, medicines, and insecticides. They make mineral-rich soil fertilizer and fodder for cattle, sheep, and pigs.

Horsetail Kelp
Laminaria digitata
To 3 feet (1 meter) long on exposed shores

Sugar Kelp
Laminaria saccharina
To 10 feet (3 meters) long on sheltered shores

Sea Colander
Agarum cribrosum
To 4 feet (1.2 meters) long on exposed shores, 6 feet (1.8 meters) long where sheltered

Edible Kelp
Alaria esculenta
To 10 feet (3 meters) long on exposed shores; blade floats because midrib is hollow

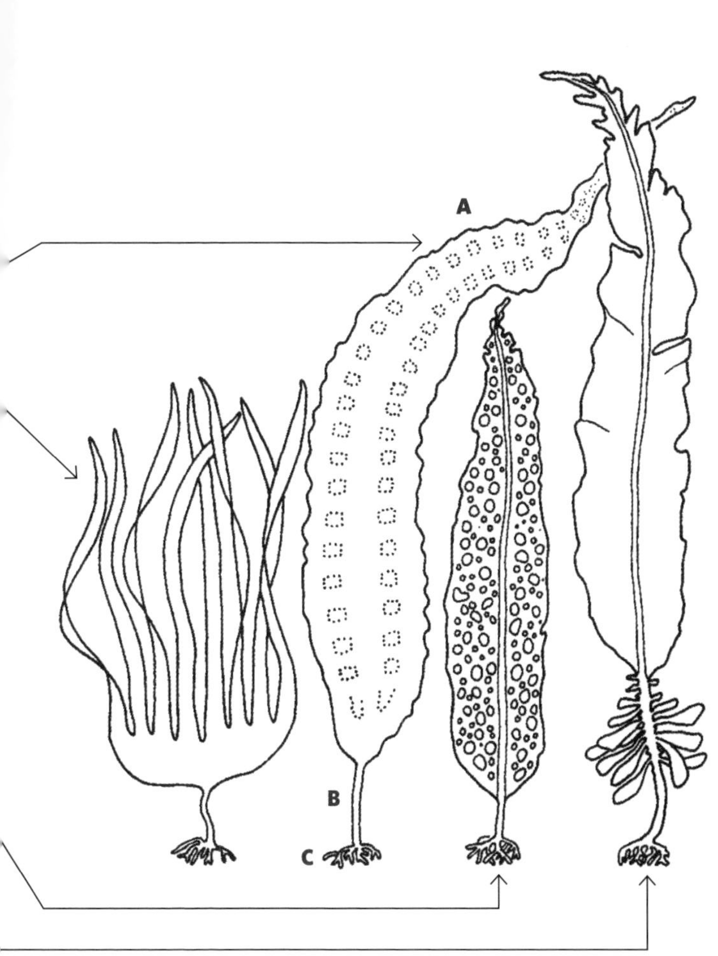

A
B
C

Identifying Sea Anemones

SEA ANEMONES ARE SOLITARY POLYPS; tentacles surround their mouths and pedal disks anchor their broad bases. With colors ranging from cream, olive green, to orange and bright pink, anemones resemble flowers. They grow on lower level rocks, pilings, and in crevices and shaded tide pools on sheltered shores. Their tentacles are armed with stinging nematocysts to immobilize prey (zooplankton and small fish), and for self-defense. When disturbed or exposed by low tide, they contract their body columns and withdraw their tentacles to form soft mounds that are easily overlooked. They can glide slowly over the substrate on pedal disks. Sea anemones look frail but subtidal individuals may live over a century. Growth is continuous throughout life. They may reproduce by budding or by producing eggs and sperm that unite to form free-swimming larvae that eventually settle; or bits of pedal disk left behind generate new, complete anemones. Sea slugs and sea spiders can eat anemones without discharging anemones' nematocysts.

Striped Anemone

Haliplanella luciae

Body column to ¾ inch (2 cm) high, ¼ inch (5 mm) wide. Overall color is brown to olive green with longitudinal stripes of cream, yellow, or orange. On lower level rocks and pilings on sheltered shores. Tolerates reduced salinity.

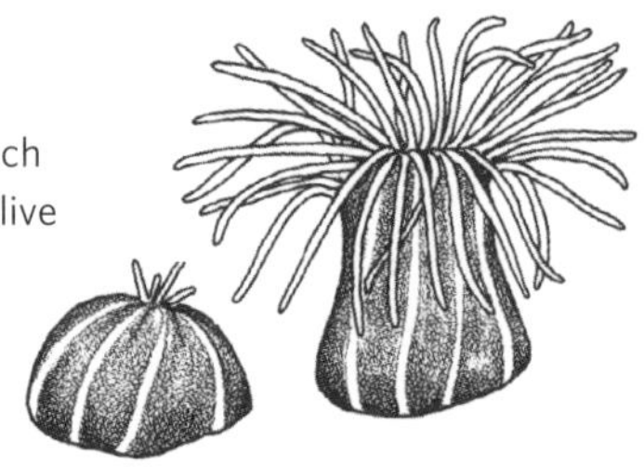

Northern Red Anemone

Tealia felina

Smooth body column to 3 inches (7.5 cm) high, 5 inches (12.5 cm) wide. Bright pink. Young attached to rocks, under rockweed, and in tide pools in lower zone from Casco Bay northward; found in subtidal zone elsewhere.

Frilled Anemone

Metridium senile

Body column to 4 inches (10 cm) high, 3 inches (7.5 cm) wide. Olive green to orange brown. Fine, cream-colored tentacles give a frilled look. Most common species in the lower zone.

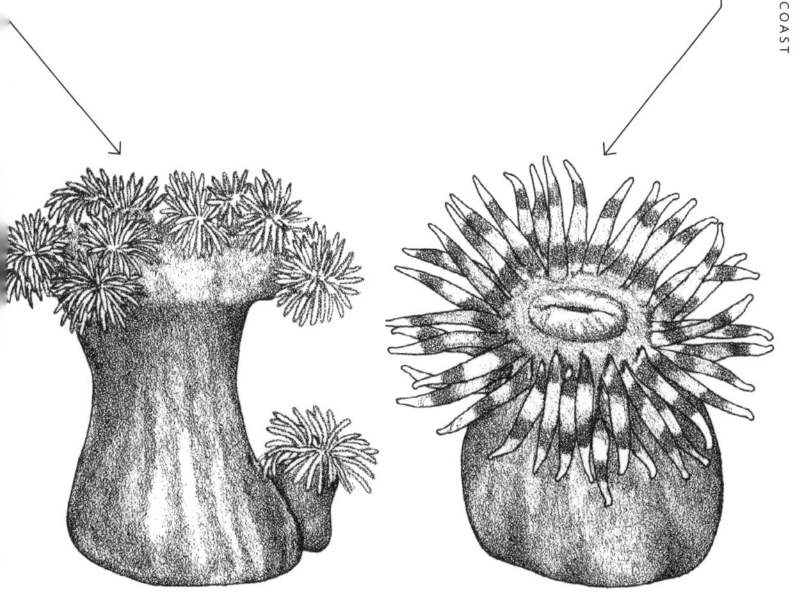

Identifying Periwinkles and Whelks

SHELLS OF THESE SNAIL-LIKE GASTROPODS grow unevenly into compact spirals instead of long, cumbersome tubes. Internal organs accommodate to twist. Common on rocks, among rockweeds, and in crevices and tide pools in all zones. Each has a well-defined head with sensory tentacles and moves by contractions of a broad muscular foot. At low tide, foot can be withdrawn inside shell and shell opening sealed by horny plate (operculum) (A). This retains moisture and protects them from predators (crabs, sea stars, and seabirds).

Rough Periwinkle
Littorina saxatilis

Shell to ½ inch (15 mm) long, gray brown with darker bands; has conspicuous spiral cords and sutures. Found grazing on *Calothrix sp.* in splash zone. Most stress-tolerant periwinkle. Withstands extreme temperature and salinity changes in upper-level tide pools. Female broods eggs within mantle cavity, and eggs hatch young periwinkles instead of larvae.

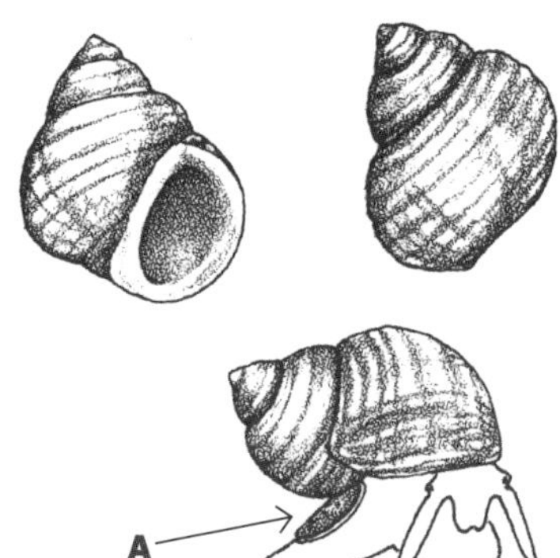

Common Periwinkle
Littorina littorea

Shell to 1¼ inches (3 cm) long,
olive brown to tan with dark bands,
cream-colored shell opening. Since its
accidental introduction to Nova Scotia
from western Europe around 1860, it has
become the most abundant species in the middle
zone. Uses radula to scrape algal film off rocks. Female
attaches eggs to rocks; a jelly coating keeps them moist
at low tide. They hatch into free-swimming larvae that
settle and mature. May live ten years.

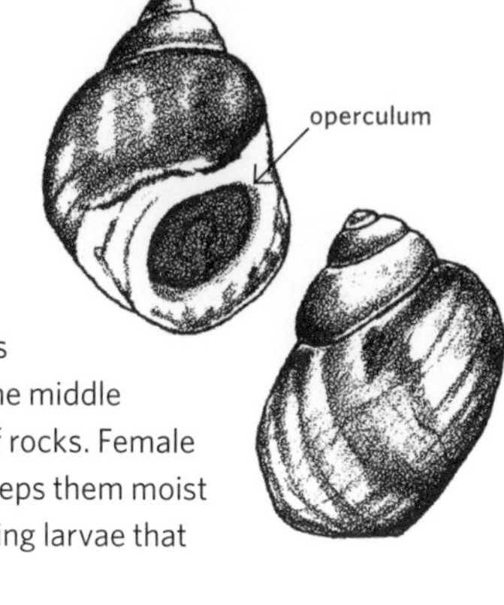

Northern Yellow Periwinkle
Littorina obtusata

Shell to ½ inch (13 mm) long, bright yellow, orange to olive green. Least
stress-tolerant periwinkle. Seeks shelter in moisture and even temperatures of rockweed in middle and lower zones. Also called "smooth"
periwinkle. Round shell mimics air bladders of knotted and bladder
wracks. Grazes on algal film that coats rockweeds.

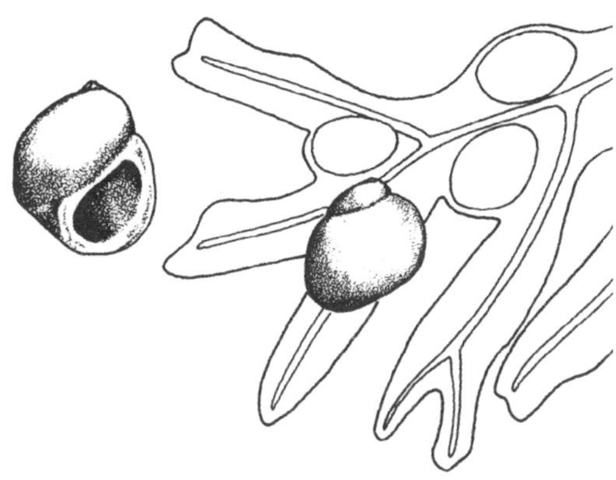

Waved Whelk

Buccinum undatum

Stout shell to 4 inches (10 cm) long, cream with pink, orange, or tan tint. Young hide under Irish sea moss, protected from herring gulls during low tide. Uses powerful foot to grasp mussel, then wedges lip of its shell into mussel's gape. May erode mussel's shell margin with its lip, prying it apart to insert proboscis and feed. Subtidal adults scavenge dead fish and invade lobster traps to steal bait. Clusters of empty egg masses wash ashore.

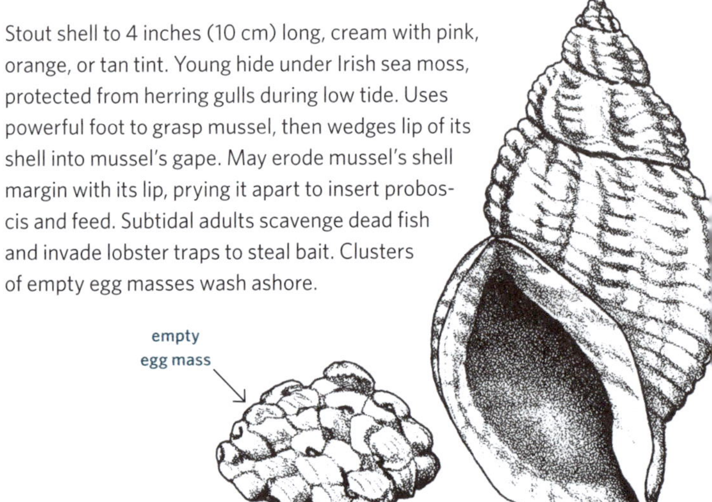

Periwinkle Problems

Periwinkles are intermediate hosts to parasitic blood fluke of seabirds. Fluke larvae hatch from eggs in seabird feces and seek periwinkle host. By boring through host's tissue, they migrate to its liver, where several generations of a second type of larvae are produced. Thousands of these swarm into the water seeking a second host type to complete their cycle. Some find and penetrate bare skin of seabirds and migrate to circulatory system. Mature flukes develop, mate, and send fertilized eggs into bird's digestive tract to repeat cycle. Sensitive humans get "swimmer's itch" from incomplete penetration of second larval type.

Dog Whelk
Thais lapillus

Rough-textured shell to 1½ inches (4 cm) long; color varies due to diet. Young dog whelks scrape coiled tubeworms from lower-level rocks. Larger, thick-shelled adults prey on either mussels or barnacles in middle zone. Eating barnacles produces cream-white shell with yellow bands; mussels give purple-brown tint. Kills barnacles by releasing toxin, then pries apart top plates to insert proboscis and feed. Drills hole in mussel shell by chemically softening shell, then scraping with radula. Extends proboscis with radula through hole to tear up victim's tissue. Attaches leathery, cream-colored egg capsules under rocks and rockweed, and in crevices. Has no larval stage. Each capsule (A) hatches several miniature whelks. Crabs and herring gulls eat them.

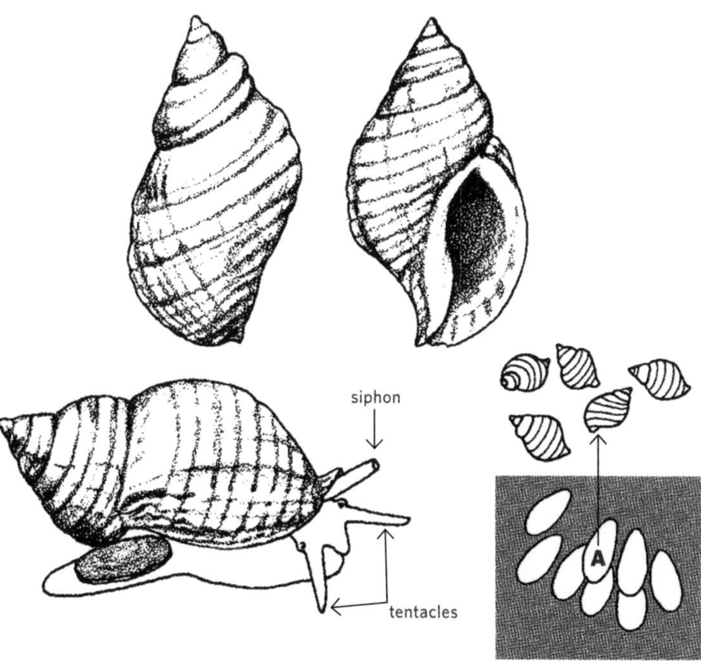

Identifying Sea Slugs or Nudibranchs

THESE SHELL-LESS GASTROPODS cling to lower level rocks, pilings, and seaweeds. Most numerous in winter and early spring. Rarely live to second year. Dorid nudibranchs have circle of retractable secondary gills on lower back; scrape barnacles and bryozoans off rocks using their radulae; retract sensory tentacles and feathery gills to become gumdrop-like lumps at low tide. Rely on camouflage to avoid enemies. Eolid nudibranchs have projections called cerata down their backs, eat hydroids and sea anemones, and have jaws for cutting prey. Ingest prey's nematocysts without discharging them and concentrate them at tips of cerata for their own defense. Brightly colored eolids warn fish, lobsters, and crabs of the poisonous nematocysts.

Rough-Mantled Nudibranch

Onchidoris bilamellata (Dorid)

Body to 1 inch (2.5 cm) long, cream, tan, to orange brown. Rough-textured back has rounded tubercles. On middle-level pilings and under barnacle-encrusted rocks. Eats acorn barnacles.

Hairy Nudibranch
Acanthodoris pilosa (Eolid)

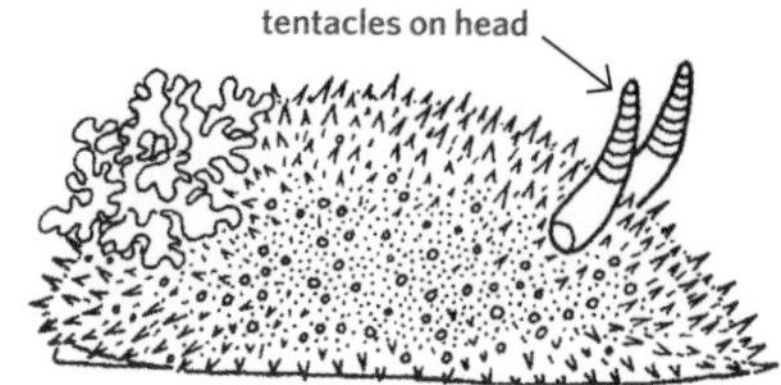

Body to 1¼ inches (3 cm) long, pale yellow, brown, to purple. Back is rough textured with tapered tubercles. On lower level rocks and seaweeds with encrusting bryozoans.

Maned Nudibranch
Aeolidia papillosa (Eolid)

Body to 4 inches (10 cm) long, gray to orange with brown and white flecks. Rows of slender cerata line back. On lower level pilings, in crevices and shaded tide pools with frilled sea anemones.

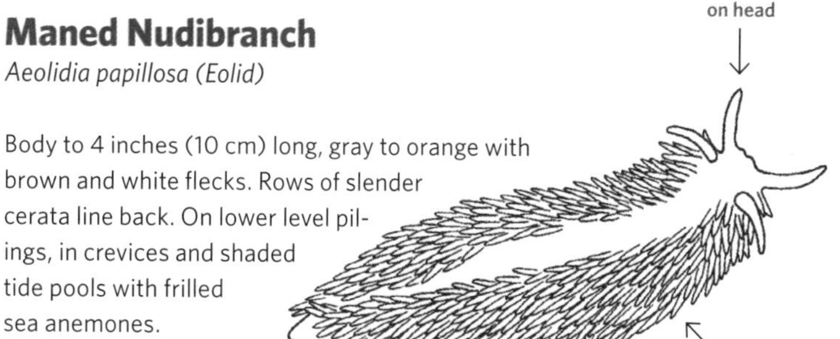

Red-Gilled Nudibranch
Coryphella sp. (Eolid)

Body to 1½ inches (3.5 cm) long, translucent white; long, tapered cerata are bright red at core. On lower-level seaweeds and hydroids, particularly Obelia and Tubularia.

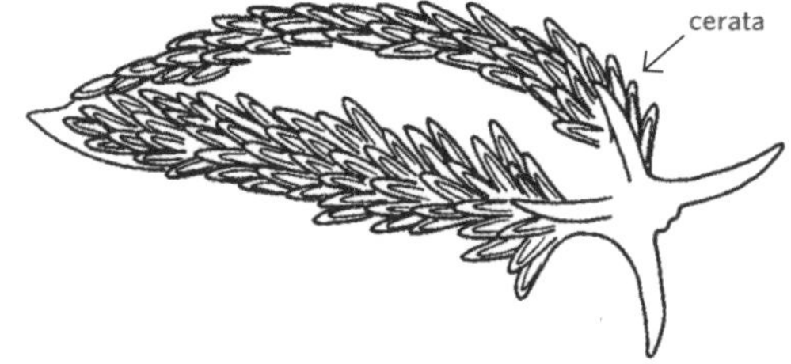

Identifying Crabs

TRUE CRABS HAVE FOUR PAIRS of walking legs and two grasping claws. They walk forward and run sideways. Abdomens are reduced to flaps folded under flattened bodies. Males have a narrow flap; females a broad one to carry fertilized eggs for six to nine months until they hatch. For about two months, larvae swim free in the plankton community, then settle. Crabs scavenge in middle and lower zones and prey on polychaete worms, mussels, periwinkles, even sea stars and sea urchins. Throughout life, growth is accomplished by successive molts that leave them temporarily soft and vulnerable to bottom-feeding fish and lobsters. Mating takes place just after females molt. Calcium carbonate extracted from surrounding seawater helps shells harden.

Green Crab
Carcinus maenas

Body to 3 inches (7.5 cm) wide. Green mottled with yellow and black on top, yellow below, but reddish orange in adult females. Young hide under rockweed, in crevices. Adults forage in midtide pools. After dark at low tide, they scurry over rocks and rockweed in search of food, fiercely brandishing front claws when startled. Tolerates estuarine conditions.

Rock Crab
Cancer irroratus

Body to 5 inches (12.5 cm) wide, smooth, yellow to olive brown with reddish-brown blotches on top, pale yellow below. Active. Runs from danger or burrows in shell debris between rocks.

Jonah Crab

Cancer borealis

Body to 6 inches (15 cm) wide, rough with tubercles, red to pinkish purple on top, cream colored below. Stands and fights with front claws or hides in crevices.

Young crabs of both species are found under rocks, among Irish sea moss, and in low tide pools. Subtidal adults scavenge invertebrates and dead fish. Both species are edible and important to the shellfish industry in the Gulf of Maine. They are caught in crab traps that are similar to lobster traps.

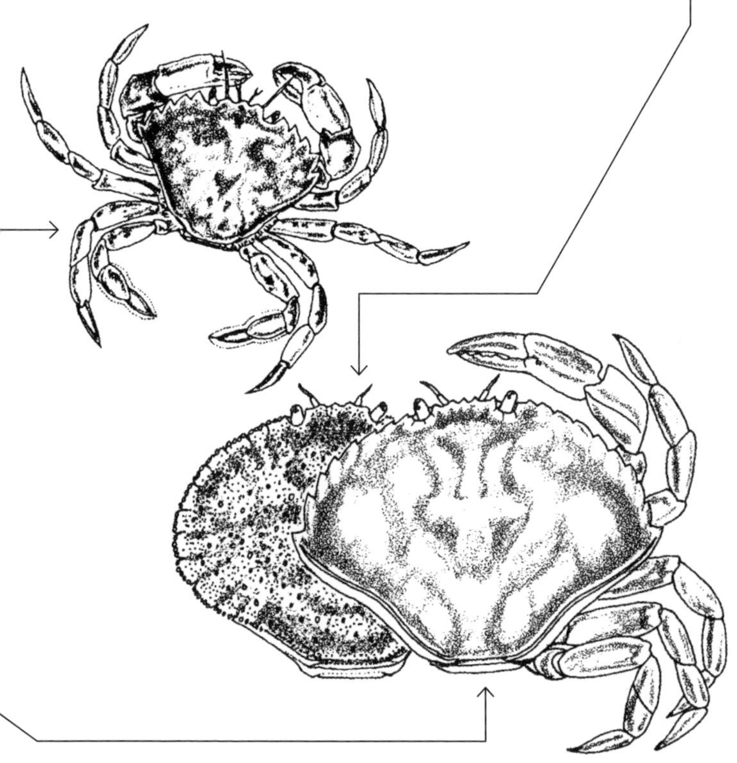

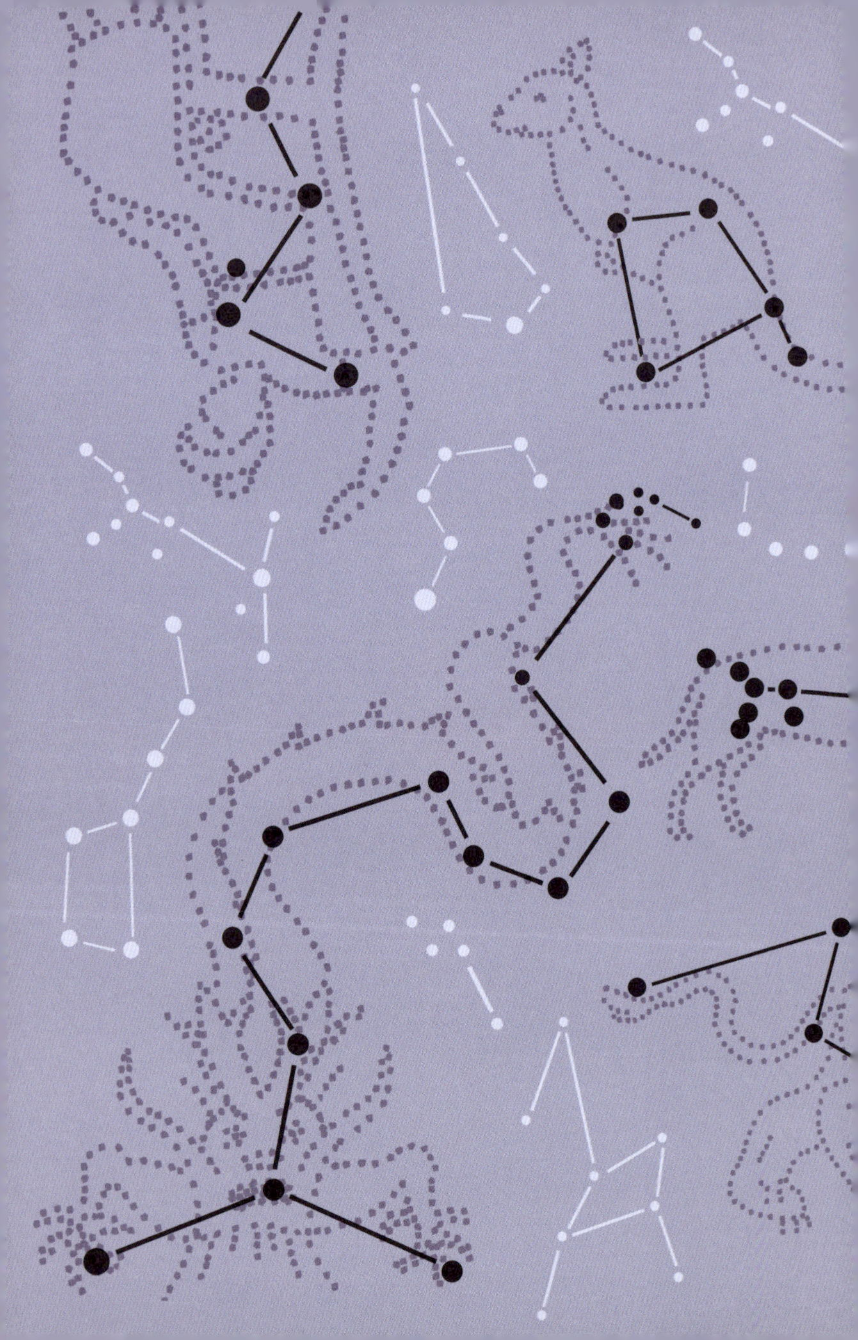

THE NIGHT SKY

Constellations

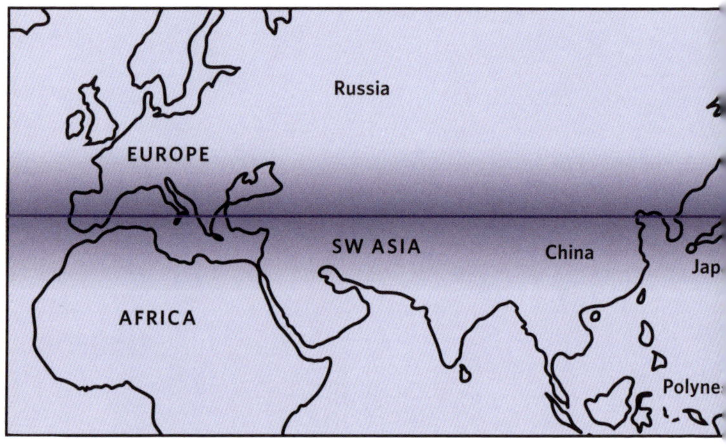

About the maps:

Maps generally show the brightest stars in each group. Some lesser stars are not shown. Lines have been drawn between stars to depict the person or animal for which the constellation is named. If you see a better way to connect the dots, that's fine, too.

On mini-maps, the five-pointed stars ★ are the very brightest in the sky.

If you see a bright "star" in the sky that's not shown on the maps, it's probably a planet. Planets change position, relative to constellations, through the year. Viewed from earth, the major planets appear to move through the zodiac, which is composed of the constellations that lie in the ecliptic, the sun's apparent annual path across the sky.

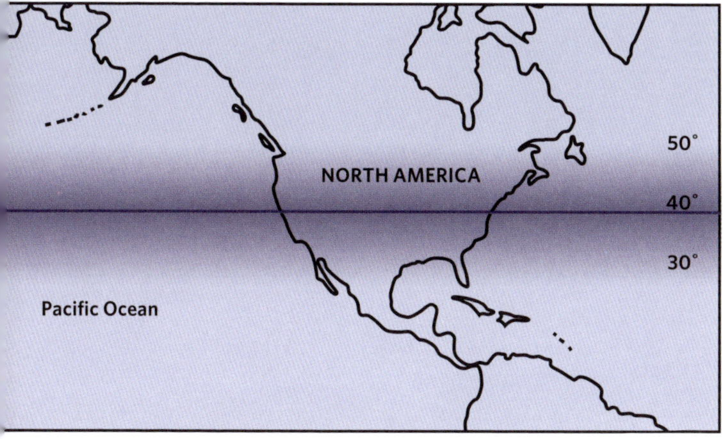

Everyone living at the same latitude sees the same stars.

The seasonal sky maps on pages 336–351 show the sky as seen at latitude 40° north, which runs through northern California, Denver, Philadelphia, Spain, Turkey, northern China, and Japan.

Sky conditions permitting, people living in a wide band (marked in blue above) north and south of the 40th parallel will be able to see most of the constellations shown in this guide.

Locating Stars from the Big Dipper

To find the **BIG DIPPER,** stand with your right shoulder in the direction of sunrise and your left shoulder in the direction of sunset. You are now facing north.

Look for a large rectangle with a string of stars off one corner. This is the asterism **BIG DIPPER,** a star group in **GREAT BEAR (URSA MAJOR),** page 352.

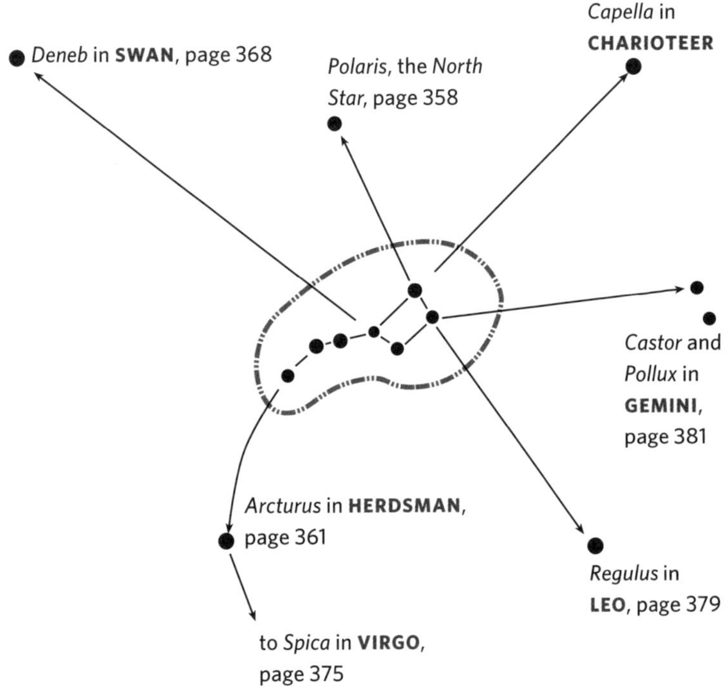

Locating Stars from ORION

ORION is visible in the night sky in fall, winter, and early spring. Look in the east, overhead (facing south), or west. **ORION** has five bright stars: a line of three closely spaced stars (Orion's belt), with reddish Betelgeuse and whitish Rigel nearby. Rigel is one of seven bright stars that form a hexagon around **ORION,** page 384.

Look for the Pleiades and Hyades, star clusters near Aldebaran in **TAURUS.**

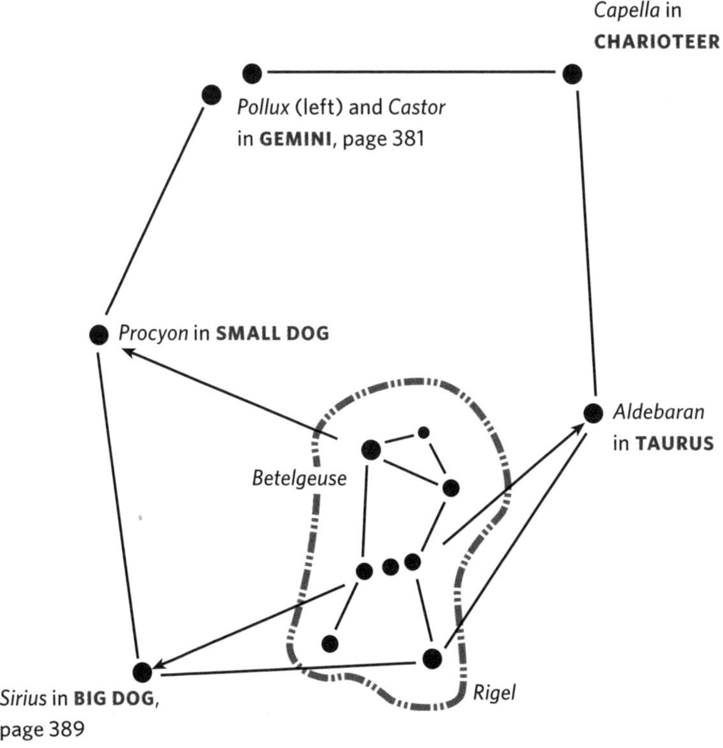

Patterns in the Evening Sky

If you see . . .

A cross
it is **Northern Cross** in **SWAN, page 368** ⟶

A semicircle
in the northern sky it is **NORTHERN CROWN, page 363** ⟶

An "M" or a "W"
it is **CASSIOPEIA, page 364** ⟶

A pair of bright stars
it is *Pollux* and *Castor* in **GEMINI, page 381**

Three stars in a row

- in the winter, it is **ORION, page 384** ⟶

- in the summer, it is **EAGLE**

A sickle
it is the **Sickle** in **LEO, page 379** ⟶

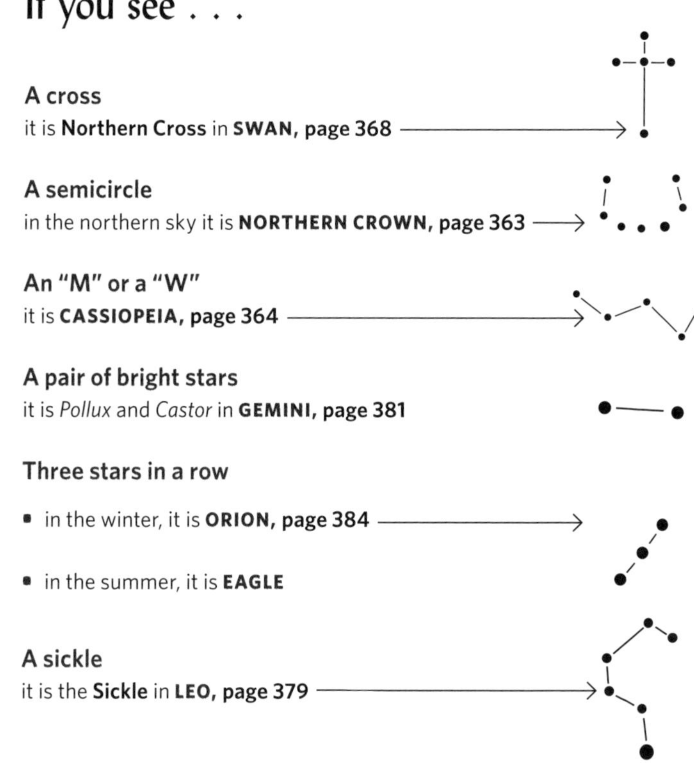

NOTE:
Patterns shown
on these pages
are not to scale.

A dipper
looking north any time of year ⟶

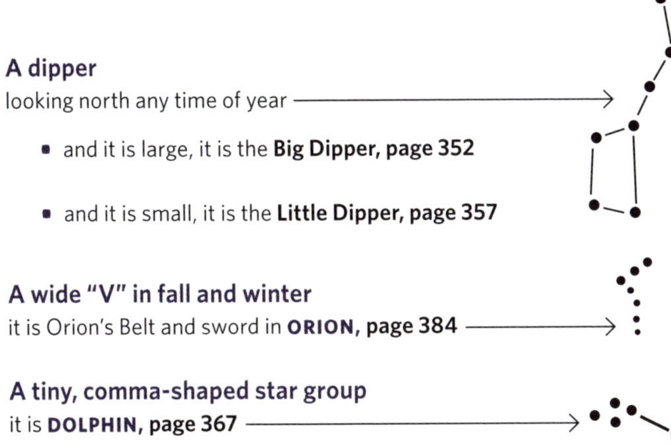

- and it is large, it is the **Big Dipper, page 352**

- and it is small, it is the **Little Dipper, page 357**

A wide "V" in fall and winter
it is Orion's Belt and sword in **ORION, page 384** ⟶

A tiny, comma-shaped star group
it is **DOLPHIN, page 367** ⟶

A bright star that is

- orangish, and it is near the **Big Dipper,** it is *Arcturus* in
  HERDSMAN, page 361

- reddish, in the fall, winter, and early spring, it is
  Betelgeuse in **ORION, page 384**

BEWARE! Reddish Mars can be found in or near **SCORPIO**
and in **ORION.**

Tips for Watching Stars

CHOOSE A NIGHT WITH NO MOON and a dark place far from bright lights. If conditions are not perfect, scan the sky anyway. You may be able to locate the brightest constellations.

If you can't quite see something, look a little to its side. Your peripheral vision can provide additional clarity.

When using this constellation finder, cover your flashlight with a piece of red plastic or a cloth bandanna to help your night vision, or use a red light.

Binoculars show you too little of the sky, so use just your eyes when learning constellations.

A single "star" that looks fuzzy may actually be a nebula, a concentrated cloud of gas and dust that shows up as a hazy area.

The Constellations

Constellations were created when people looked up from earth and saw patterns in the arrangement of stars in the night sky. They gave these patterns names and told stories about them. Most of the constellations familiar to English speakers come from astronomers in Classical Greece. A few are more modern. Charts in this guide use English counterparts of Latin names for the classical constellations, except for the best-known Latin names.

Cultures around the world have their own constellations and star stories. See, for example, the many interpretations of the group of seven bright stars we call the Big Dipper, starting on page 352.

Some well-known star groups lie within constellations. They are called asterisms. The Big Dipper is an asterism in **GREAT BEAR (URSA MAJOR)**, page 352.

Using the Seasonal Sky Maps

The maps on pages 336–351 show the constellations in the sky on:

March 1, 9–11 p.m. (also November 1, dawn), **pages 336–339**
June 1, 9–11 p.m. (also January 15, dawn), **pages 340–343**
September 1, 9–11 p.m. (also June 1, dawn), **pages 344–347**
December 1, 9–11 p.m. (also August 16, dawn), **pages 348–351**

Choose the date and time closest to when you want to view the sky, and the direction—north, east, south, or west—and find the map for that season and direction. Times given are standard time. For daylight savings time, subtract one hour.

The maps are drawn so that you can hold them up to the sky. When you face north, east will be to your right and west will be to your left. When you face south, east will be on your left and west will be on your right.

Remember that stars, like the moon and sun, rise in the east and set in the west.

If you are viewing before 9 p.m., after 11 p.m., or at a time other than March, June, September, or December, the location of the constellations will vary from the positions shown in the seasonal sky maps. Some figures may not have risen and others may have already sunk below the horizon.

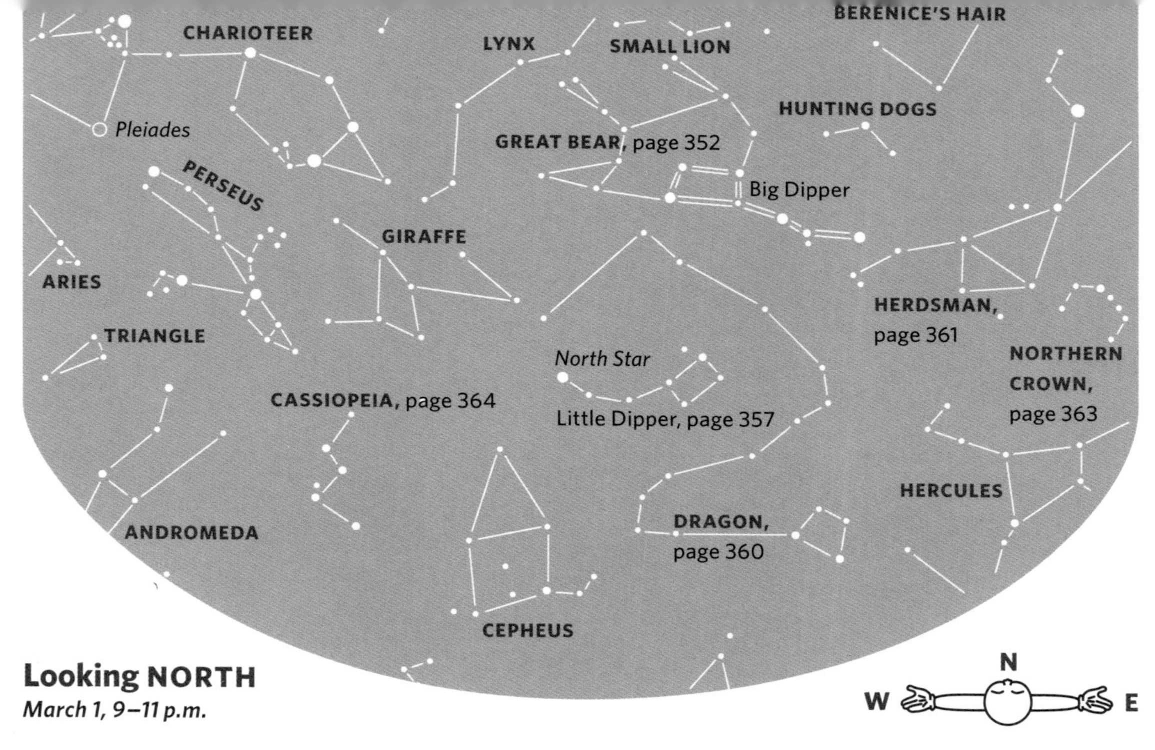

Looking NORTH
March 1, 9–11 p.m.

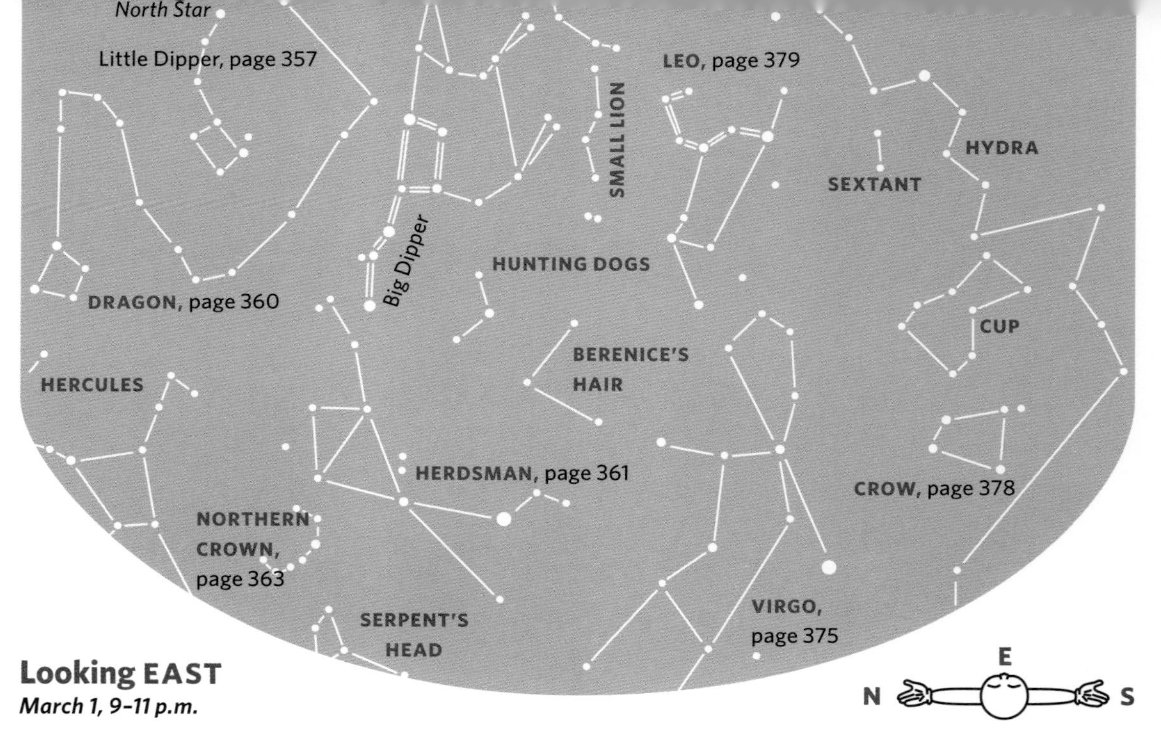

Looking EAST
March 1, 9–11 p.m.

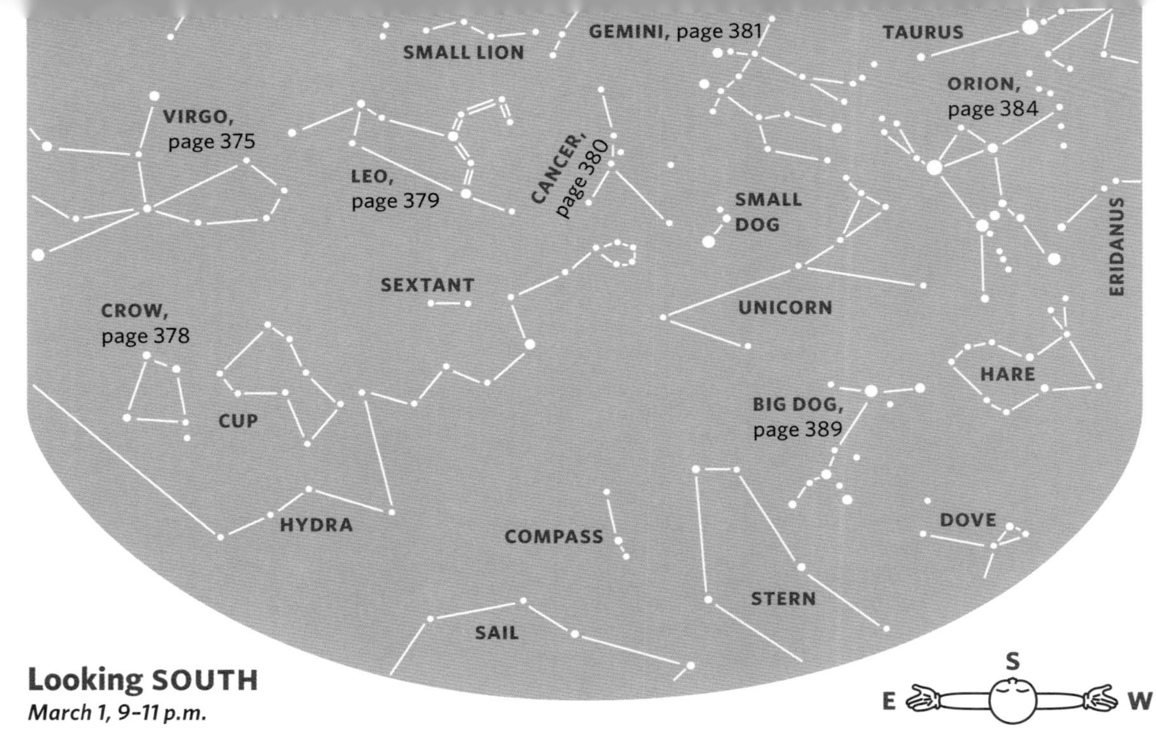

Looking SOUTH

March 1, 9–11 p.m.

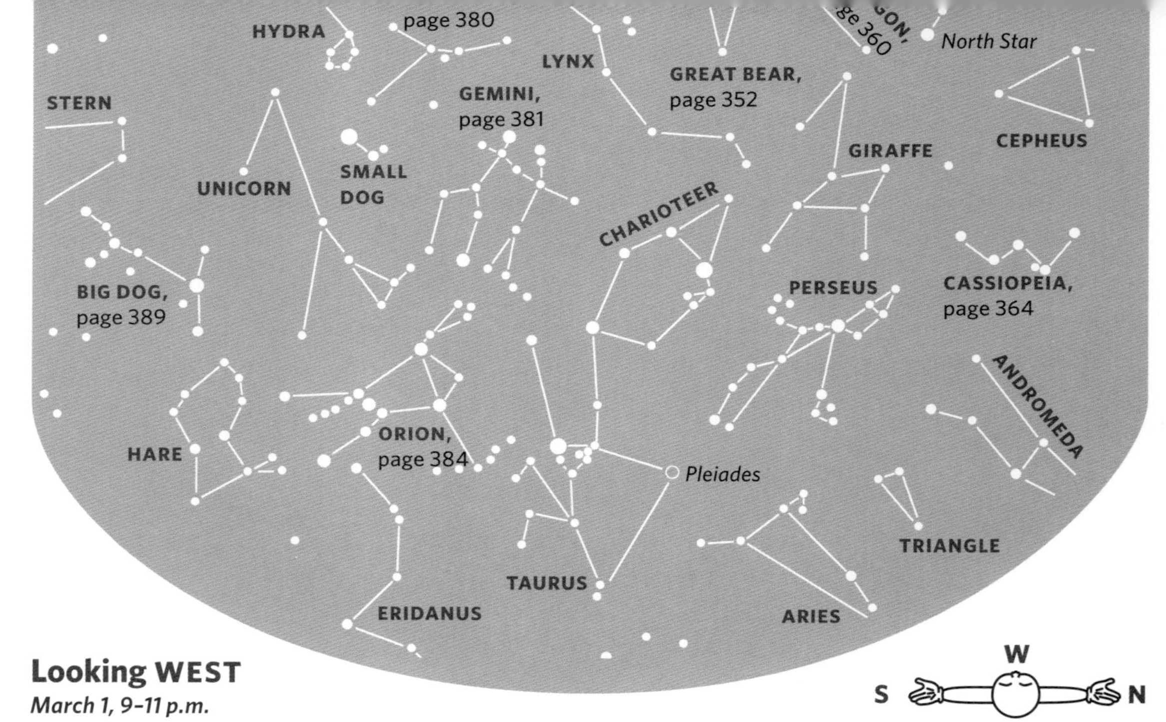

Looking WEST
March 1, 9–11 p.m.

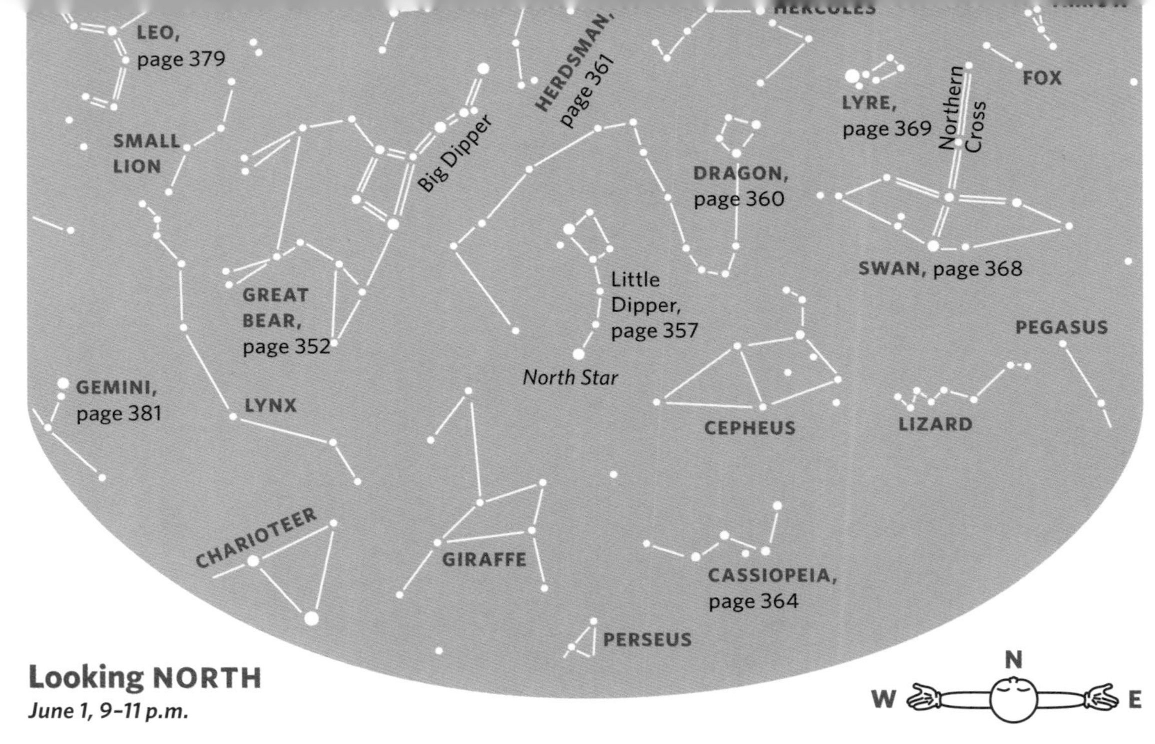

Looking NORTH
June 1, 9–11 p.m.

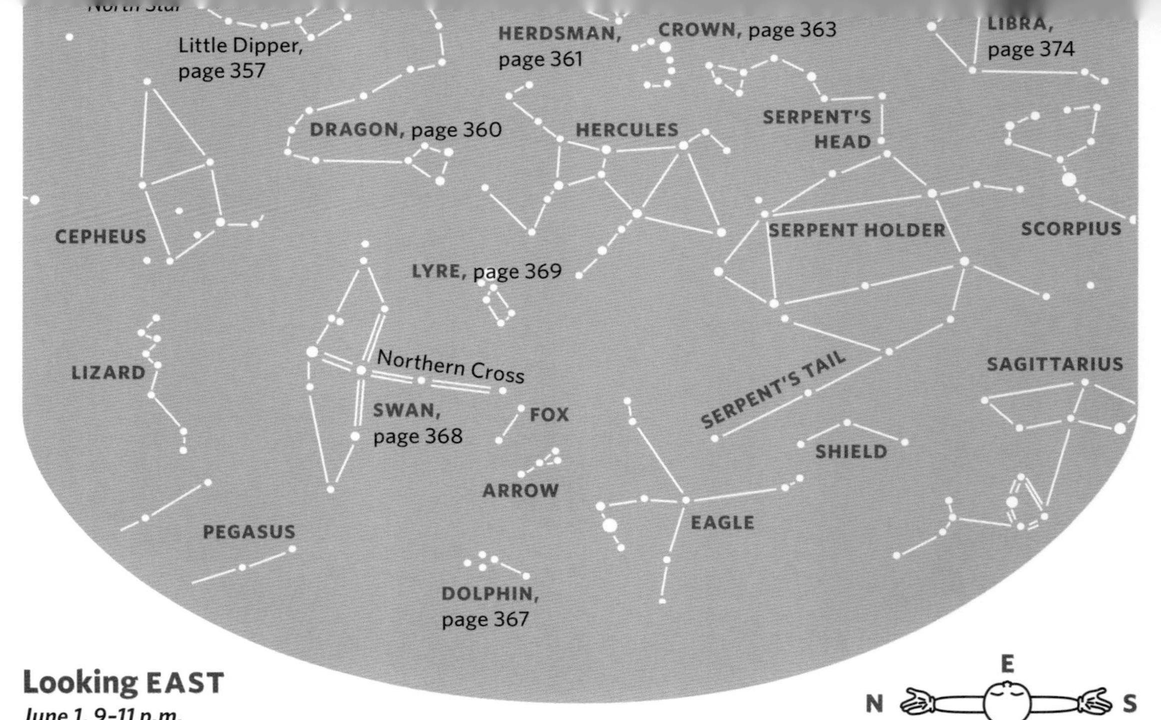

Looking EAST
June 1, 9–11 p.m.

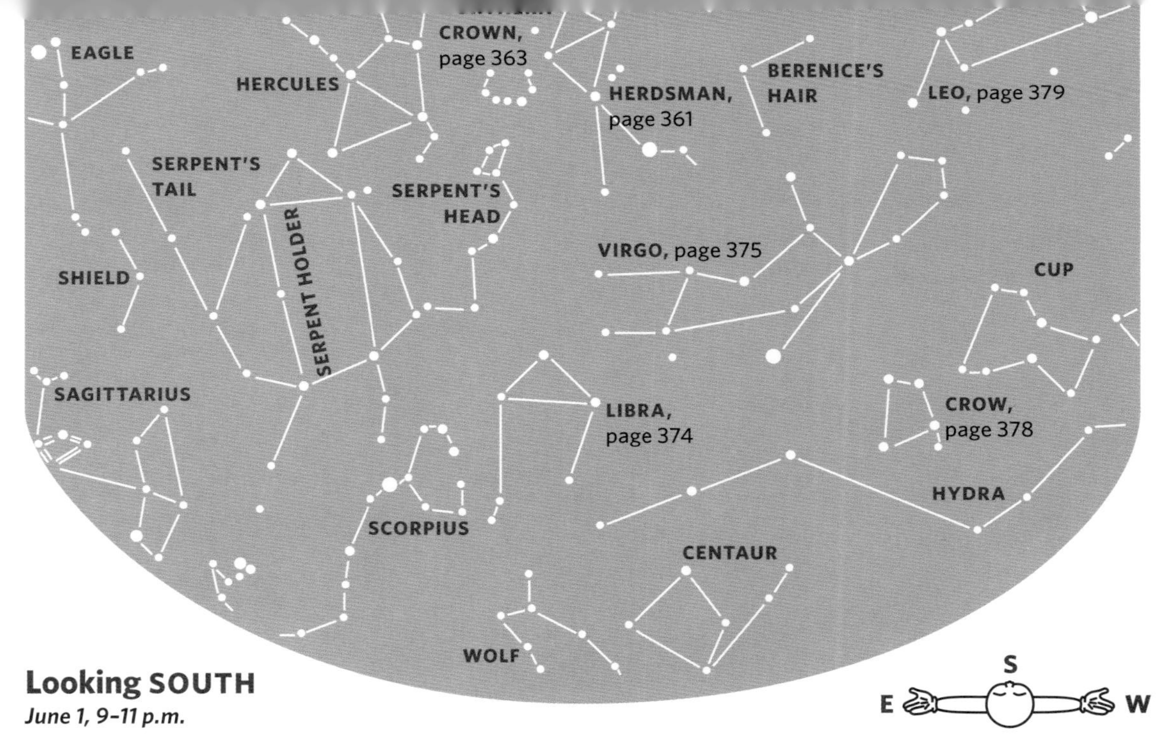

EAGLE
CROWN, page 363
HERCULES
BERENICE'S HAIR
LEO, page 379
HERDSMAN, page 361
SERPENT'S TAIL
SERPENT'S HEAD
SERPENT HOLDER
VIRGO, page 375
SHIELD
CUP
SAGITTARIUS
LIBRA, page 374
CROW, page 378
SCORPIUS
HYDRA
CENTAUR
WOLF
Looking SOUTH
June 1, 9–11 p.m.
S
E
W

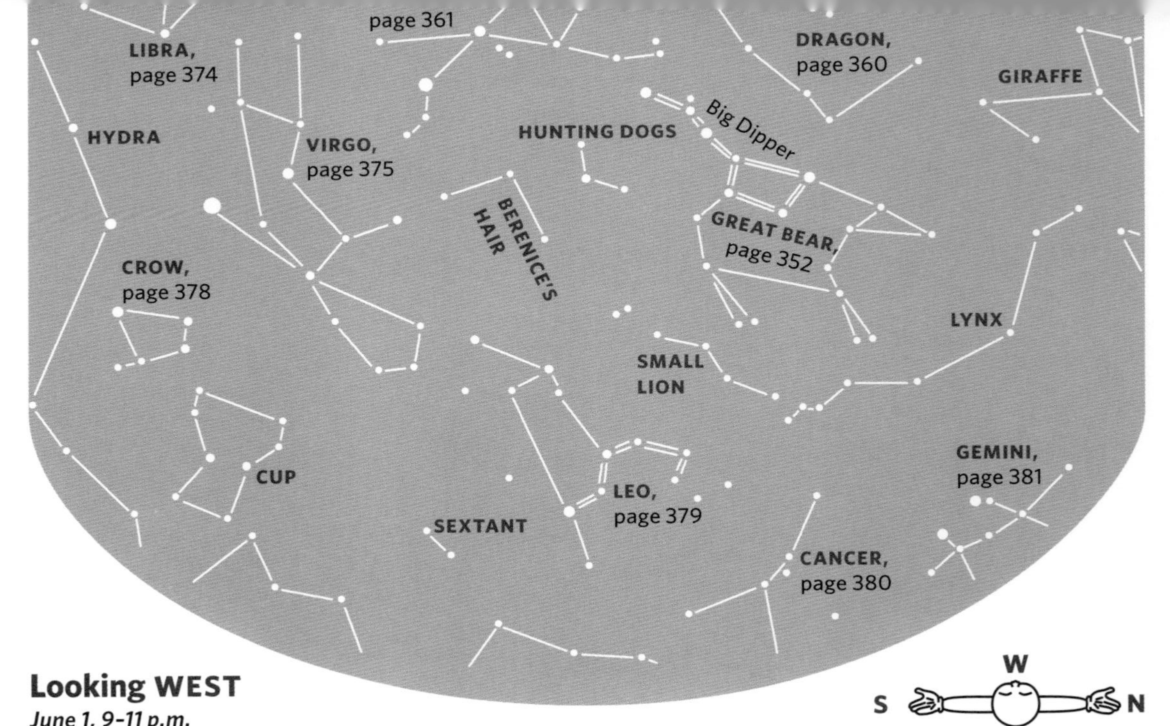

Looking WEST

June 1, 9–11 p.m.

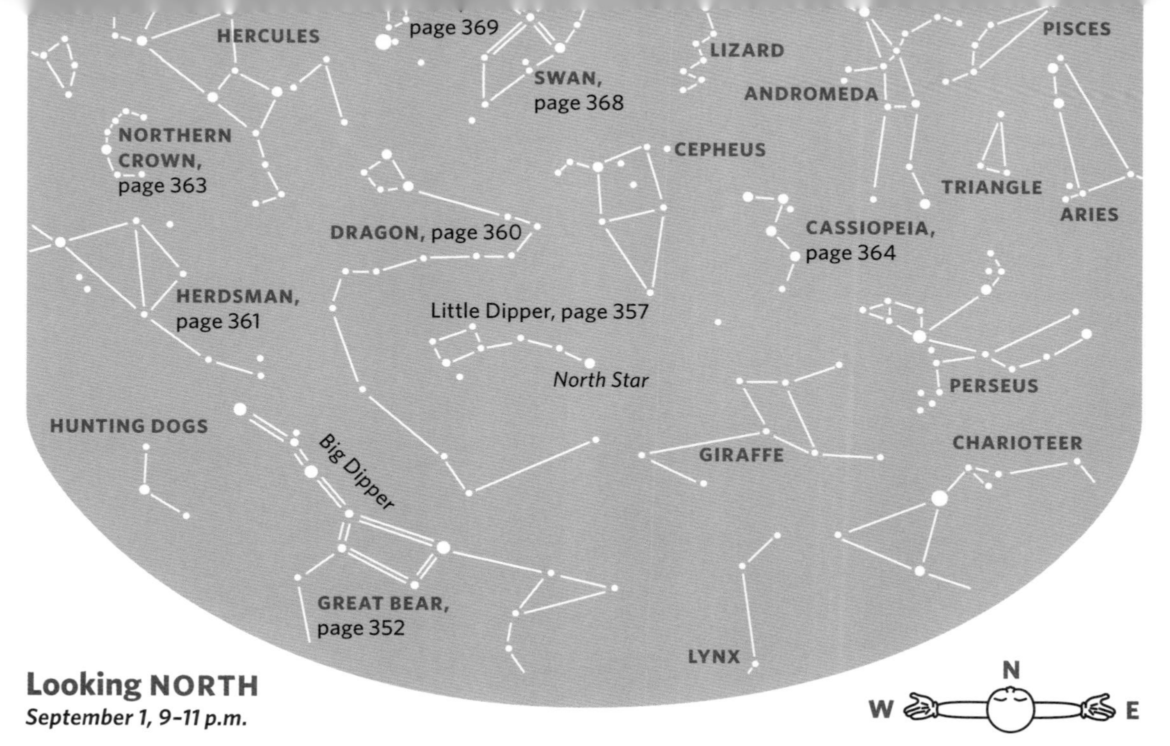

Looking NORTH
September 1, 9–11 p.m.

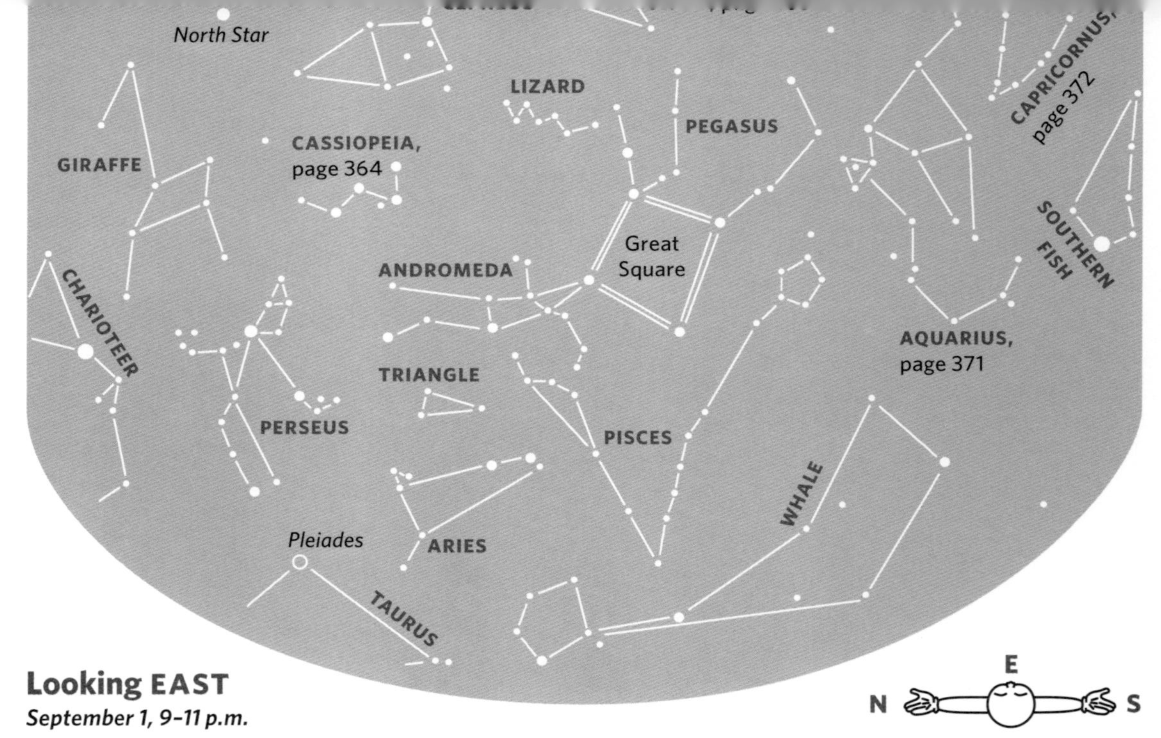

Looking EAST
September 1, 9–11 p.m.

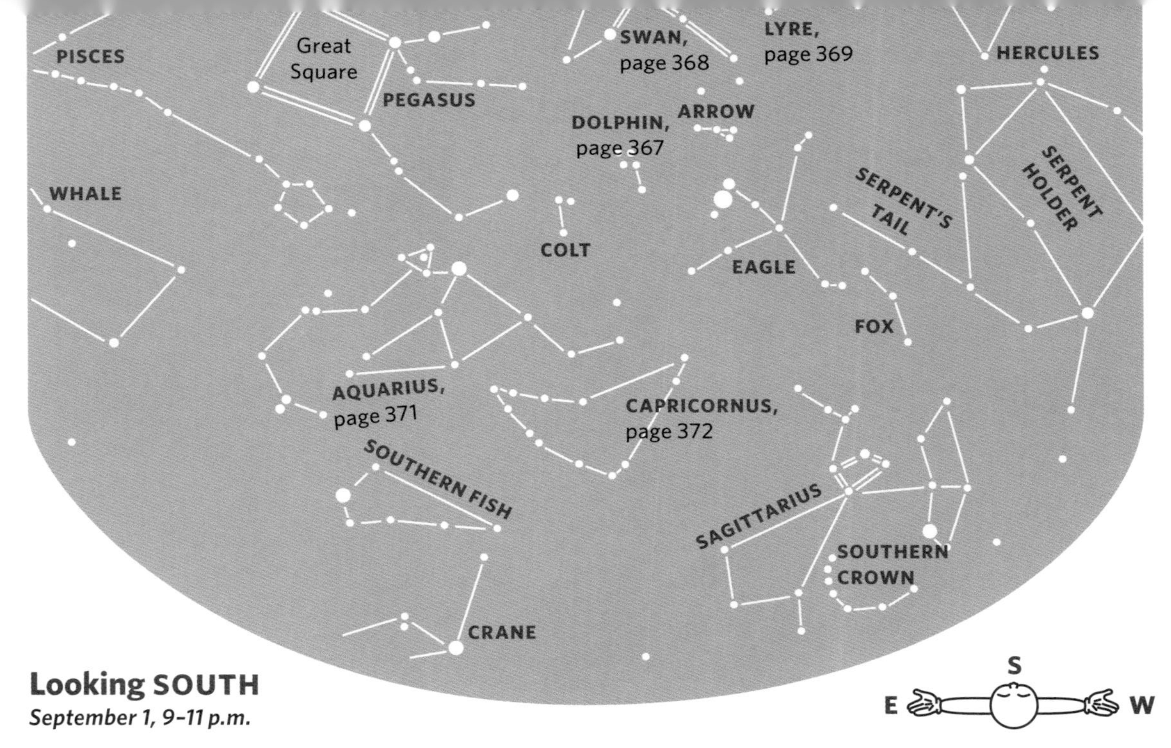

Looking SOUTH
September 1, 9–11 p.m.

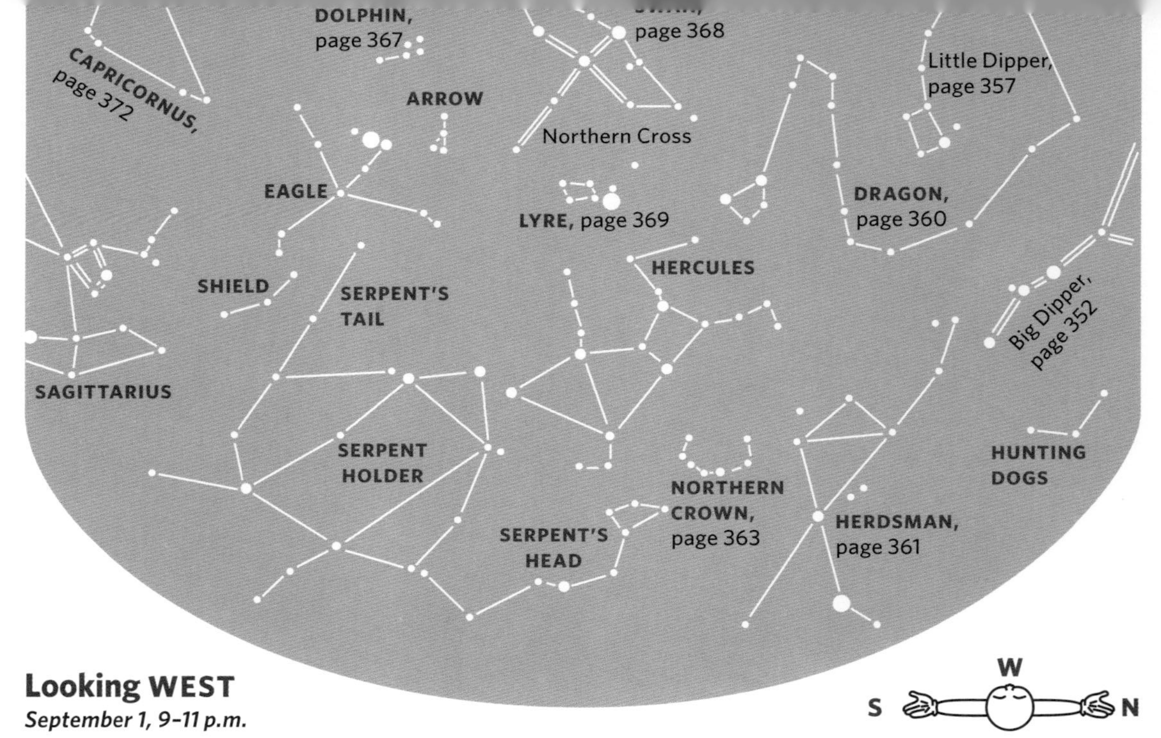

Looking WEST
September 1, 9–11 p.m.

Looking NORTH
December 1, 9–11 p.m.

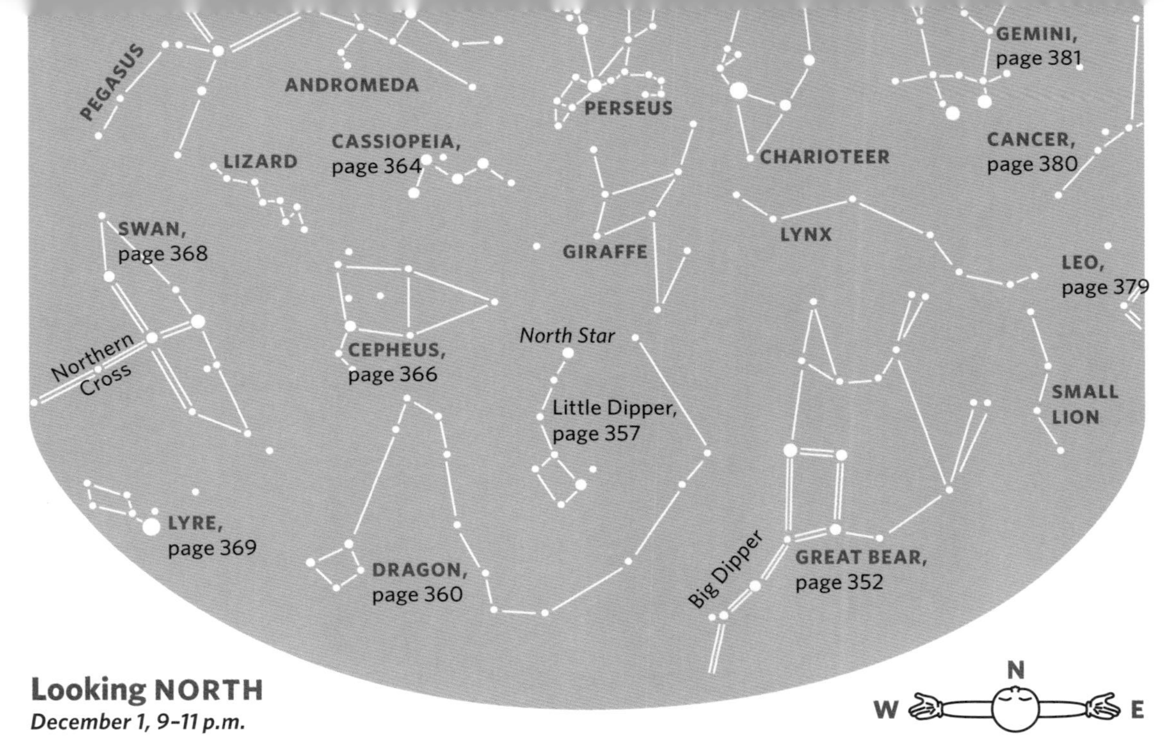

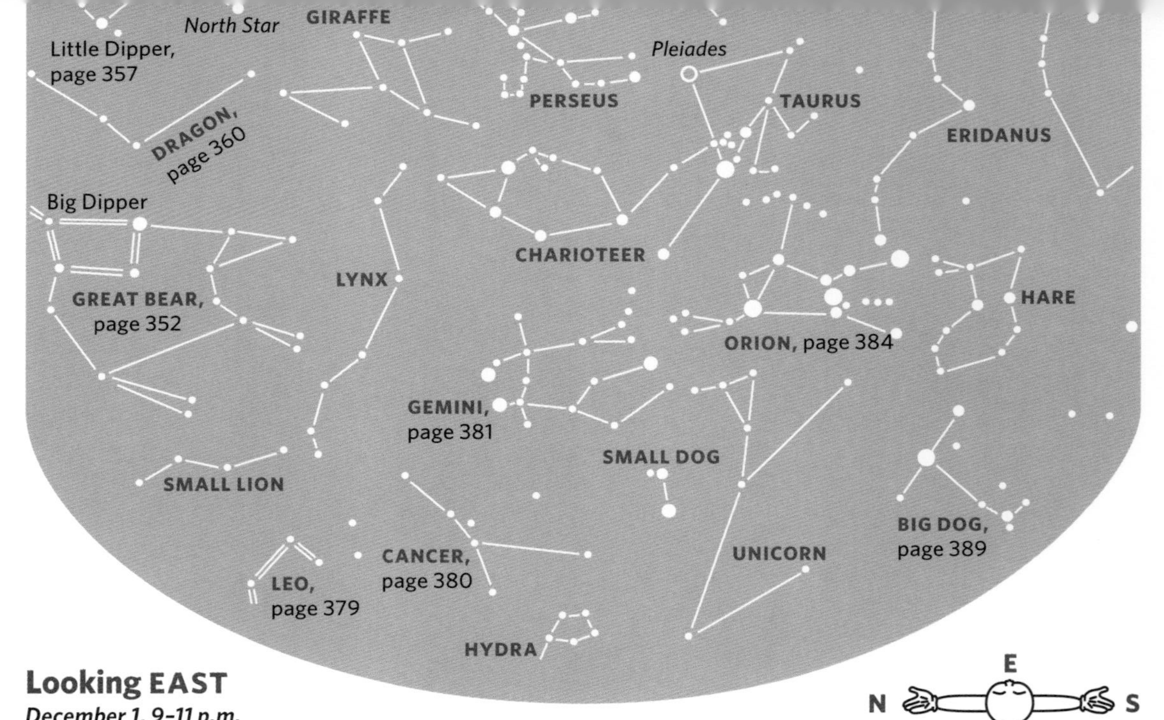

Looking EAST
December 1, 9–11 p.m.

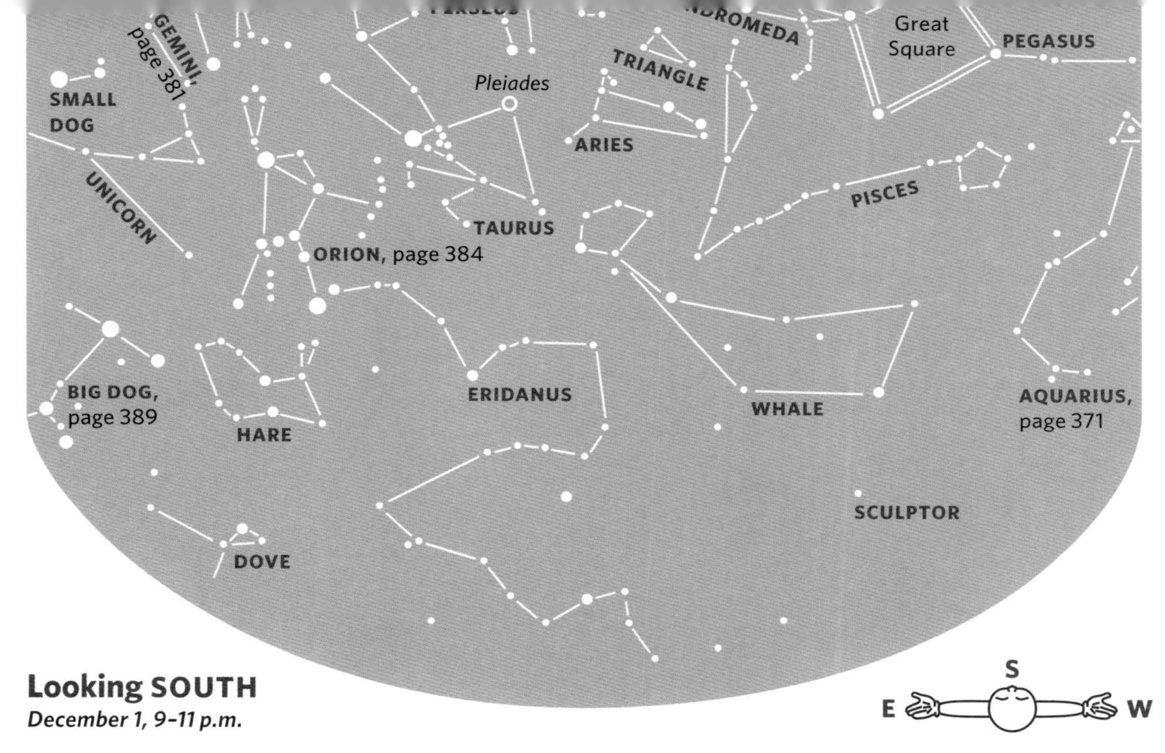

Looking SOUTH
December 1, 9–11 p.m.

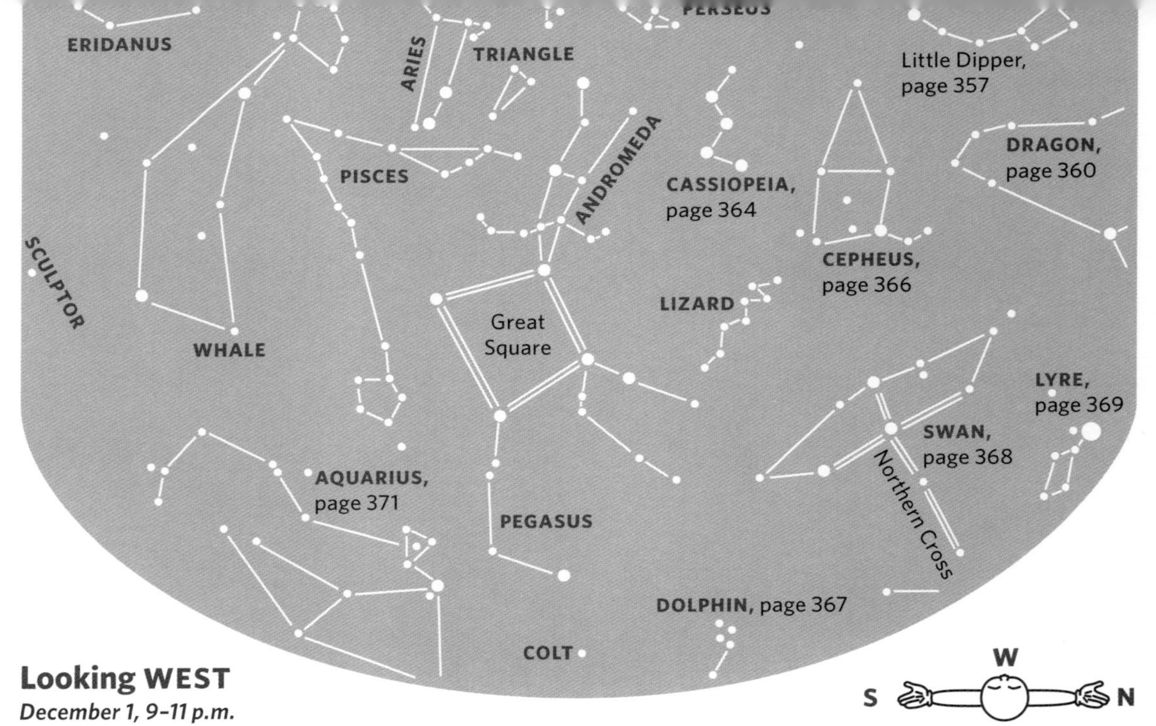

Looking WEST
December 1, 9–11 p.m.

Great Bear/Ursa Major (UR-sa MA-jor)
Asterism: *Big Dipper*
Best viewing: *All year*

Best known for the seven bright stars that form a large dipper in the north sky. The **BIG DIPPER** has many other names, especially among peoples of the far north.

BEAR: Greek. Zeus loved a mortal woman, Callisto, who bore him a son. Zeus's angry wife turned Callisto into a bear. Years later, Callisto saw her grown son and tried to embrace him. As the son drew his weapon to kill the attacking bear, Zeus raised them to the sky, mother as **Great Bear** and son as **Little Bear.** Star names in the Great Bear:

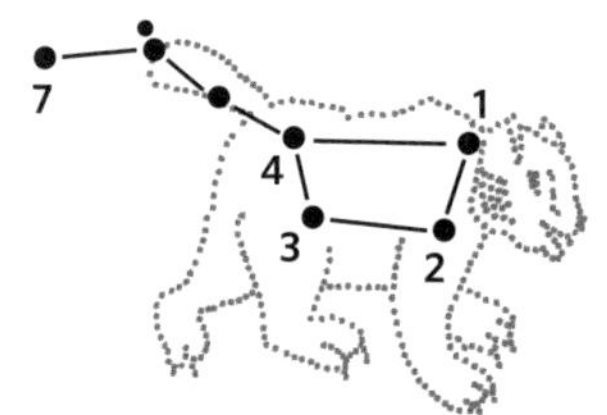

1 *Dubhe,* the bear
2 *Merak,* loins of the bear
3 *Phecda,* thigh
4 *Megrez,* root of the tail
7 *Alkaid* or *Benetnash,* end of the tail

BEARS AND HUNTERS: Chukchi in Siberia; Micmac and other tribes in eastern North America. The bowl of the **Dipper** is the bear, the three stars of the handle are hunters, and the semicircle of stars to the south (**Northern Crown**) is the bear's den.

STAG, DEER: Many Siberian groups

ELK: Saami, northern Siberia

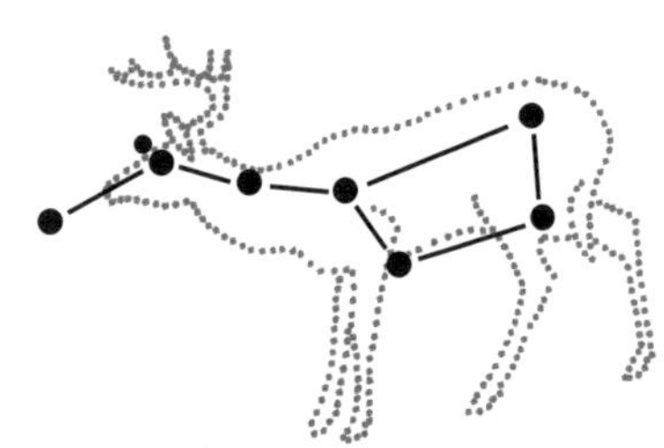

HORSE: Teleuts of east Siberia

CARIBOU: Inuit, Labrador to Alaska

LOONS: Klamath of western United States

CAMEL: Berbers (Imazighen) of North Africa

SEVEN OXEN: Roman

SEVEN SAGES: Hindu

SEVEN BLACKSMITHS: Buryats of southern Siberia

EMPEROR AND HIS COURT: Chinese

BULL'S THIGH: Egyptian. The thigh represents immortality because these stars never disappear. Priests used a dipper-shaped tool and a bull's thigh to "open the mouth" of a mummy so it could proceed to the afterlife. These stars are identified with Set, who attacked and killed Osirus (**see ORION, page 384**).

SEVEN THIEVES: Siberian. Seven Burkhans stole a girl (from *Pleiades*) who now stays with them (as ★8). ———→

FISHER: East Cree, Ojibwa. The land of the north was always cold and snowy because the people of the south kept summer to themselves. Fisher helped free summer, but as he escaped an arrow (★8) hit his tail.

DIPPER: European, east Slavic, Chinese, tribes of Pacific Northwest

DRINKING GOURD: Southern United States. Escaping slaves in pre-Civil War United States used this prominent constellation to locate the *North Star* (**page 358**).

STICK FOR GATHERING SAGUARO FRUIT: Hualapai, Tohono O'odham of southwest United States

BASE FOR LAMP: Inuit of Greenland

NET FOR CATCHING RABBITS: Northern Paiute

WAGON AND (sometimes) **THREE HORSES:** Norse (Thor's Wagon), Irish (King David's Chariot), and other cultures of Europe and Scandinavia

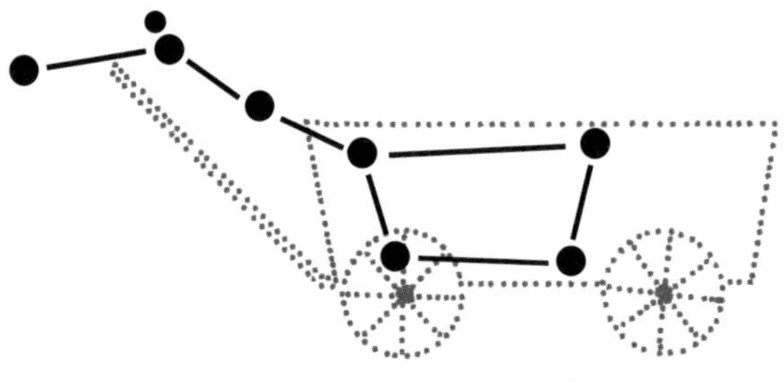

BOAT, SKIFF: Creek, Seminole, Alabama, Caribbee in Caribbean

CANOE: Malay

WINNOWING SHOVEL: Chinese

BED: Tungus of Siberia

CRADLE: Isleta Pueblo

SEAT: Inuit of northern Quebec, Labrador

BARBECUE GRILL: Taulipang of Brazil

SAUCEPAN: French peasantry

PLOW: British

PILLAR: Tahitian

TEZCATLIPOCA, LORD OF THE NIGHT:
Aztec. One foot is missing because the
Earth ate it (part of the **Big Dipper** goes
below the horizon).

SEVEN MACAW: Maya. Seven Macaw (a
scarlet macaw) had sparkling teeth and
bright metal eye ornaments. He pretended
he was the sun and moon. Hunahpu shot
him with a blow dart that loosened his
teeth; two curers tricked him into giving
up his finery. The real sun and moon came
into the world. Seven Macaw is now hum-
bled; he remains in the sky only part of the
night, then falls below the horizon.

CORNER: Japanese. In Japan, north is a
"bad" direction, so facing the corner—
which is in the north—means having to
deal with great strife.

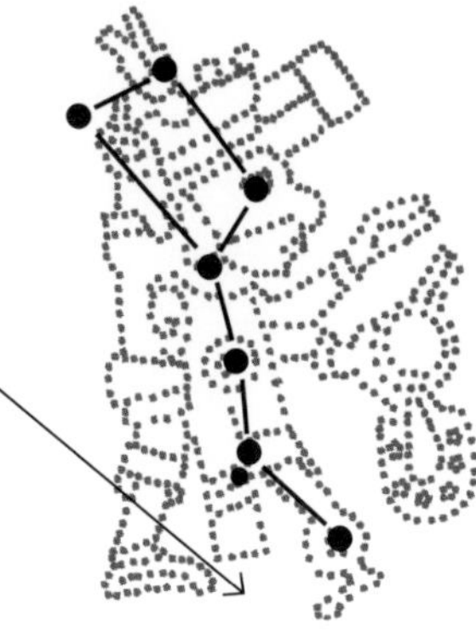

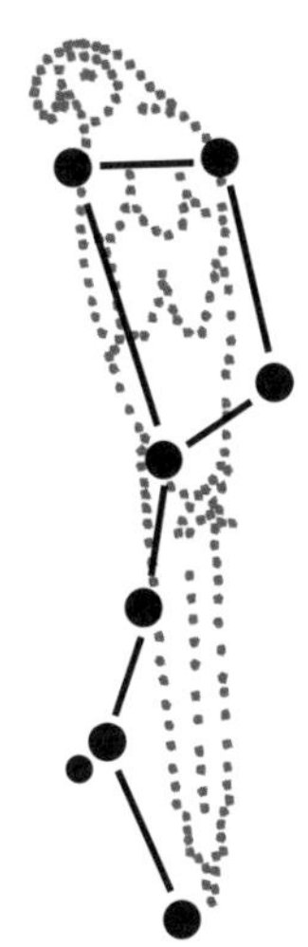

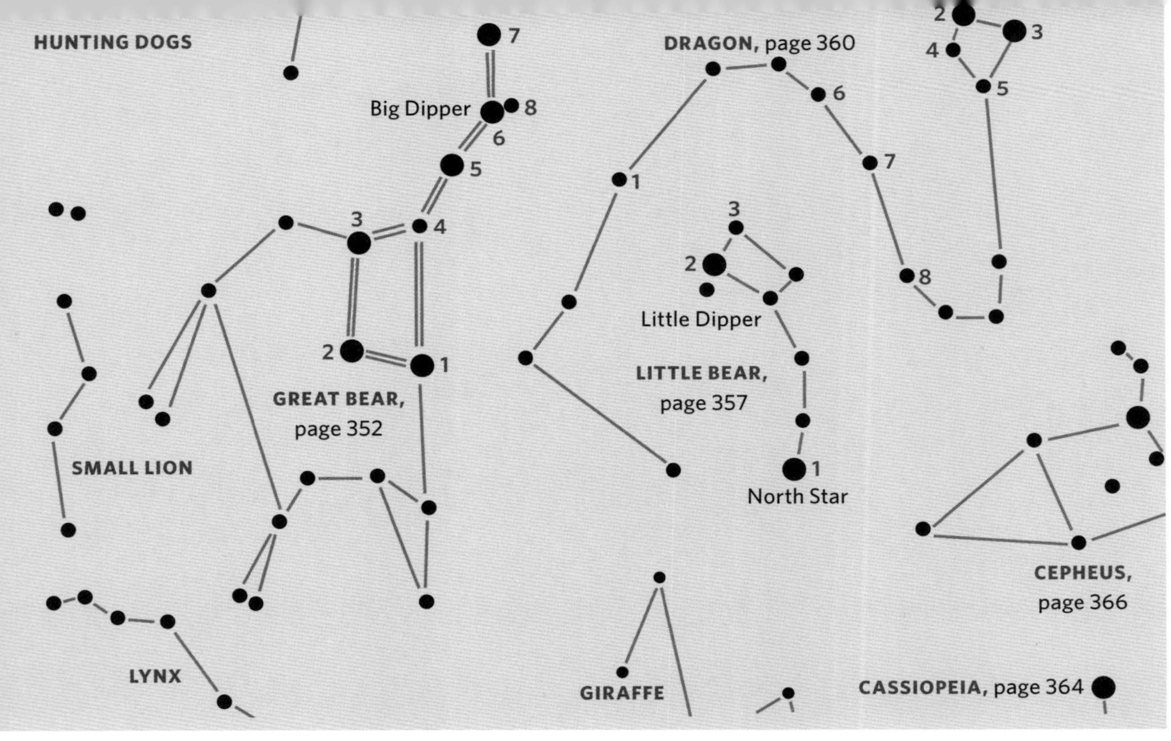

HUNTING DOGS
Big Dipper
7
8
6
5
3
4
2
1
GREAT BEAR,
page 352
SMALL LION
LYNX
DRAGON, page 360
6
7
8
1
Little Dipper
3
2
LITTLE BEAR,
page 357
1
North Star
2
3
4
5
6
7
8
CEPHEUS,
page 366
GIRAFFE
CASSIOPEIA, page 364

Little Bear/Ursa Minor (UR-sa MI-nor)

Asterism: *Little Dipper*
Best viewing: *All year*

The Little Dipper hangs opposite the Big Dipper in the northern sky.

LITTLE BEAR: Arab, Greek

LESSER WAGON: Babylonian, widely in Europe

SMALL DIPPER: Chinese

BEAVER THAT SPREADS ITS SKIN: Seneca

SEVEN MICE: Mortlocks, South Pacific

TWO HORSES TIED TO A STAKE: Kyrgyz of northern Asia. If wolves
(stars in the **Big Dipper**) kill the horses, the world will end.

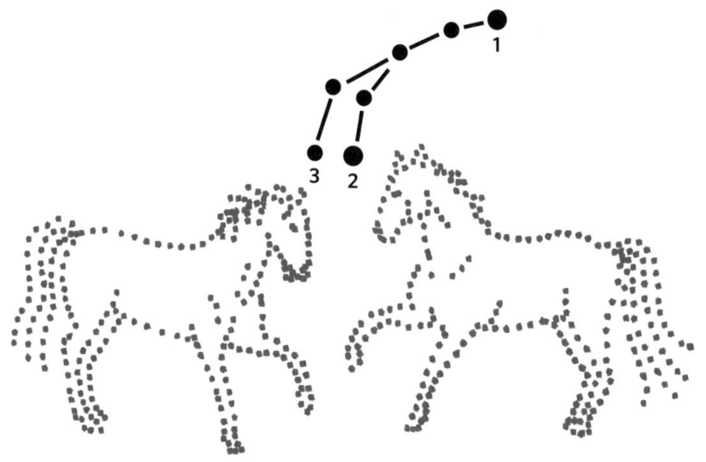

HUNTING HORN: Spanish, Portuguese

WAR CLUB: Samoa (includes ★1 of **Big Dipper**)

THE GUARD: Chinese

SEVEN BROTHERS: Kurds

CHIMALMAT: Maya. Wife of Seven Macaw (see page 355).

FRUIT OF ALMOND AND CHERRY PLUM TREES: Arab, Persian

GREAT RULERS: Chinese. ★1 Great Imperial Ruler of Heaven, ★2 Emperor, ★3 Crown Prince

CAMEL CALF: Berbers (Imazighen)

FISH: Arab

DOG'S TAIL: Greek

YOKE: Tewa of SW southwestern United States

The bright star (★1) at the end of the Little Dipper's handle is Polaris, the North Star, used for ages by people in the Northern Hemisphere to find the direction north. To find the North Star, draw a line through the two stars that mark the outer corners of the Big Dipper (see page 356). Other constellations appear to rotate around the North Star in the sky.

★1 NORTH STAR

NORTH STAR OR GUIDE OF MERCHANTS: Yucatec Maya

SHIP STAR: Anglo-Saxons

STEERING STAR: English

GUIDE OF THE PEOPLE: East Cree

STAR THAT STANDS STILL: many tribes in North America

PIVOT OF PLANETS: early northern India

NAIL OF THE SKY/NORTH: Siberian, Saami, Finn, Estonian, Norse, Persian

GRIZZLY BEAR: Kootenai

EYE OF THE CREATOR: Pomo of California

Dragon/Draco
Best viewing: *All year*

These dim stars wind around the Little Dipper.

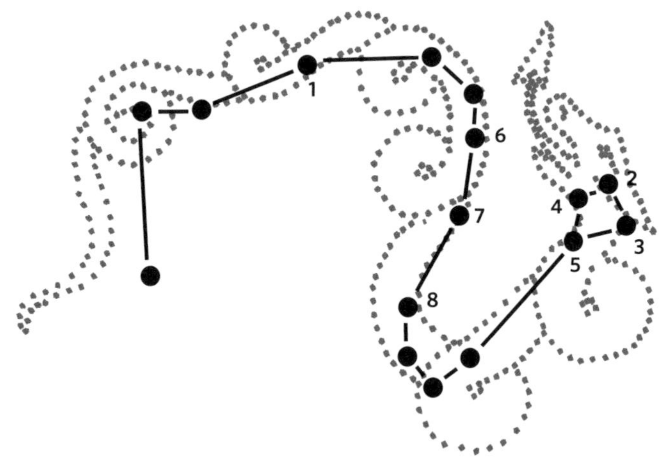

DRAGON: Greek. ★1 is *Thuban*, dragon; ★2 *Rastaban;* and ★3 *Eltanin*, dragon's head.

CAMELS AND HYENAS: Arab. ★6–7 are hyenas stalking a camel foal (not shown), who is protected by four female camels (★2–5, the stars of **DRAGON**'s head); ★8 is a desert nomad's tent.

Herdsman/Boötes (boo-OH-teez)

Best viewing: *April to August*

Follow the arc of the Big Dipper's handle to find Arcturus, the brightest star in the **HERDSMAN**. (See following page for star number references.)

BEAR WATCHER or **OX DRIVER:** Greek. ★1 is *Arcturus*, bear watcher; ★5 *Izar*, the girdle; ★7 *Alkalurops*, the herdsman's staff.

ICARIUS: Greek. Icarius, the first person to make wine, shared it with shepherds who got drunk and killed him. His daughter Virgo killed herself out of grief. Their dog (*Procyon* in **SMALL DOG**) joined them in the sky.

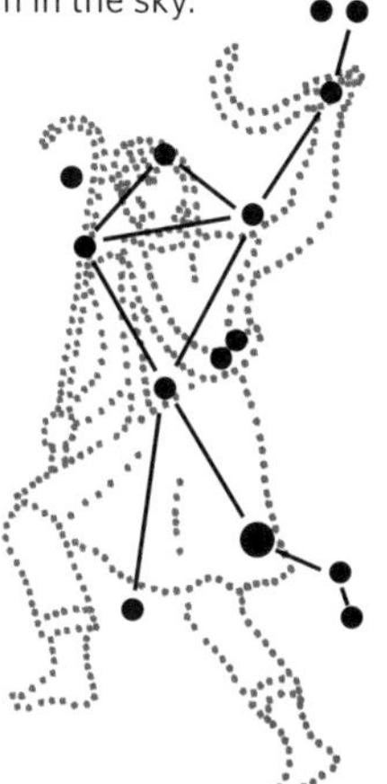

HERDSMAN: early Arab. ★1 is a shepherd and nearby stars are animals. ★2, 3, 4, 7 are wolves; in the **Little Dipper**, ★1 is a goat, 2 and 3 are calves; stars in **DRAGON (page 360)** are hyenas and camels; stars in **CEPHEUS (page 366)** are sheep.

PIRANHA: Kobeua of Brazil (★1, 6, 8 tail; 2, 3, 4 head)

TWO IN FRONT: Inuit. When an orphan revealed that an old man had killed his brother-in-law, the man (★1) chased the boy (★6) into the sky; the boy's grandmother (Vega in **LYRA**) was too far behind to rescue the child.

OTHER NAMES FOR ★1 ARCTURUS:
Great Horn: Chinese
Coral Bead, Gem, Pearl: India
Parakeet: Tribes of the Gran Chaco in Argentina, Bolivia, and Paraguay
Eye of the Rainmaster: Guajiro, northern Colombia

A small nearby constellation is Hunting Dogs, see following page.

Hunting Dogs/Canes Venatici
(Dogs of the **HERDSMAN***)*

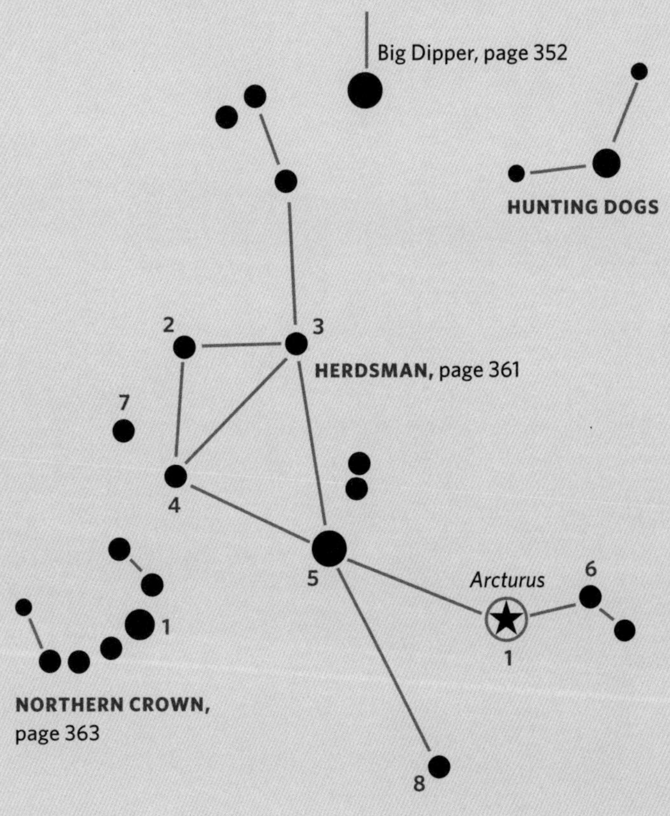

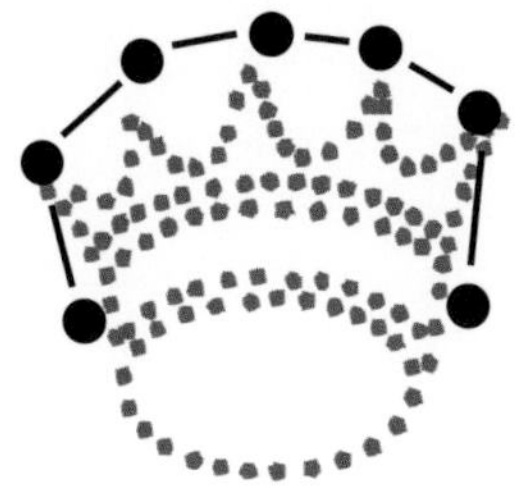

Northern Crown/Corona Borealis (Kor-RO-na Bor-ee-AL-is)

Best viewing: *April to August*

Semicircle near head of **HERDSMAN**

CROWN: Greek. Ariadne, daughter of King Minos of Crete, helped Theseus kill the Minotaur and escape from its maze. She married Dionysus, who gave her this crown. The brightest star ★1 is *Gemma*, gem.

DISH, BROKEN DISH: Arab, Persian

POLAR BEAR'S PAW: Siberian

COUNCIL OF CHIEFS: Pawnee

FLOWER GARDEN: Lithuanian

STRING WITH COINS: Chinese

BOOMERANG: Aborigines, Australia

ARMADILLO: Tribes of Colombia

DEN OF GREAT BEAR (see page 352): Greek

Cassiopeia

Best viewing: *August to January, but visible all year*

From the Big Dipper, find the North Star and continue on to W- or M-shaped constellation. (See page 366 for star number references.)

WOMAN IN A CHAIR: Arab

QUEEN OF ETHIOPIA, SHE OF THE CHAIR: Greek. ★1 *Schedar*, the breast; ★2 *Caph*, hand. A vain queen boasted of her beauty until sea nymphs convinced Poseidon, god of the sea, to send a monster to destroy her country. Her husband, King **CEPHEUS,** chained their daughter **ANDROMEDA** to a rock as a sacrifice to appease Poseidon. **PERSEUS** arrived, carrying the head of Medusa, whom he had recently killed. He freed Andromeda, slew the monster (either **WHALE** or **PISCES**), and asked for her hand in marriage. The queen, on her throne, must spend half the time with her head down (as she rotates around the *North Star*) to humble her as punishment for bragging.

FEMALE ONE WHO REVOLVES: Navajo

SPIDER: Maricopa and Pima, southwest United States

RABBIT: Pawnee

STAR ZIGZAG: Zuni

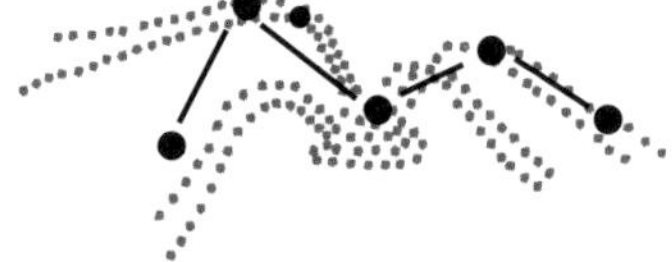

BELT: Lithuanian

HAND STAINED WITH HENNA: Arab. ★1–5 are fingertips stained with henna, a plant used to make dye.

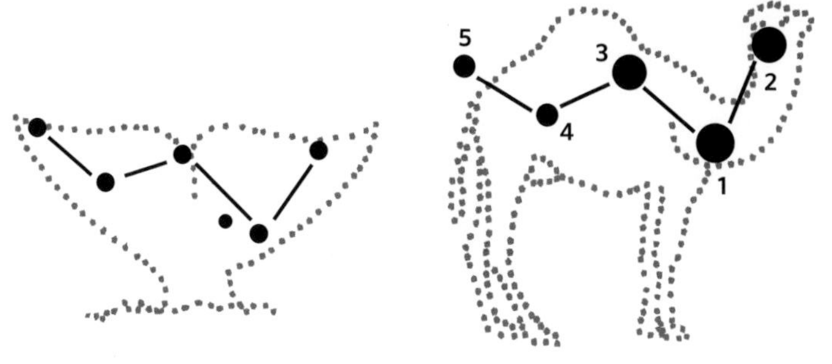

TAIL OF FISH, PORPOISE, OR WHALE: Carolines, Nukuoro, Truk in South Pacific

ELK SKIN: Quileute, Pacific Northwest

CAMEL: Middle East

LAMP STAND: Greenland Inuit. Three stones or posts support a seal oil lamp (★1–3).

SOD CUTTER: Alaskan Inuit. A triangular tool used to cut turf for winter houses (★1–3).

CHARIOT: Chinese (★1, 2 as wheels and body; ★3 as whip)

SEAL-SKIN OIL CONTAINER: Inuit

DON'S COURT: Celtic. Don is king of the fairies; his daughter Arionrod lives in the **NORTHERN CROWN, page 363.**

WHITE BEAR, FIVE REINDEER STAGS: Chukchi in Siberia

MOOSE ANTLER: Saami

DEER: Sumerian (★2–5)

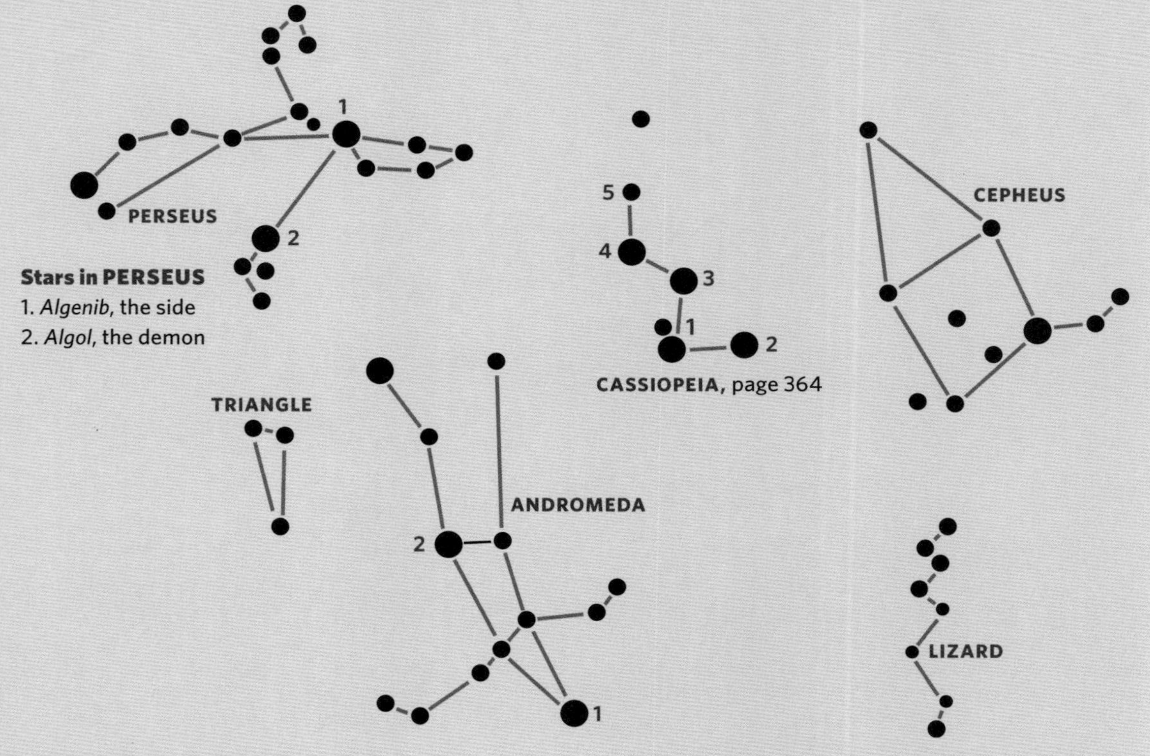
PERSEUS
1
2
Stars in PERSEUS
1. Algenib, the side
2. Algol, the demon
TRIANGLE
CASSIOPEIA, page 364
5
4
3
1
2
ANDROMEDA
2
1
CEPHEUS
LIZARD

Dolphin/Delphinus (del-FINE-us)

Best viewing: *July to November*

Small, discrete constellation near Northern Cross

PORPOISE: Greek, Arabic, Hindu

GOURD: Chinese, Polynesian

WOODEN BOWL: Marshall Islanders and Mortlocks in South Pacific

BIRD CAGE: Jaina of India

TRUMPET SHELL: Torres Strait Islanders, Australia

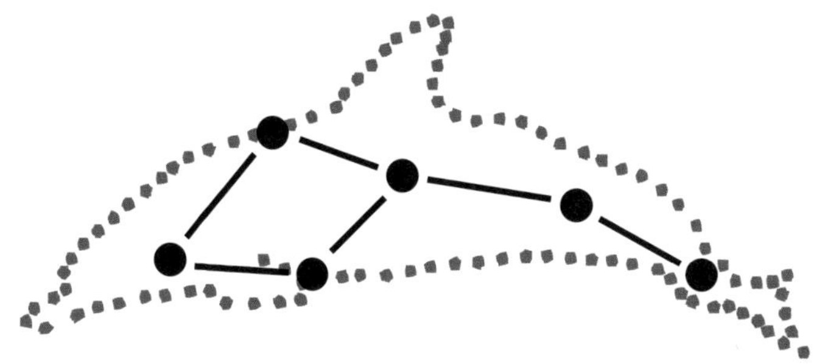

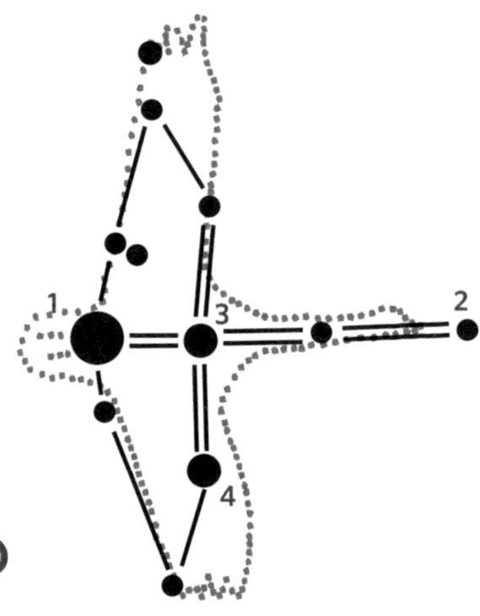

Swan/Cygnus (SIG-nus)

Asterism: *Northern Cross*
Best viewing: *June to November*

In summer and fall, this cross rises in the northeast, travels overhead, and sets in the northwest. ★1 is Deneb, the tail; ★2 Albireo; ★3 Sadr, hen's breast; ★4 Gienah, the wing.

HEN: Arab

BIRD: Greek

SWAN: Roman

WHITE SEA-SWALLOW: Society Islanders in South Pacific

TURKEY: Warao in Venezuela

CROSS: Christian

MAGPIES: Chinese. Birds form a bridge so that two lovers may unite (see **WEAVING PRINCESS, page 369**).

Lyre/Lyra (LI-ra)

Best viewing: *May to November*

Look for ★1 Vega (eagle), a very bright, bluish-white star near the Northern Cross.

LYRE (HARP): Arab and Greek. Arion played his lyre as shipboard robbers threw him into the sea to die. A dolphin, attracted by the beauty of Arion's music, carried him to safety.

TALYN ARTHUR (ARTHUR'S HARP): Early Briton

SWOOPING STONE EAGLE OF THE DESERT: Arab (★1–3).

TORTOISE: Persian. ★2 *Sheliak* and ★3 *Sulaphat* both mean tortoise.

GRANDMOTHER: Inuit (see **TWO IN FRONT, page 361**)

RAM AT HEAD OF FLOCK: Peruvian

SCORPION WOMAN'S STINGER: Chumash, California

WEAVING PRINCESS: Chinese. A sky princess (★1 *Vega*) and a cowherd (*Altair,* ★1 in **EAGLE**) fell in love. They married and were so happy they neglected their duties, and were ordered to opposite sides of a river (the Milky Way). They may unite only on the seventh night of the seventh month of the Chinese calendar each year, when magpies form a bridge (**SWAN, page 368**) so the lovers can be together.

BUZZARD'S RIGHT HAND: Luiseño, California

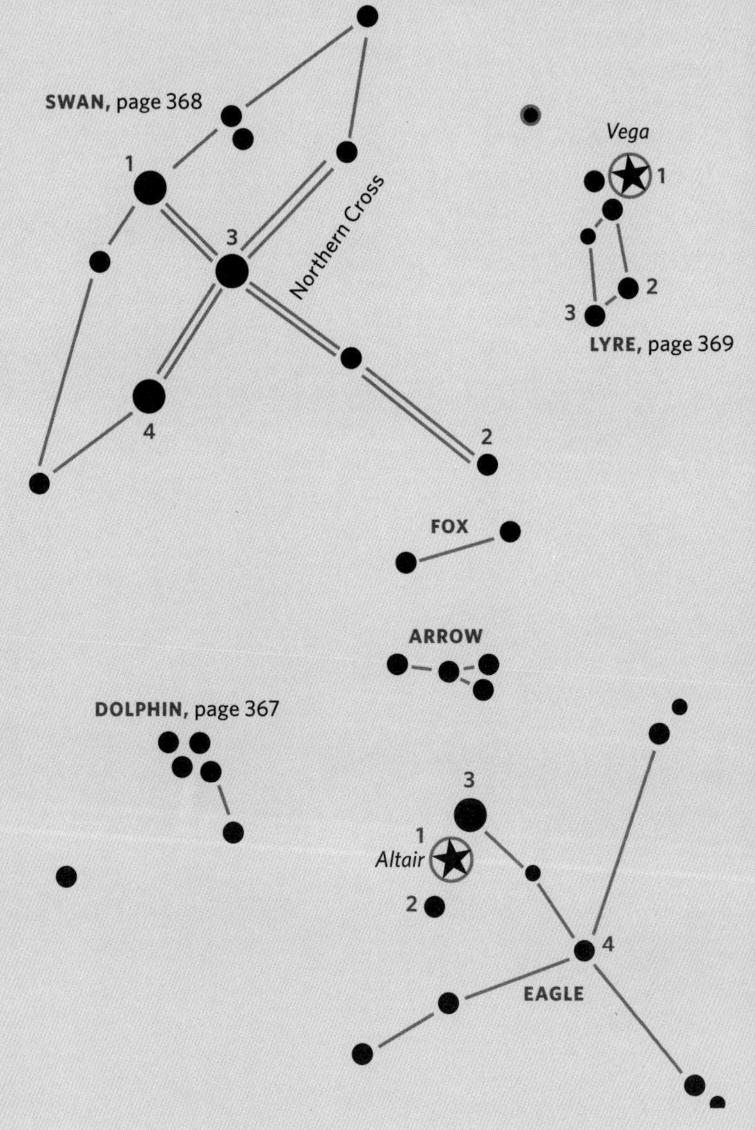
SWAN, page 368
Northern Cross
1
3
4
Vega
1
2
3
LYRE, page 369
2
FOX
ARROW
DOLPHIN, page 367
3
1
Altair
2
4
EAGLE

Aquarius (a-QUAIR-e-us)/Water Carrier

Best viewing: *August to October*

Intricate, dim stars south of **Great Square**. Look for the urn, water, and (apparently headless) body.

WATER BUCKET: Arab, Persian, Hebrew, Syrian, Turkish

WATER CARRIER: Greek: Ganymede, a handsome mortal, serves drinks to the gods in Olympus. Egyptian: The god who controls the Nile pours its water from a jug.

FUNERAL MOUND AND TOMBS: Chinese

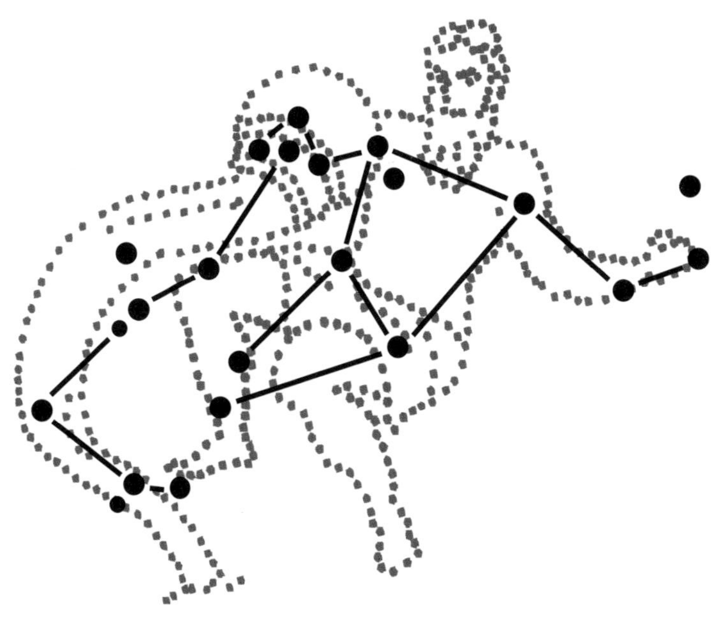

Capricornus/Sea Goat

Best viewing: *August to October*

The row of three stars in **EAGLE** points south to this faint group.

IBEX: Sumerian (left, carrying star in horns)

GOAT: Persian, Turkish, Syrian, Arab, Hindu. Star names: ★1 *Giedi*, goat; ★2 *Deneb Algedi*, tail of the goat.

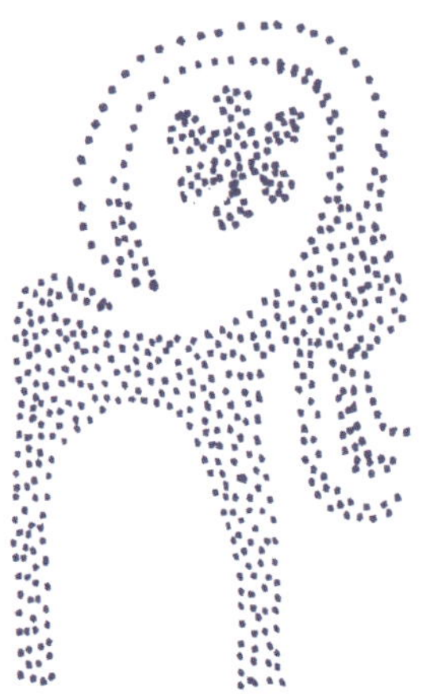

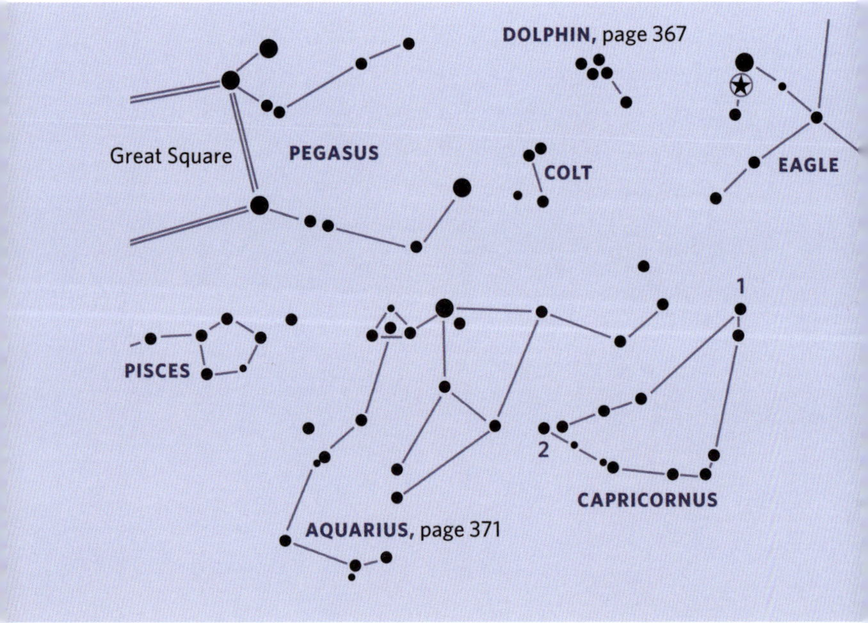

GOAT-FISH: Babylonian; Greek: Escaping a giant, the god Pan jumped into the Nile, becoming half goat, half fish. Egyptian: Associated with annual flood of the Nile.

HORNED FISH: Aztec

OX: early Chinese

ANTELOPE: Hindu

GATE OF THE GODS: Chaldean: Souls proceeded to heaven through this gate.

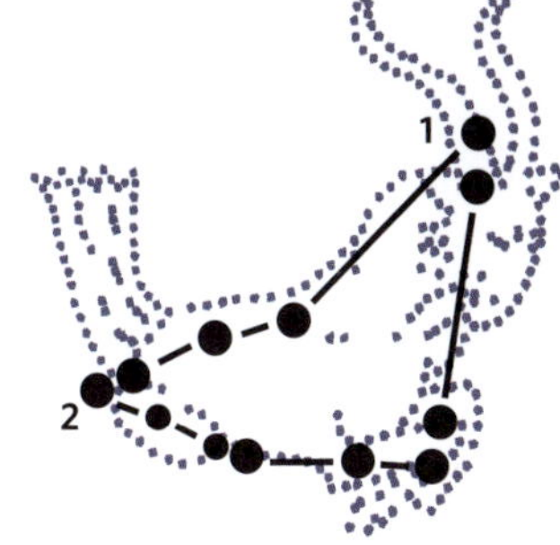

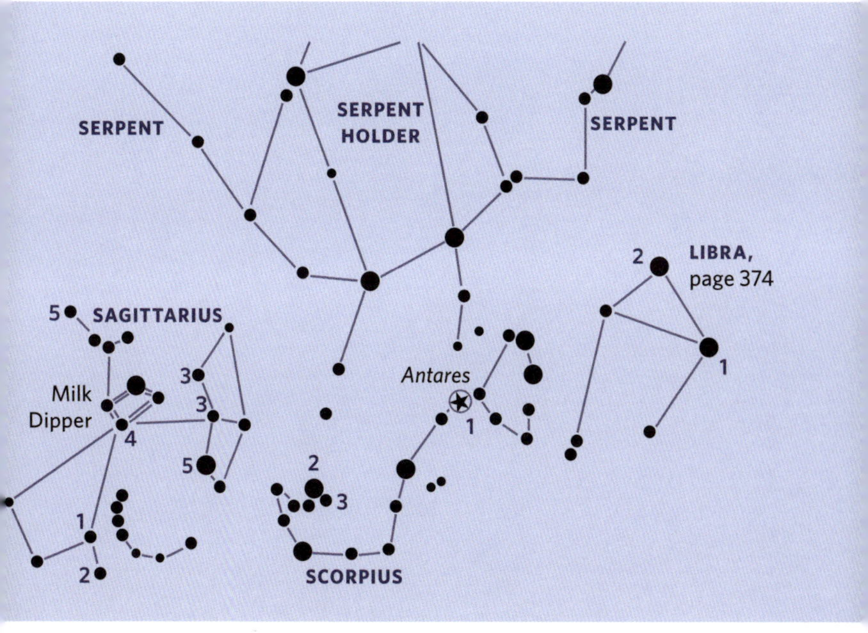

Libra (LEE-bra)/Scales

Best viewing: *June to July*

In some cultures, these faint stars were part of **SCORPIUS**.

BALANCE, SCALE BEAM: Babylonian, Persian, Syrian, Egyptian; Roman. Marks approach of autumn, when day and night are of equal length.

SCORPION'S CLAWS: Greek, Arabian. ★1 *Zubenelgenubi*, southern claw; ★2 *Zubeneschamali*, northern claw.

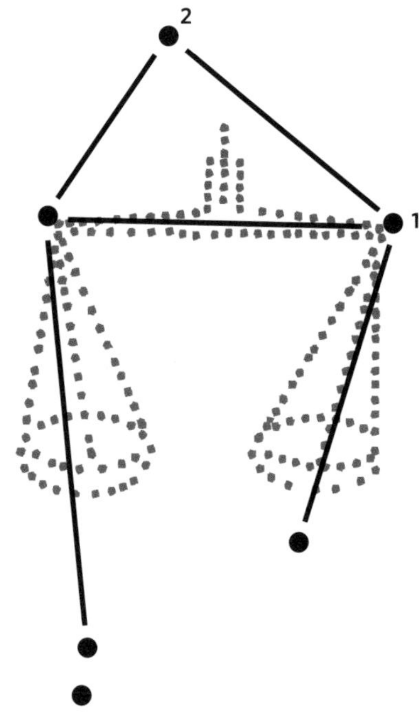

Virgo (VUR-go)/Virgin

Best viewing: *April to June*

Sweep from the **Big Dipper**'s handle through *Arcturus* to *Spica*.
VIRGO's other stars are faint.

CELESTIAL DOORWAY: Chinese

KENNEL: (★3 *Zaniah*) early Arab

LAMP, PEARL: (★1) Hindu

MAIDEN, WHEAT-BEARING MAIDEN: Sumerian, Greek, Egyptian; Tamil;
European; Sinhalese; later Arab. ★1 *Spica*, ear of wheat; ★2 *Vindemiatrix*,
grape gatherer; ★4 *Syrma*, hem of the robe.

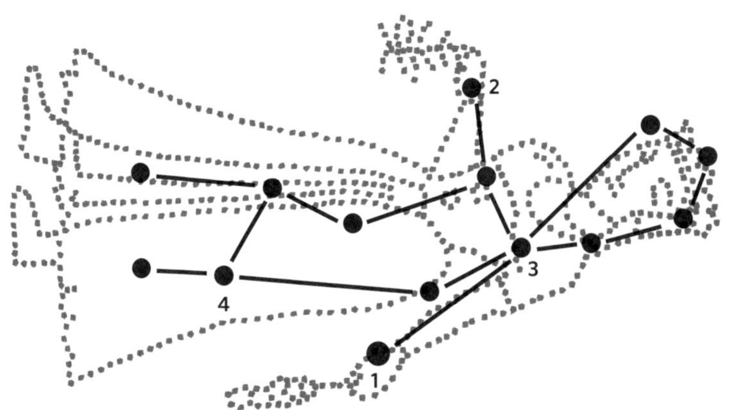

Stars in LEO

1 *Regulus*, little king
2 *Denebola*, tail of the lion
3 *Algieba*, the lion's mane
4 *Zosma*, girdle of the lion
5 *Rasales*, the eyebrows

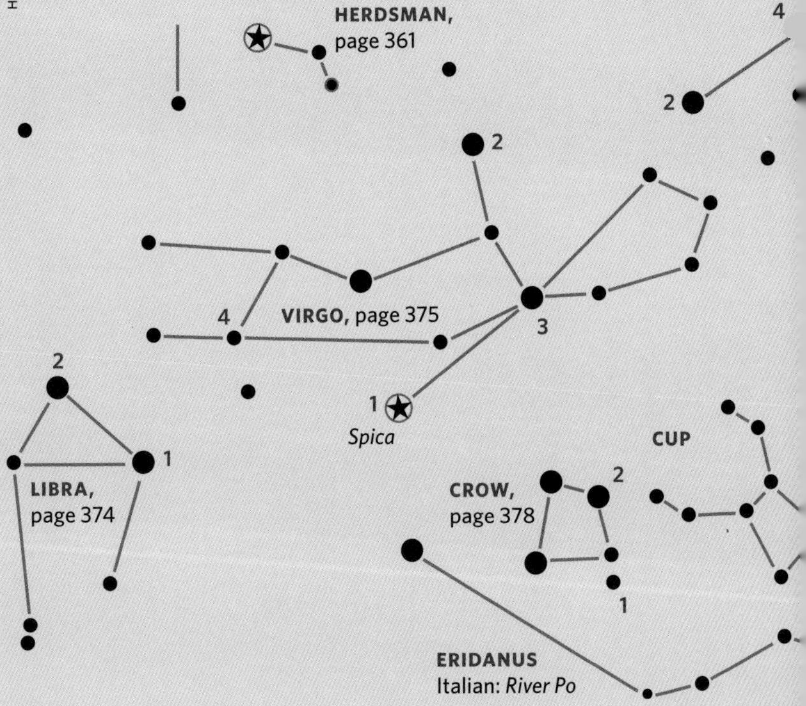

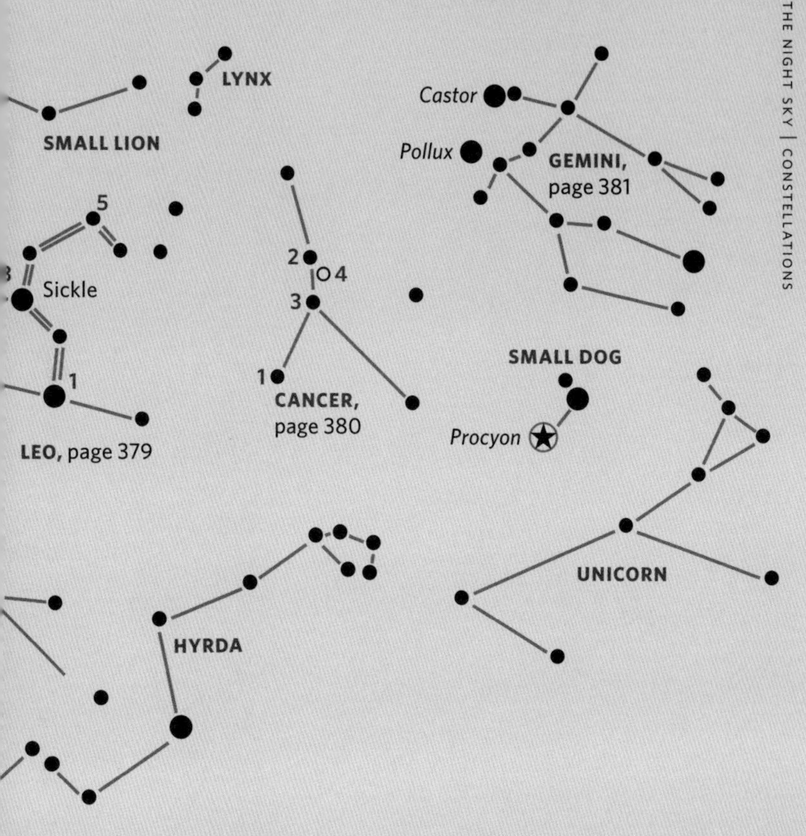
LYNX
SMALL LION
Castor
Pollux
GEMINI,
page 381
5
B
Sickle
2
O 4
3
1
1
CANCER,
page 380
LEO, page 379
SMALL DOG
Procyon
UNICORN
HYRDA

Crow/Corvus (KOR-vus)
Best viewing: *April to June*

From *Spica* in **VIRGO**, look south for a modified square. Stars ★1 *Alchiba*, the tent; ★2 *Gienah*, right wing of the raven. Just beyond **CROW** is the constellation **CUP**.

CROW/RAVEN and **CUP**: Roman; Greek: Apollo punished the disobedient Crow by placing it just far enough from a cup so that it could never satisfy its thirst.

TENT: Early Arab

KANGAROO: Wailwum, Australia

CELESTIAL CHART: Chinese

THRONE OF THE UNARMED ONE: Early Arab. (*Spica* in **VIRGO** is Unarmed One.)

TURTLE: Bororo of Brazil

ELEPHANT: Khmer of Cambodia

BIRD: Pukapuka of South Pacific

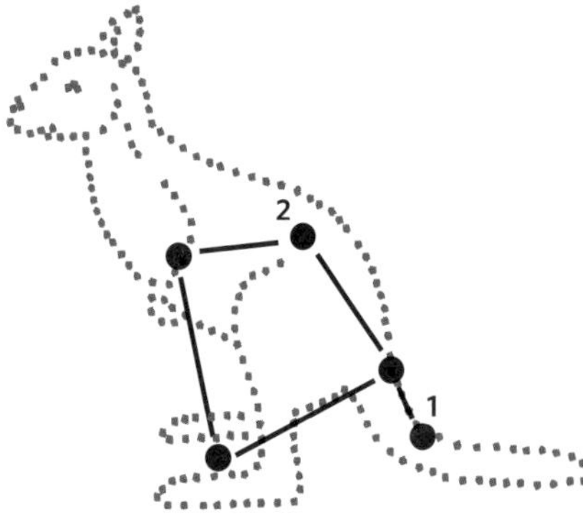

Leo (LEE-oh)/Lion

Asterism: *Sickle*
Best viewing: *March to June*

From Big Dipper (see page 352), find dazzling, blue-white ★1 Regulus set in the handle of the Sickle.

LION: Babylonian, Persian, Syrian, Hebrew, Egyptian, European; Greek: Hercules killed the Nemean lion as the first of his twelve labors.

SCIMITAR/CURVED WEAPON (THE SICKLE): Akkadian, Khorasmian, Sogdian of southwest Asia.

THE FIREPLACE (REGULUS): Lakota

GOD OF THUNDER: Taulipang of northern Brazil

CRAYFISH: Tukano, Kobeua, and Siusi of northern Brazil

SLEEPING WOMAN: Chukchi in Siberia (forepart of **LEO**)

RAIN DRAGON: Chinese

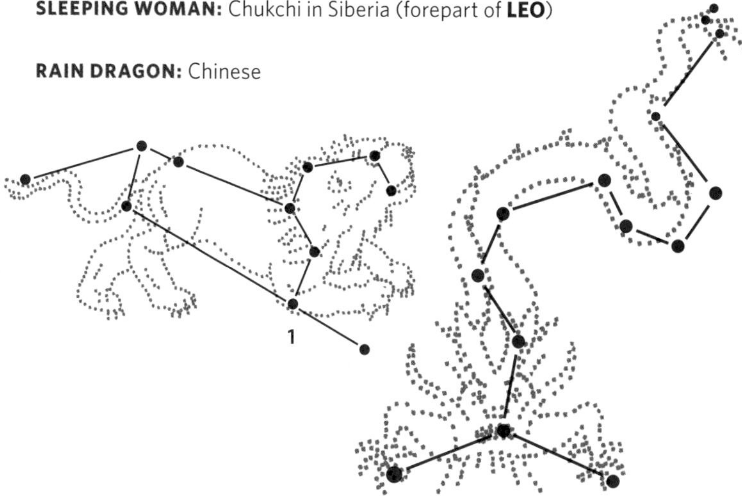

Cancer (CAN-cer)/Crab

Best viewing: *January to May*

Faint stars located between Regulus in **LEO** and Castor in **GEMINI**.

CRAB: Persian, Hebrew, Syrian, Arabian; Greek: The Crab attacked Hercules; the hero crushed it. Goddess Juno (Roman: Hera) placed it in the sky. ★1 *Acubens*, claws.

TORTOISE: Babylonian, Egyptian

FROG: Tibetan

SCARAB BEETLE: Egyptian

MANGER: Arab: Manger (O4 *Praesepe*, a star cluster) lies between two asses (★2 and ★3).

SPIRITS (PRAESEPE): Chinese

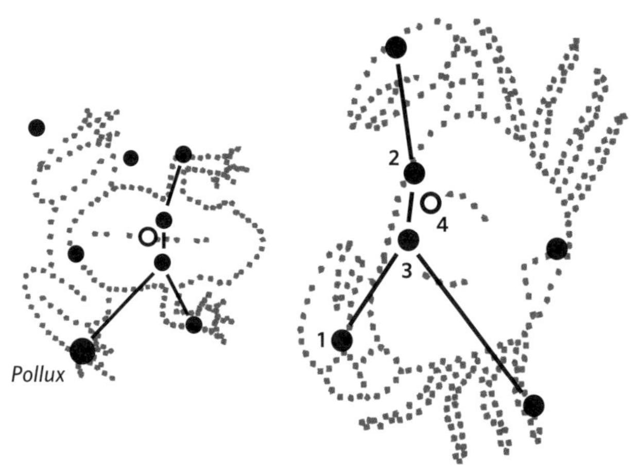

Gemini (JE-mi-neye)/Twins

Best viewing: *December to May*

Look diagonally across the Big Dipper's bowl (see page 352) to a pair of bright stars. (Orion's Belt lies just beyond.)

★1 **CASTOR** and ★2 **POLLUX**

TWINS: Babylonian, Sumerian, Arabian, Egyptian (right), Hawaiian, Klamath; Greek: Leda bore two sons—Castor with the Spartan king, and Pollux with Zeus. As a mortal, Castor went to Hades when he died; as a son of a god, Pollux was destined for Olympus. Zeus allowed them to stay together forever, spending half their time in Olympus and half in Hades.

TWO MEN: Samoan

HORSEMEN: India

PEACOCKS: Arab

TWO BUTTERFLIES: Javanese, Japanese, Chinese

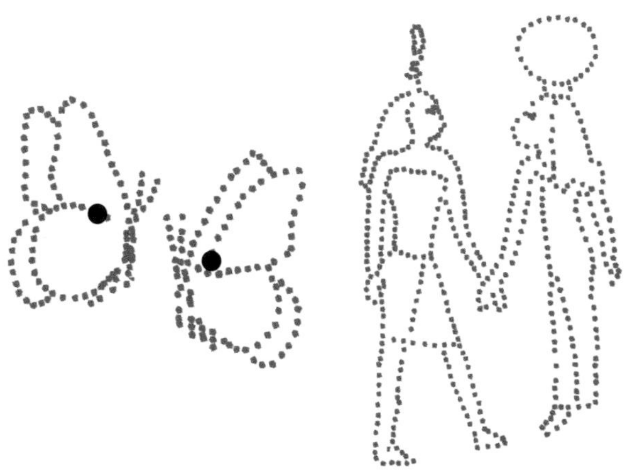

TWO FIGHTING STARS: Jicarilla Apache

STAR IN THE MIDDLE OF THE ISLANDS AND STAR IN THE MIDDLE OF THE HEAVENS: Tuamotu in South Pacific

Part of **RACE TRACK** (a hexagon around **ORION**; see page 384): Lakota

PE-HO, NORTHERN RIVER: Chinese

COLLARBONE: Inuit, pan-Arctic (★1 and ★2 in **CHARIOTEER** form the other bone.)

BEAR'S LODGE: (with other stars) Lakota

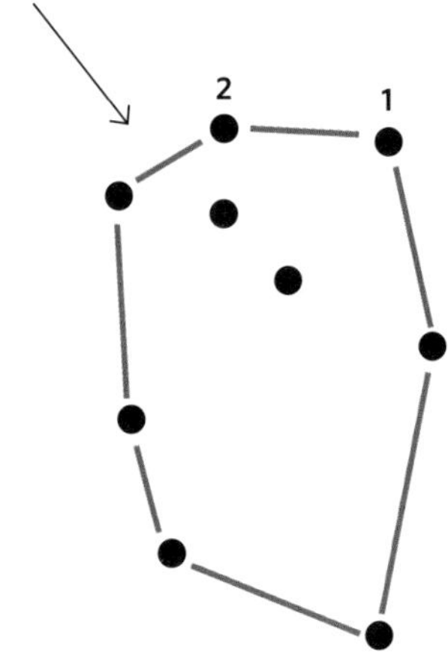

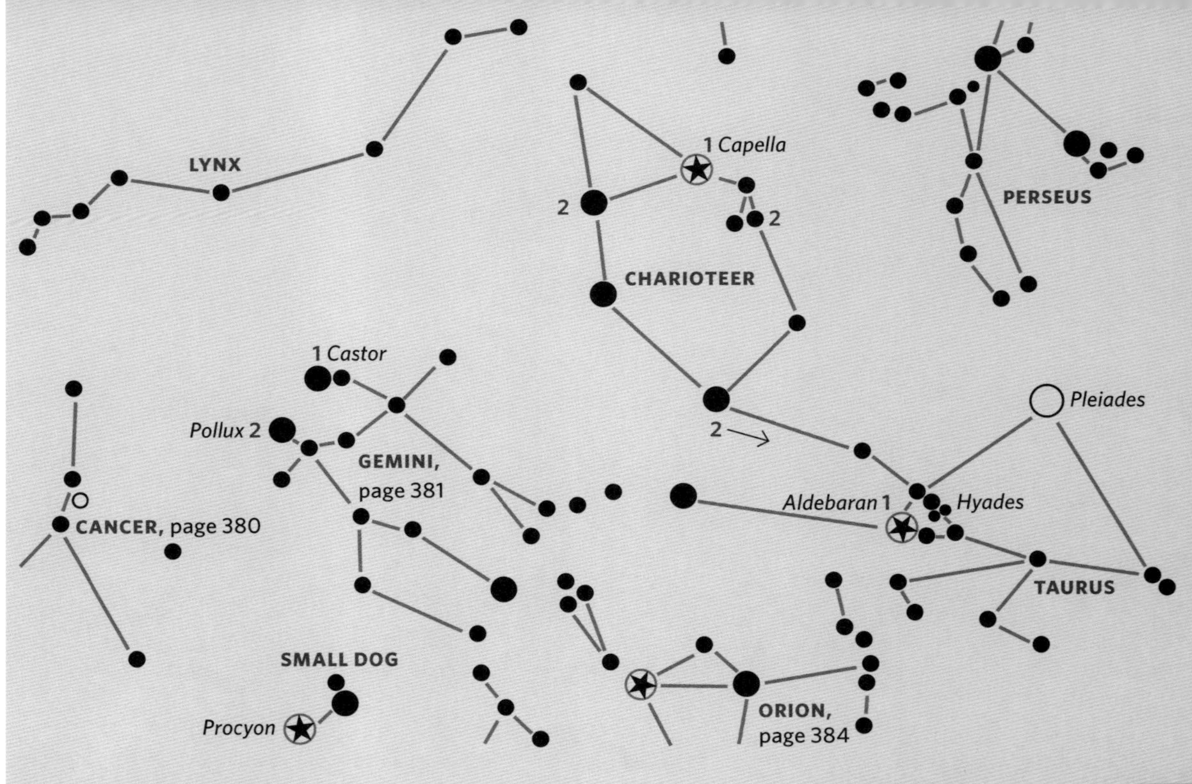

LYNX
1 Capella
2
2
CHARIOTEER
PERSEUS
1 Castor
Pollux 2
GEMINI, page 381
2
Pleiades
Aldebaran 1
Hyades
CANCER, page 380
TAURUS
SMALL DOG
Procyon
ORION, page 384

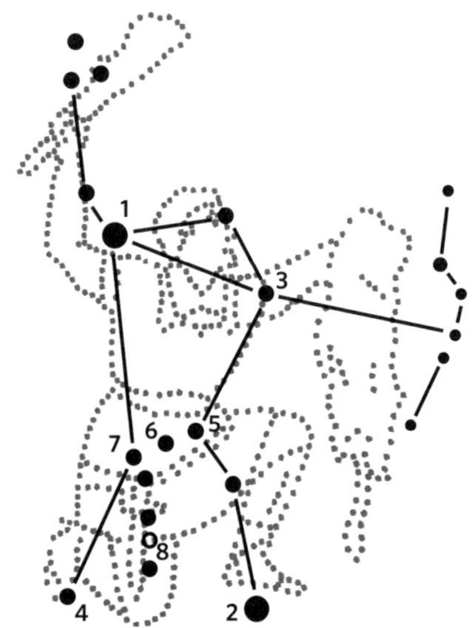

Orion (oh-RI-un)

Best viewing: *November to March*

Dominates the evening sky in winter. Look for a row of three evenly spaced stars (Orion's Belt) with reddish Betelgeuse nearby.

HUNTER: Greek. A scorpion (**SCORPIUS**) killed this hunter; Artemis placed him in the sky so that he and the scorpion could not be seen at the same time. Alternately, Orion chased some women (*Pleiades*); Zeus put them in the sky where Orion cannot reach them.

OSIRIS: Egyptian. Osiris represents the endless cycle of life, death, and rebirth. He was killed by his brother, but Isis caused Osiris's spirit to return and he rose to the sky. The reappearance of Isis (*Sirius*) in the dawn sky coincides with the life-giving floods of the Nile.

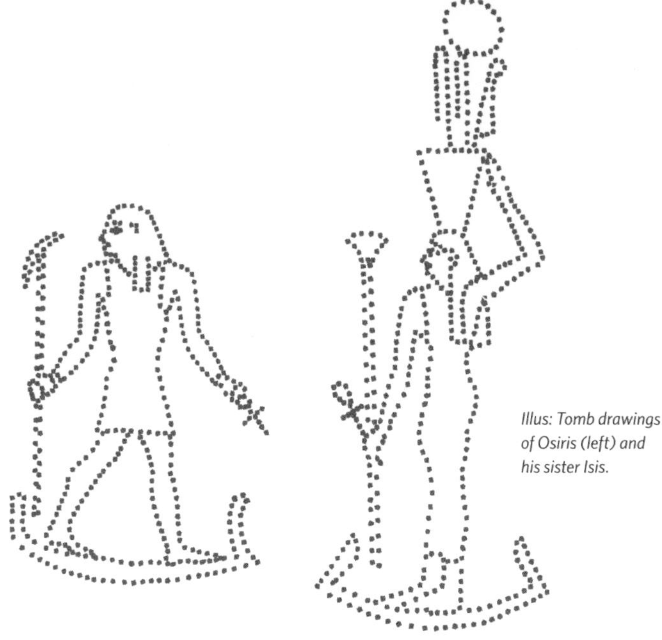

Illus: Tomb drawings of Osiris (left) and his sister Isis.

GIANT: Syrian, Arab

ORWANDIL, THE GIANT: Norse. Orwandil's big toe froze; Thor threw it (★8 in Big Dipper) into the sky next to Thor's Wagon **(page 354)**.

SUPREME COMMANDER: Chinese

HARPOONER OF HEAVEN or **PADDLER:** Kwakiutl. He hunts an otter (*Pleiades*) that escaped to the sky.

HEARTHSTONE: Quiché, Central America (stone, ★2, 4, 7; smoky fire, ★8)

DRUM: Japanese. ★5–7 secure strings from drumheads (★1, 3; 2, 4).

CANOES AND SHED: Kapingamarangi, South Pacific. Three canoes (★5–7) are stored in a shed (★1–4).

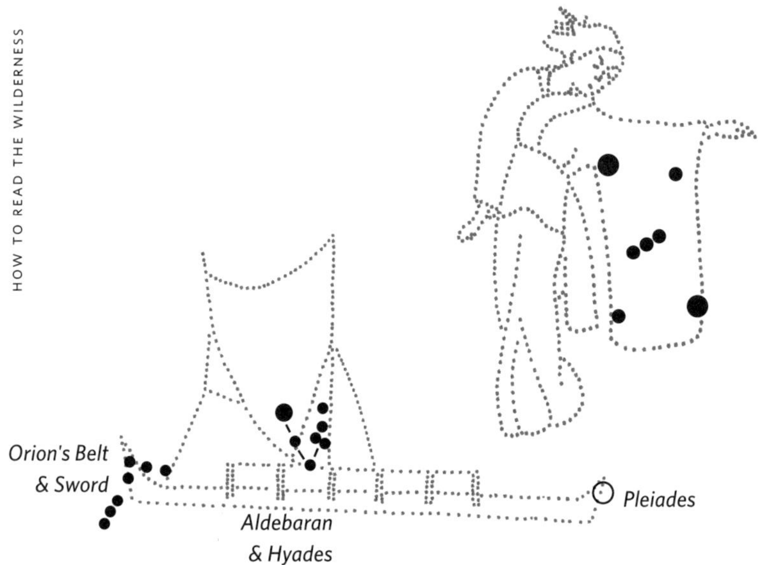

GREAT CANOE: Maori. ★5–7 (Orion's Belt) form the stern; *Pleiades*, the bow; *Hyades*, sail; Orion's sword (line of stars at right angle to belt starting at ★7) is the cable.

KIMONO SLEEVE STARS: Japanese

TURTLE: Bororo, Brazil (★2, 4–7)

DEER: Hindu. See **Deer-slayer, page 389.**

★5, 6, 7 (**ORION'S BELT**) **THREE FISHERMEN**: North Australia. These men, husbands of the *Pleiades*, caught fish (Orion's sword) that were taboo.

THREE STAGS: Siberian groups. Other stars in **ORION** are horses, dogs, hunters, and arrows.

THREE RUNNERS: Pan-Arctic. Three men and their dogs (*Hyades*) chase a polar bear (*Aldebaran*).

MORE NAMES FOR ORION'S BELT:
Three Mowers or Reapers: Early Germanic. Men stand in a row in a meadow.
Frigg's (Freya's) Distaff: Scandinavian. Frigg is the wife of Odin, mother of Thor; her staff holds flax or wool for spinning.
Three Prongs of a Fishing Spear: Japanese
Arrow and Three Zebras: Bushman. A god was hunting but his arrow (Orion's sword) fell short of the zebras; when the three stars set, three zebras step down onto earth.

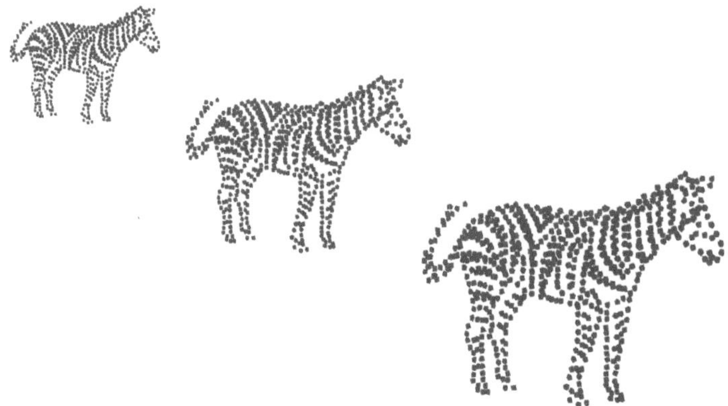

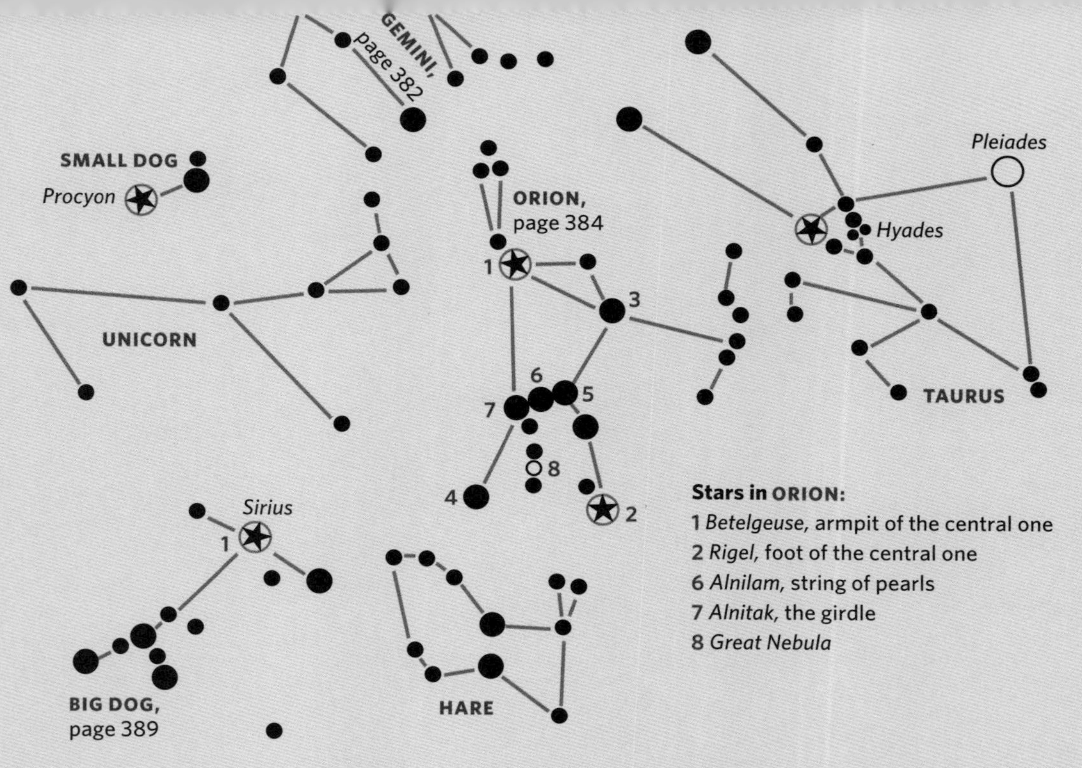
SMALL DOG
Procyon
GEMINI,
page 382
ORION,
page 384
Pleiades
Hyades
UNICORN
TAURUS
Sirius
Stars in ORION:
1 Betelgeuse, armpit of the central one
2 Rigel, foot of the central one
6 Alnilam, string of pearls
7 Alnitak, the girdle
8 Great Nebula
BIG DOG,
page 389
HARE

Big Dog/Canis Major (KAY-nis MAY-jor)
Best viewing: *January to March*

Orion's Belt points to Sirius, the brightest star in the sky.

Names for ★1 Sirius, the Dog Star:

DOG: Greek, Arab, Lithuanian

JACKAL: Chinese

BARKER: Phoenician

RED FOX AND WHITE FOX: Inuit. Various factors in the Arctic cause *Sirius* to flicker. The foxes are said to be changing places.

DEER-SLAYER: Hindu: A father (a stag, **ORION**) pursued his daughter (an antelope, *Aldebaran*) in an unfatherly way. Deer-slayer (*Sirius*) shot an arrow (Orion's Belt) and stopped him.

SMALL FISH: Landak Dayak in Borneo

ISIS: Egyptian. Sister and consort of Osiris **(page 384)**

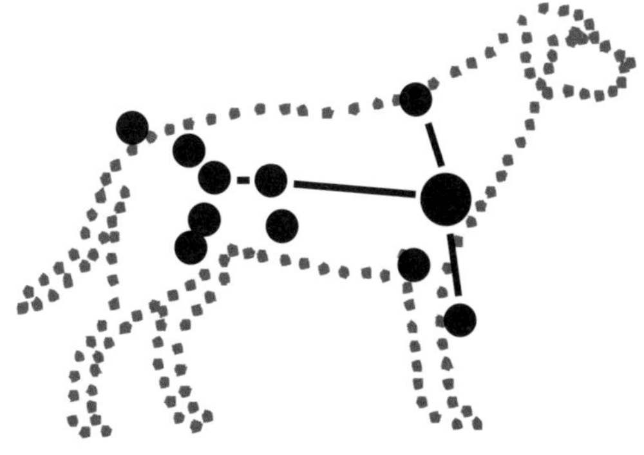

Sources

The contents of this book were curated from the following pocket guides published by the Nature Study Guild.

Pacific Coast Tree Finder, by Tom Watts

Desert Tree Finder, by May Theilgaard Watts and Tom Watts

Rocky Mountain Tree Finder, by Tom Watts and Bridget Watts

Winter Tree Finder, by May Theilgaard Watts and Tom Watts

Tree Finder, by May Theilgaard Watts

Redwood Region Flower Finder, by Phoebe Watts, with illustrations by Sarah Ellen Watts

Rocky Mountain Flower Finder, by Janet L. Wingate, Ph.D.

Flower Finder, by May Theilgaard Watts

Winter Weed Finder, by Dorcas S. Miller, with illustrations by Ellen Amendolara

Berry Finder, by Dorcas S. Miller, with illustrations by Cherie Hunter Day

Fern Finder, by Anne C. Hallowell and Barbara G. Hallowell

Pacific Coast Mammals, by Ron Russo and Pam Olhausen

Mountain State Mammals, by Ron Russo, with illustrations by Barbara Downs

Track Finder, by Dorcas S. Miller, with illustrations by Cherie Hunter Day

Pacific Coast Bird Finder, by Roger J. Lederer, with illustrations by Jacquelyn S. Guiffré

Bird Finder, by Roger J. Lederer, with illustrations by Roger C. Franke

Pacific Intertidal Life, by Ron Russo and Pam Olhausen

Life on Intertidal Rocks, by Cherie Hunter Day

Constellation Finder, by Dorcas S. Miller

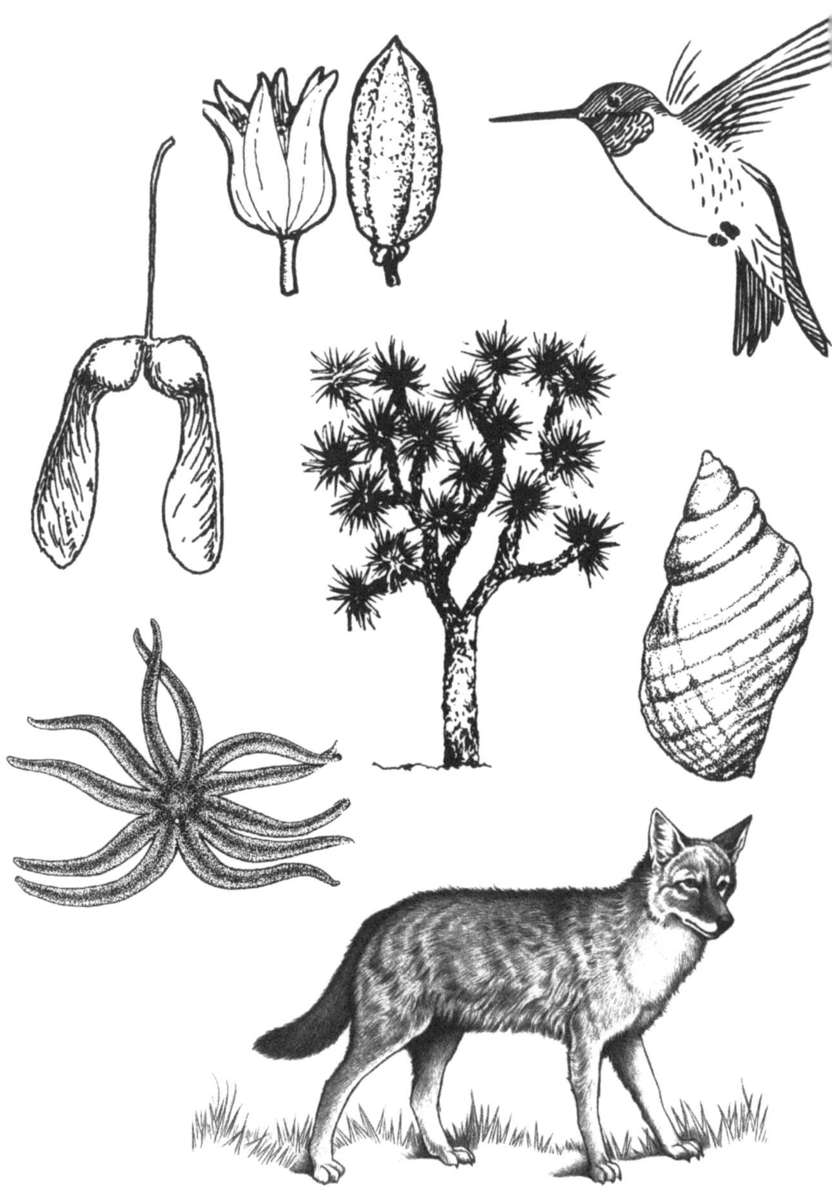

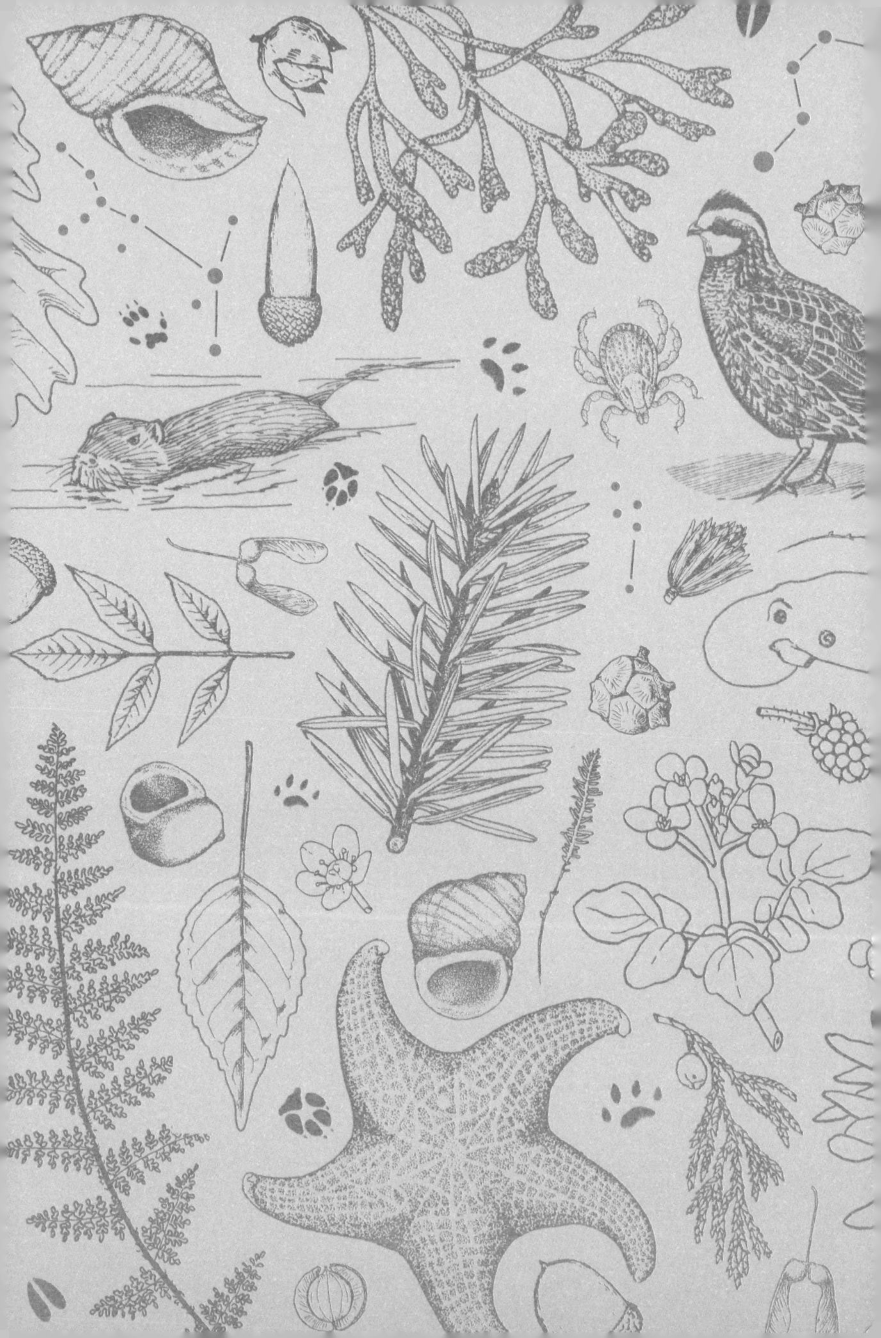